W0038491

Geodynamics of Azores-Tunisia

Edited by
E. Buforn
J. Martín-Dávila
A. Udías

2004

Springer Basel AG

Reprint from Pure and Applied Geophysics
(PAGEOPH), Volume 161 (2004), No. 3

Editors

E. Buforn
Depto. de Geofísica y Meteorología
Universidad Complutense
28040 Madrid
Spain

e-mail: ebufornp@fis.ucm.es

A. Udías
Depto. de Geofísica y Meteorología
Universidad Complutense
28040 Madrid
Spain

e-mail: audiasva@ucm.es

J. Martín-Dávila
Real Instituto y Observatorio de la Armada
San Fernando
Cadiz
Spain

e-mail: mdavila@roa.es

A CIP catalogue record for this book is available from the Library of Congress, Washington D.C., USA

Bibliographic information published by Die Deutsche Bibliothek
Die Deutsche Bibliothek lists this publication in the Deutsche
Nationalbibliographie; detailed bibliographic data is available in the
Internet at http://dnb.ddb.de

ISBN 978-3-7643-7043-5   ISBN 978-3-0348-7899-9 (eBook)
DOI 10.1007/978-3-0348-7899-9

Printed on acid-free paper produced from chlorine-free pulp

9 8 7 6 5 4 3 2 1

Contents

Pure appl. geophys. 161 (2004) 473–476
0033–4553/04/030473–4
DOI 10.1007/s00024-003-2458-1

❙ Pure and Applied Geophysics

Introduction

E. Buforn, J. Martín-Dávila, and A. Udías

Azores-Tunisia Geodynamics

The Azores-Tunisia region is formed by the western part of the plate boundary between Eurasia and Africa. This plate boundary presents a complex nature due to its proximity to the pole of rotation of the African plate. This situation produces crustal extensions and normal faulting at the Azores archipelago, transcurrent motion with strike-slip faulting at the center part of the Azores-Gibraltar fault and at the eastern end, from the Gulf of Cadiz to Tunisia, plate convergence with reverse faulting. In this last part, the collision of Iberia with northern Morocco produces complex phenomena with intermediate depth and deep earthquakes and an extensional regime at the Alboran Sea. Recently, new evidence has been gathered in this region, based on observations from geology, geodesy, mainly through GPS measurements, seismology, especially with the installation of broad-band stations, and other fields of geophysics, such as paleomagnetism and gravimetry. The synthesis of these new observations is contributing to a new understanding of the complex geodynamics of this region.

On May 31 to June 2, 2001, a Workshop was held in San Fernando (Cadiz, Spain) organized by the Real Instituto y Observatorio de la Armada (San Fernando) and the Universidad Complutense (Madrid) with the title "Geodynamics of the Western Part of the Eurasia-Africa Plate Boundary (Azores-Tunisia)". Five invited lectures and 77 papers were presented at the workshop covering many aspects of the geology, seismology, geodesy, paleomagnetism and other geophysical observations of the region. Attendance brought together 105 participants from Europe, Northern Africa and America. New observations in the different fields were presented and new models of the geodynamics of the region were proposed. The present volume is a selection of the papers presented at the workshop. The 12 papers of the volume include 5 papers pertaining to geological and tectonic structure of the regions and parts thereof, 4 papers focus on seismology and seismotectonics, 2 address geodetic observation and one paleomagnetism. These papers are representative of those presented at the workshop.

The volume opens with a paper by Zeck which presents a new model for the orogenic evolution of the central part of the region, the Alpine Betic-Rif belt. The model proposes for the evolution of this orogenic belt two main stages: A continental collision with formation of primary nappes and high-pressure metamorphic parageneses, and a subsequent tectonic extrusion with vertically directed and extensional tectonics. The model may also offer important constraints to the reconstruction of the tectonic evolution of other convergent terranes characterized by high exhumation rates. The same Betic-Rif orogenic belt is analyzed in the paper by Chalouan and Michard which presents a different scheme in which the structure is the result of a subduction-subduction-transform fault triple junction. The older subduction is part of the NW-dipping subduction extending from the northern Apennines to the Maghrebides belts and the younger is produced as a result of incipient blocking of the older. Both subductions meet at the Betic-Rif, west of Alboran, together with the Azores-Gibraltar transform fault, to form some sort of a triple junction. Seismic and paleomagnetic evidence is brought in support of this model. A more restricted area, the frontal part of the Rif cordillera and the Saïs basin, is studied in the paper by Bargach *et al.* Recent tectonic deformations in this area are explained, in the general context of compressive regional stress field resulting from plate motion. The active structures of the Rif notably disturb these stresses, producing local deformations with alternating trends of compression and extension. Evidence of the relationship between local deformations and general plate motion is found at the Prerif Ridges.

The northern areas of the Betics and the passive margin of Iberia are dealt with in the next two papers. Ruano *et al.* present a study of the recent tectonic structures at the Central Betic Cordillera. The main structures of the Cordillera are folds trending predominantly E-W to NE-SW with crustal thickening and uplift which are related and are compatible with the Eurasian-African plate convergence. At the Internal Zones normal faults developed in the upper crust associated to the development of asymmetrical basins like the Granada depression. The model presented for this region is compatible with the presence of an active subduction in the Alboran Sea and contrasts with the models based on the activity of transcurrent faults. The paper by Zitellini *et al.* presents the results of two multi-channel seismic surveys offshore of the southern coast of Iberia west of Cape San Vicente. The seismic data, according to the authors show evidence of a compressional deformation involving both oceanic and continental crust extending from the southern border of the Tagus Abyssal Plain to the Seine Abyssal Plain on the continental margin of SW Portugal. The most active structures are located where present seismic activity is higher. The deformation is expressed mainly by blind thrusts, thrust faults which do not reach the surface. One of them which reaches the surface constitutes the Marquês de Pombal structure. This fault is considered to be related with the source region of the 1755 Lisbon earthquake.

The following four papers deal with the seismicity and seismotectonic of the region. Carrilho *et al.* present the first results of GEOALGAR, a project initiated in 2000 to monitor the seismic activity in the Algarve region (southern Portugal). In this paper results of the relocation of epicenters and determination of fault plane solutions are presented. The new epicentral locations show a more organized spatial distribution which could indicate a possible correlation with some known tectonic features. Fault plane solutions are predominantly of strike-slip motion consistent with a horizontal compression in the NW-SE to NNW-SSE direction. The paper by Yelles-Chaouche *et al.* presents a detailed study of the 22 December, 1999 earthquake at Ain Temouchent (northwest Algeria) of magnitude 5.7. The earthquake caused serious damage in the town of Ain Teemouchent with 25 casualties and 25000 people left homeless. Intensity map, surface features and the focal mechanism, based on wave form analysis, are shown. The mechanism corresponds to reverse fault motion with planes striking NNE-SSW resulting from horizontal compression in the NW-SE direction. This corresponds to the general mechanism found for Algeria earthquakes. Buforn *et al.* present a study of the characteristics of the plate boundary between Africa and Iberia, from west of Cape San Vicente to Algeria, using seismicity and source mechanism data. The region is divided into three areas which manifest different characteristics. The central area (Alboran and Betics) show a different seismotectonic character in comparison with those of the west (Gulf of Cadiz) and the east (Algeria). A special study is presented of the intermediate depth and deep earthquakes present in the region. Finally a seismotectonic model is presented which integrates the different observations. The last paper treating seismicity is by Góngora and Carrilho, which presents the calibration of the local magnitudes in the Azores Archipelago based on digital data from seven stations for the period of 1998–2000.

The following two papers deal with geodetic measurements. Nocquet and Calais present the results of the use of an efficient and rigorous combination procedure to integrate several geodetic solutions into a single and consistent reference frame. They stress that the most reliable results are obtained using continuous GPS data. Among the results presented in the paper, the authors find that geodetic data consistently indicate an Africa-stable Europe convergence rate slower than the prediction of geological models, and that the oblique convergence in the Western Mediterranean is mostly accommodated by transpressional deformation in north Africa and southern Spain. In the second paper Fernandes *et al.* present recent geodetic results in the Azores region. The GPS observations have been carried out between 1993 and 2000. The observations confirm that the western group of the Azores islands belongs to the North American plate and that, in the vicinity of the central and eastern groups of islands, the relative motion between Africa and Eurasia is very small (4 mm/yr). The analysis also confirms the existence of a diffuse boundary zone from Graciosa to Santa Maria islands. The last paper of this issue by Osete *et al.* deals with results from paleomagnetic observations in the Betic Cordillera. Using data from the Jurassic formations of the External Betics the authors are able to constrain the timing of block rotations in the Subbetics. These are large clockwise rotations (35°–140°) indicated by paleomagnetic declinations. Rotations in the central Subbetics took place before the remagnetization event during the Neogene, while rotations in the western Subbetics occurred after this event.

In conclusion, this collection of papers presents the analysis of recent geological, seismological, geodetic and paleomagnetic observations of the region from the Azores Islands to Tunisia. Based on these observations new models for the geodynamics of this complex region are presented.

Editors:

E. Buforn
Depto. De Geofísica y Meteorología
Universidad Complutense
Madrid
ebufornp@fis.ucm.es

J. Martín-Dávila
Real Instituto y Observatorio de la Armada
San Fernando
Cadiz
mdavila@roa.es

A. Udías
Depto. de Geofísica y Meteorología
Universidad Complutense
Madrid
audiasva@ucm.es

 To access this journal online:
http://www.birkhauser.ch

Pure appl. geophys. 161 (2004) 477–487
0033–4553/04/030477–11
DOI 10.1007/s00024-003-2459-0

❘ Pure and Applied Geophysics

Rapid Exhumation in the Alpine Belt of the Betic-Rif (W Mediterranean): Tectonic Extrusion

H. P. Zeck[1]

Abstract — Extreme cooling rates (~500 °C/m.y.) during the late stage, 22–18 Ma, orogenic evolution of the Alpine Betic-Rif belt are suggested to result from rapid exhumation caused by tectonic extrusion and concomitant extensional tectonics. The extrusional/extensional tectonic setting is controlled by the SW-NE trending break-off scar left in the lithosphere of the Alborán Sea and SE Spain after detachment of a lithospheric slab. The extruded material represents the collisional crustal nappe pile (together with fragments of underlying mantle, such as the Ronda peridotites) and the cause of the extrusion is the thermal softening within the crustal section during and after collision. The extrusion/extension took place under the influence of a NW-SE directed compressive regime, perpendicular to the collisional belt. At the same time the sub-lithospheric mantle still showed the E-W compressive regime of the collisional stage. The Alpine tectono-metamorphic evolution of the Betic-Rif belt in the W Mediterranean thus comprises two main stages: (1) continental collision with formation of primary nappes and high-pressure metamorphic parageneses, (2) tectonic extrusion with vertically directed tectonics (high pressure, very rapid decompression) and extensional tectonics with roughly horizontal, lateral transport and final emplacement of the extruded mélange in the form of a stack of detachment sheets (low pressure, very rapid cooling). This model for the Betic-Rif may offer important constraints to all rapidly exhumed convergent terranes.

Key words: Tectonic extrusion, rapid exhumation, Alpine Betic-Rif, W Mediterranean.

Introduction

The Betic-Rif mountain chain (Fig. 1) forms the southwestern termination of the Alpine orogenic belt in W Europe – N Africa. Most current tectonic models for the Betic-Rif rely on N-S Africa-Iberia/Europe convergence with the formation of a ~250 km wide E-W trending collisional prism and lithospheric root. A central element in many of these models is the late stage convective removal of the thickened lithospheric root which is thought to have resulted in regional topographic uplift and late stage gravitational collapse of the Alborán area (e.g., PLATT and VISSERS, 1989; GARCÍA-CASCO and TORRES-ROLDÁN, 1999). However, seismicity and seismic tomography studies in the region do not support

[1] Tectonics Special Research Centre, University of Western Australia, Perth, Australia. Present address: Geological Museum, Copenhagen University; E-mail: zeck.frappier@wanadoo.fr

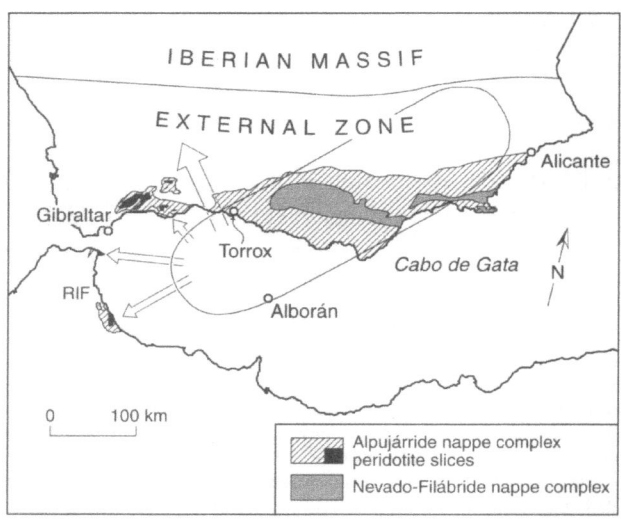

Figure 1

Geological sketch map showing distribution of the two major collisional nappe complexes of the internal zone in the Alpine Betic-Rif belt comprising the Betic Cordilleras, between Alicante and Gibraltar in S Spain, and the Rif in N Africa (after ZECK, 1999). Late stage extrusional/extensional tectonics have reworked the collisional nappe complexes resulting in dislodgment of secondary allochthons onto the Iberian and African forelands (open arrows; cf. Figs. 3b,c). The SW-NE trending ellipsoid is the vertical projection of the sinking lithospheric slab (Fig. 3a), detached after subduction terminated (BLANCO and SPAKMAN, 1993; ZECK, 1997; CALVERT et al., 2000).

an E-W oriented subduction zone (e.g., GRIMISON and CHEN, 1986; BLANCO and SPAKMAN, 1993; data in SEBER et al., 1996, re-interpreted by ZECK, 1997; CALVERT et al., 2000). Neither does the small magnitude of the Africa-Iberia convergence (150–200 km; DEWEY et al., 1989; SRIVASTAVA et al., 1990) support these models.

To meet these geophysical and geological objections, the orogeny in the Betic-Rif was recently explained by a sinking slab model (ZECK, 1996, 1997, 1999). For the early and major stage of the orogeny this model suggests a development under the influence of E-W convergence within a SW-NE striking subduction system dipping westward under eastward drifting Iberia. Subduction came to an end with the introduction of continental lithosphere into the subduction zone, leading to nappe piling and HP metamorphism (e.g., NIJHUIS, 1964; GOFFÉ et al., 1989; TUBÍA and IBARGUCHI, 1991; AZAÑON and GOFFÉ, 1997). The timing of this collisional event is not well established; perhaps it occurred within the period of 50–30 Ma (ZECK, 1997, 1999). After the collision the subducted oceanic lithospheric slab broke off (BLANCO and SPAKMAN, 1993) some time before 22 Ma (ZECK, 1996, 1997) and sank rapidly into the asthenosphere (Fig. 3a) – the detached slab still produces ~600 km deep earthquakes reflecting the E-W directed stress field of the subduction stage (CHUNG and KANAMORI, 1976).

After slab break-off, during the age interval 22–18 Ma, events were controlled by extensional tectonics associated with very rapid exhumation and cooling at rates of ~500 °C/m.y. in rock complexes of the internal zone (Fig. 2). A combination of isostatic surface/topographic uplift and erosion or tectonic de-roofing cannot explain the high rock uplift rates demanded by these extreme cooling rates. Very large and rapid tectonic rock uplift must have played a considerable role. However, the question of which mechanism could produce these rapid rock uplifts has remained enigmatic. The present paper introduces to the arena the tectonic extrusion model recently suggested by THOMPSON et al. (1997a,b) and discusses its relevance to the rapidly exhumed Betic-Rif tectonic belt.

Extrusional Tectonics

Widely accepted subduction and collision scenarios currently explain formation of orogenic nappe piles and prograde regional metamorphic regimes. Models

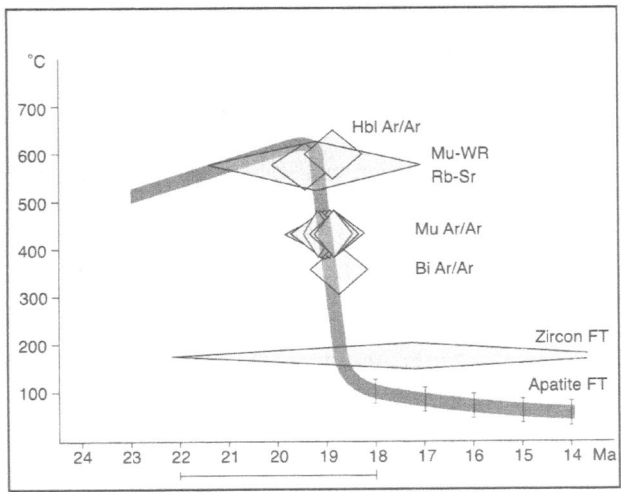

Figure 2

Cooling constraints for a small area close to the village of Torrox (Fig. 1) in the central Betic Cordilleras, after ZECK (1996). Stratigraphy, tectonic setting and regional geological correlations are given in ZECK and WHITEHOUSE (1999). Results of six thermochronometers with widely varying closure temperatures are compiled, defining a cooling trajectory which constrains a cooling rate of not less than 500 °C/m.y. within the time interval 19.5–18.5 Ma. The apatite cooling constraint brackets reflect best-fit thermal histories based on fission track length distribution patterns (ANDRIESSEN and ZECK, 1996). Similar cooling events in other parts of the Betic-Rif may have occurred slightly earlier, nonetheless all within the interval 22–18 Ma (ZECK, 1999).

to explain the rebound of such collisional tectono-metamorphic complexes are less convincing. Commonly it is assumed that isostatic surface uplift and concomitant erosion or tectonic de-roofing of the thickened continental crustal root complex explain its exhumation (e.g., THOMPSON and ENGLAND, 1984; PLATT, 1993).

THOMPSON et al. (1997a,b) have suggested an alternative model which explains how collisional prisms between rigid, colliding lithospheric plates may become thermally softened and rapidly extruded upward "like tooth paste in a tube." A basic assumption is that the thickened, siliceous, continental crust in the collisional root complex of a subduction zone is heated by adjacent mantle during and after collision. The model is based on numerical homogeneous particle flow simulations (using quartz dislocation flow boundary rates at 350 °C and 500 °C of SIBSON, 1977), which support calculation of particle paths in (x,y,z), P, T, t dimensions. Within the given thermal-tectonic setting two further factors exert control over the rate of rock elevation: (1) *the obliquity of convergence* – rapid rock elevation is only obtained in (near-)orthogonal convergent settings, (2) *the width of the collisional belt* – the narrower the belt, the faster the resulting rock uplift.

Tectonic Extrusion Applied to the Betic-Rif

Position of the Crustal Root in the Sinking Slab Setting of the Betic-Rif

The sinking slab model outlined above for the Betic-Rif observes the basic requirements of the tectonic extrusion model: its collisional belt is located between two rigid lithospheric plates. Towards the northwest it borders upon the Iberian lithospheric block. Its southeastern edge follows the Almería-Murcia Iberian coast line (Fig. 1; ZECK, 1999, Fig. 4) and is contiguous to an offshore lithospheric fragment suggested to represent old, Mesozoic, and therefore cold and rigid Tethyan oceanic lithosphere. Preservation of this older lithospheric fragment, which is terminated to the northeast by the Alicante transform fault zone, is explained by the absence of rollback activity in this southern sector of the Betic-Ligurian subduction zone system (ZECK, 1999).

Obliquity Factor

After westward subduction terminated and the subducted lithospheric slab was detached, NW-SE directed Africa-Iberia compression became dominant as indicated by seismic analysis of recent shallow and intermediate earthquakes in the Iberia-Alborán lithosphere (GRIMISON and CHEN, 1986; MORALES et al., 1999). This is in agreement with the hinged, anti-clockwise movement of the African plate since ~50 Ma (DEWEY et al., 1989; MUELLER, 1989; SRIVASTAVA et al., 1990) under the influence of higher spreading rates in the S Atlantic compared to the N Atlantic. The regional compression over the Betic-Rif crustal root belt thus was (approximately) perpendicular to its length direction, which according to the tectonic extrusion model would give maximum velocity of upward rock transport.

Width Factor – Influence of Lithological Variation within the Extruding Mélange

For a 100-km wide belt with a plate convergence rate of 10 km/m.y. the calculations of THOMPSON *et al.* (1997b) suggest a rock uplift rate of c. 6 km/m.y. Total collisional belt width for the Betic-Rif belt is in the order of 120–150 km (Fig. 3a), difficult to reconstruct more precisely, as the primary collisional root complex, the internal zone, was tectonically reworked during the subsequent extensional stage. Total Africa-Iberia convergence within the period ~50–20 Ma is estimated at ~150 km (SRIVASTAVA *et al.*, 1990), an average of ~5 km/m.y. As rock uplift is slower for wider belts and lower plate convergence rates, the model predicts rock uplift for the Betic-Rif at rates much slower than 6 km/m.y. This is too slow to explain the large rock uplift rates required by *P-T-t* path reconstruction for the Betic-Rif (~20 km/m.y.; Fig. 4). Even though the location of the 19.5 Ma age point on the *PT* curve (Fig. 4) is a rough estimate, and might be positioned at lower pressure – providing for an earlier start of the fast tectonic rock uplift, and thus slower uplift rates – it seems clear that rock uplift rates in the Betic-Rif were considerably faster than predicted by the calculations by THOMPSON *et al.* (1997a,b).

However, these calculations were based on total belt width, which indeed is the only input readily available for a homogeneous particle flow model. When considering natural lithological complexes, total root belt width is hardly the essential factor, though. The Alpine nappe stratigraphy in the Betic-Rif is lithologically highly variable, harbouring pronounced mechanical heterogeneities. As a result, the deformation has been heterogeneous and tectonic transport was concentrated within incompetent rock bodies. This has reduced the effective width of ascending units very considerably (Figs. 3b,c), implying the feasibility of extremely rapid upward rock transports. Examples of such incompetent rock horizons are the Permo-Triassic, gypsum-bearing phyllites/schists which occur between massive carbonate rocks and coarsely crystalline Hercynian basement complexes (e.g., SANZ DE GALDEANO, 1997; ZECK and WHITEHOUSE, 1999). Another example of pertinent lithological heterogeneity is given by the leucogranitic rocks, often found associated with the Ronda peridotite bodies (e.g., PRIEM *et al.*, 1979), but also further east in the Betic Cordilleras (ZECK *et al.*, 1989). These very small intrusive bodies have massive structure and form discrete dykes and veins with sharp contacts indicating that their magmatic crystallization post-dates the tectono-metamorphic effects seen in their country rocks. Probably the leucogranitic melts yielding these rocks were formed by local partial melting of quartzo-feldspathic lithologies at deeper crustal levels and acted as lubricants to

assist rock ascent, notably of mantle fragments, in the extrusional mélange
(Figs. 3b,c); their crystallization taking place after tectonic extrusion. Their
estimated intrusion age based on data from ZECK *et al.* (1989) is 19.5 ± 1 Ma,
which is in good agreement with the suggested model.

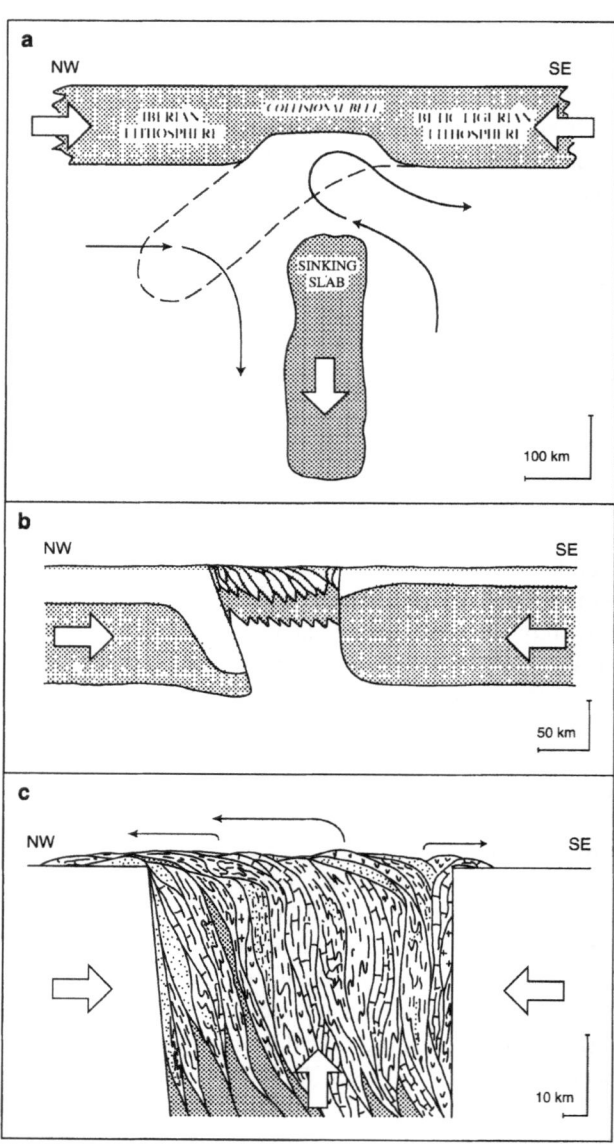

Thermal Factor

The tectonic extrusion model is dependent on a period of heating during and after collision to weaken the root complex (thermal softening). Figure 4 summarizes the thermobarometric constraints for a central part of the Betic Cordilleras and presents a *P-T*-time trajectory for the tectonic extrusion scenario. Collisional culmination was estimated at c.12–13 Kb and 450–500 °C (based on stability relations of carpholite, chloritoid, chlorite and kyanite; AZAÑON and GOFFÉ, 1997) and was followed by internal heating during decompression on the collisional rebound. It is suggested that the flow of HT-asthenosphere into the scar left by the detached slab (Fig. 3a) may have been an important heating factor which diverted the *PT*-loop towards higher temperatures (Fig. 4). Reconstructing the precise configuration of the *PT*-trajectory at this stage is complicated by the weakly defined timing of the collision (?50–30 Ma?) which makes it difficult to estimate how long the collisional prism was heated before slab break-off. Whatever the details of the early thermal history have been, temperatures at the start of the root elevation were estimated at ∼600 °C (Muscovite Rb-Sr and Hornblende $^{40}Ar/^{39}Ar$ thermochronometers; ZECK et al., 1992; MOINIÉ et al., 1994), well within the range required by the THOMPSON et al. (1997a,b) model. Rock complexes at the bottom of the root pile may have experienced *PT* conditions of up to ∼8-9 Kb / 650 °C (Fig. 4), whereas rock complexes that started their ascent from the top of the collisional pile have undergone considerably lower pressures and temperatures.

Extrusion and Extension

Geometric considerations imply that the vertically directed extrusional tectonic regime must be combined with a higher level lateral, extensional regime during which the extrusional mélange reaches its final emplacement in the form of a stack of

◄

Figure 3

Cartoon outlining the exhumation model for the Betic-Rif. The three frames illustrate the present state drawn at different scales; only the lightly shaded westward dipping slab in (a) refers to an earlier stage. Cross-section direction is perpendicular to the SW-NE trending ellipsoid in Figure 1. (a) In light shading, the subducted Tethyan oceanic lithospheric plate at the final stage of subduction. After subduction, the slab broke off and sank rapidly into the sub-lithospheric mantle (CHUNG and KANAMORI, 1976; BLANCO and SPAKMAN, 1993; ZECK, 1996, 1997). High-*T*, low density asthenosphere (white) flowed into the widening gap above the sinking slab. Evidence for the break-off scar, that is for the thinner lithosphere above the sinking slab, is given by multichannel seismic reflection and well data (WATTS et al., 1993) and seismicity compilations (SEEBER et al., 1996; cf. ZECK, 1997). (b) Close-up of (a). Asthenosphere in white, lithospheric mantle in dark shading, crustal section in lighter shading. Iberian crust dipping south-easterly under collisional section is after MORALES et al. (1999). (c) Close-up of collisional section in (b); horizontal and vertical scales are different. Thermal softening within the collisional prism trapped between the two convergent lithospheric plates caused mechanical failure along incompetent horizons producing a mélange of tectonic units – mantle fragments in dark shading. These units were deformed and in part recrystallized in the strongly anisotropic, vertical tectonic regime (HT, very rapid decompression) of the tectonic extrusion. Final allochthon emplacement took place in a horizontally directed extensional regime (*LP*, very rapid cooling) with emphasis on more brittle deformation concentrated in discrete detachment horizons.

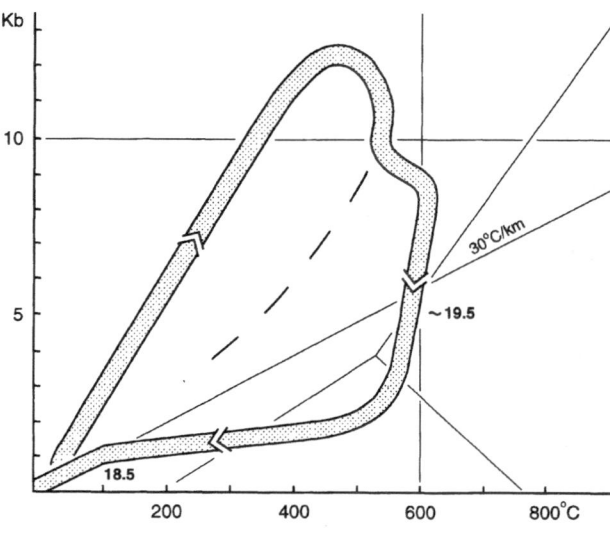

Figure 4

Alpine *P-T*-time path for the Torrox (Fig. 1) gneiss complex in the central part of the Betic-Rif, after ZECK and WHITEHOUSE (1999). Stability fields for Al-silicates (HOLDAWAY, 1971), 30 °C/km geothermal gradient, isothermal 600 °C line and isobaric 10 Kb line are given for calibration. Bold figures on the *PT*-curve indicate approximate timing in Ma. Alpine orogeny started in the Betic-Ligurian subduction system (Fig. 3a) resulting in collisional nappe piling and a *PT* path approximately along a 10 °C/km geothermal gradient which culminated at 12–13 Kb and 450–500 °C (AZAÑON and GOFFÉ, 1997). After subduction the collisional prism experienced internal heating during decompression on the collisional rebound. After slab break-off, high-*T* asthenospheric underplating (Fig. 3a) caused further heating which diverted the *PT*-loop towards the 600 °C isotherm. This resulted in thermal softening and tectonic extrusion (Fig. 3c) along a steep *PT*-trajectory (very rapid decompression), followed by extensional tectonics and very rapid cooling (approximately horizontal *PT*-trajectory). This rapid cooling took place in the time interval 19.5–18.5 Ma (Fig. 2) where after cooling rates were reduced to an average of c. 5 °C/m.y. (ANDRIESSEN and ZECK, 1996). Dashed curve outlines the hypothetical rebound trajectory of the collisional rock pile, which would have been followed if heating resulting in thermal softening and tectonic extrusion had not occurred.

extensional detachment sheets (Figs. 3b,c; cf. DEWEY, 1988). Although essentially continuous, the tectono-metamorphic character of the extrusional regime is rather different from that of the extensional regime.

During tectonic extrusion, rock uplift has been so fast that isotherms were uplifted and rocks did not cool appreciably until arriving at near-surface conditions. Tectonic extrusion thus was characterized by HT conditions and very rapid decompression. The extensional regime was typically *LP* and, because elevated temperatures were inherited from the extrusional stage, characterized by very rapid cooling (Fig. 4). The overlap in temperature between the two tectonic stages makes it difficult to reconstruct which part of the tectono-metamorphic features observed in presently exposed rock complexes was formed there. This is the more so because both stages are controlled by strongly anisotropic stress fields. It might seem feasible, though, to suggest that the major part of the pronounced schistose and linear crystalloblastic structures was formed during the vertical anisotropic regime of the tectonic extrusion as its *PT* conditions are more conducive of interpenetrative recrystallization structures. During the high-level, horizontal extensional stage emphasis may have been on more brittle deformation concentrated in discrete detachment horizons.

MORALES *et al.* (1999) carried out detailed geophysical investigations (seismic tomography, gravimetry, seismic wave analysis) in an area straddling the boundary between the lithospheric break-off scar and the Iberian block. It was shown that the Iberian crustal section extends to depths of c. 100 km, dipping southeastward under the asthenospheric mantle (Fig. 3b) with a very steep angle (the shallower angle in a cross section given by these authors is an artefact caused by the oblique orientation of their cross section). This study offers an elegant explanation for the asymmetric character of the extensional belt, with extensional allochthons concentrated at the northwestern side of the break-off scar. Its normal fault setting combines well with the considerable decrease in belt width implied by the tectonic extrusion model. The study also confirms the anomalous character of the mantle filling the break-off scar (integrated density 3.12 g/cm^3 compared to 3.32 g/cm^3 for the standard mantle of the Iberian lithosphere), cf. Figure 3.

Conclusions

The extremely rapid cooling ($\sim$500 °C/m.yr) which took place within the period $\sim$22–18 Ma in the Alpine Betic-Rif belt may be explained along the lines of the tectonic extrusion model suggested by THOMPSON *et al.* (1997a,b). The cooling is suggested to be related to the extrusional/extensional tectonics, which represent the final stage of the Alpine orogeny in the Betic-Rif belt. It consists of two essentially continuous stages: a) the extrusional stage characterized by a vertical tectonic regime, high temperature and very rapid decompression, and b) the extensional stage with a horizontal stress regime, low pressure and very rapid cooling. The model here developed for the Betic-Rif may offer important constraints to the reconstruction of the tectonic evolution of other convergent terranes characterized by high exhumation rates.

Acknowledgements

The study was supported by Carlsberg Foundation (grant 45/10-800), Danish Research Council (SNF grant 9800918), Gledden Foundation (University of Western Australia) and the Japanese Ministry of Education (Monbusho fellowship). ISEI directors Ikuo Kushiro and Masaru Kono, and TSRC director John Dodson are thanked for their hospitality and gracious support at the Misasa and Perth institutes, respectively; John Dewey (Davis) and Allan Thompson (Zürich) for inspiring reviews. This is Tectonics Special Research Center publication number 237.

REFERENCES

ANDRIESSEN, P. A. M. and ZECK, H. P. (1996), *Fission Track Constraints on Timing of Alpine Nappe Emplacement and Rates of Cooling and Exhumation, Torrox Area, Betic Cordilleras, S Spain*. Chemical Geology, Isotope Geoscience *131*, 199–206.

AZAÑÓN J. M. and GOFFÉ, B. (1997), *Ferro- and Magnesiocarpholite Assemblages as Record of High-P Low-T Metamorphism in the Central Alpujarrides, Betic Cordillera (SE Spain)*, European Journal of Mineralogy *9*, 1035–1051.

BLANCO, M. J. and SPAKMAN, W. (1993). *The P-wave Velocity Structure of the Mantle below the Iberian Peninsula: Evidence for Subducted Lithosphere below Southern Spain*, Tectonophysics *221*, 13–34.

CALVERT, A., SANDVOL, E., SEBER, D., BARAZANGI, M., ROECKER, S., MOURABIT, T., VIDAL, F., ALGUACIL, G. and JABOUR, N. (2000), *Geodynamic Evolution of the Lithosphere and Upper Mantle beneath the Alboran Region of the Western Mediterranean: Constraints from Travel-Time Tomography*. J. Geophys. Res., *105*, **B5**, 10871–10898.

CHUNG, W. -Y. and KANAMORI, H. (1976), *Source Processes and Tectonic Implications of the Spanish Deep-focus Earthquake of March 29, 1954*, Phy. Earth and Plan. Interiors *13*, 85–96.

DEWEY, J. F. (1988), *Extensional Collapse of Orogens*, Tectonics *7*, 1123–1139.

DEWEY, J. F., HELMAN, M. L., TURCO, E., HUTTON, D. H. W. and KNOTT, S. D. (1989), *Kinematics of the Western Mediterranean*, Geolog. Soc. London, Special Publication *45*, 265–284.

GARCÍA-CASCO, A. and TORRES-ROLDÁN, R. L. (1999), *Natural Metastable Reactions Involving Garnet, Staurolite and Cordierite: Implications for Petrogenetic Grids and the Extensional Collapse of the Betic-Rif Belt*, Contributions to Mineralogy and Petrology *136*, 131–153.

GOFFÉ, B., MICHARD, A., GARCÍA-DUEÑAS, V., GONZÁLEZ-LODEIRO, F., MONIÉ, P., CAMPOS, J., GALINDO-ZALDIVAR, J., JABALOY, A., MARTÍNEZ-MARTÍNEZ, J. M. and SIMANCAS, J. F. (1989), *First Evidence of HP, LT Metamorphism in the Alpujárride Nappes, Betic Cordilleras (SE Spain)*, European Journal of Mineralogy *1*, 139–142.

GRIMISON, N. L. and CHEN, W. P. (1986), *The Azores-Gibraltar Plate Boundary: Focal Mechanisms, Depths of Earthquakes, and their Tectonic Implications*, J. Geophys. Res. *91*, 2029–2047.

HOLDAWAY, M. H. (1971), *Stability of Andalusite and the Aluminum Silicate Phase Diagram*, Am. Sci. *271*, 97–13.

MONIÉ, P., TORRES-ROLDÁN, R. L. and GARCÍA-CASCO, A. (1994), *Cooling and Exhumation of the Western Betic Cordilleras, $^{40}Ar/^{39}Ar$ Thermochronological Constraints on a Collapsed Terrane*, Tectonophysics *238*, 353–379.

MORALES, J., SERRANO, I., JABALOY, A., GALINDO-ZALDIVAR, J., ZHAO, D., TORCAL, F., VIDAL, F. and GONZÁLEZ-LODEIRO, F. (1999), *Active Continental Subduction Beneath the Betic Cordillera and the Alboran Sea*, Geology *27*, 735–738.

MUELLER, St. (1989), *Deep-reaching Geodynamic Processes in the Alps*, Geolog. Soc. London, Special Publication *45*, 303–328.

NIJHUIS, M. (1964), *Plurifacial Alpine Metamorphism in the SE Sierra de los Filabres S of Lubrín, SE Spain*, Proefschrift, Amsterdam University, 1–151.

PLATT, J P. (1993), *Exhumation of High-pressure Rocks; A Review of Concepts and Processes*, Terra Nova *5*, 119–133.

PLATT, J. P. and VISSERS, R. L. M. (1989), *Extensional Collapse of Thickened Continental Lithosphere: A Working Hypothesis for the Alboran Sea and Gibraltar Arc*, Geology *17*, 540–543.

PRIEM, H. N. A., BOELRIJK, N. A. I. M., HEBEDA, E. H., OEN, I. S., VERDURMEN, E. A. T. and VERSCHURE, R. H. (1979), *Isotopic Dating of the Emplacement of the Ultramafic Masses in the Serrania de Ronda, Southern Spain*, Contributions to Mineralogy and Petrology *70*, 103–109.

SANZ DE GALDEANO, C. (1997), *La Zona Interna Betica-Refeña*, Universidad de Granada Colleciones Monografias Tierra Sur, 1–316.

SEBER, D., BARAZANGI, M., IBENBRAHIM, A. and DEMNATI, A. (1996), *Geophysical Evidence for Lithospheric Delamination beneath the Alboran Sea and Betic-Rif Mountains*, Nature *379*, 785–790.

SIBSON, R. H. (1977), *Fault Rocks and Fault Mechanisms*, J. Geolog. Soc. London *133*, 191–213.

SRIVASTAVA, S. P., ROEST, W. R., KOVACS, G., LÉVESQUE, S., VERHOEF, J. and MACNAB, R. (1990), *Motion of Iberia since the Late Jurassic: Results from Detailed Aeromagnetic Measurements in the Newfoundland Basin*, Tectonophysics *184*, 229–260.

THOMPSON, A. B., ENGLAND, P. C. (1984), *Pressure-temperature-time Paths of Regional Metamorphism II. Their Inference and Interpretation Using Mineral Assemblages in Metamorphic Rocks*, J. Petrology *25*, 929–955.

THOMPSON, A. B., SCHULMANN, K. and JEZEK, J. (1997a), *Thermal Evolution and Exhumation in Obliquely Convergent (Transpressive) Orogens*, Tectonophysics *280*, 171–184.

THOMPSON, A. B., SCHULMAN, K. and JEZEK, J. (1997b), *Extrusion Tectonics and Elevation of Lower Crustal Metamorphic Rocks in Convergent Orogens*, Geology *25*, 491–494.

TUBÍA, J. M. and IBARGUCHI, J. I. G. (1991), *Eclogites of the Ojén Nappe: A Record of Subduction in the Alpujárride Complex (Betic Cordilleras, Southern Spain)*, J. Geolog. Soc., London *148*, 801–804.

WATTS, A. B., PLATT, J. P. and BUHL, P. (1993), *Tectonic Evolution of the Alboran Sea Basin*, Basin Research *5*, 153–177.

ZECK, H. P. (1996), *Betic-Rif Orogeny: Subduction of Mesozoic Tethys under E-ward Drifting Iberia, Slab Detachment Shortly before 22 Ma and Subsequent Uplift and Extensional Tectonics*, Tectonophysics *254*, 1–16.

ZECK, H. P. (1997), *Mantle Peridotites Outlining the Gibraltar Arc — Centrifugal Extensional Allochthons Derived from the Earlier Alpine Westward Subducted Nappe Pile*, Tectonophysics *281*, 195–207.

ZECK, H. P. (1999), *Alpine Plate Kinematics in the Western Mediterranean — a W-ward Directed Subduction Regime Followed by Slab Roll-back and Slab Detachment*, Geolog. Soc. London, Special Publications *156*, 109–120.

ZECK, H. P., ALBAT, F., HANSEN, B. T., TORRES-ROLDÁN, R. L. and GARCÍA-CASCO, A. (1989), *Alpine Tourmaline-bearing Muscovite Leucogranites, Intrusion Age and Petrogenesis, Betic Cordilleras, SE Spain*, Neues Jahrbuch für Mineralogie, Monatshefte, 513–520.

ZECK, H. P., MONIÉ, P., VILLA, I. M. and HANSEN, B. T. (1992), *Very High Rates of Cooling and Uplift in the Alpine Belt of the Betic Cordilleras Southern Spain*, Geology *20*, 79–82.

ZECK, H. P. and WHITEHOUSE, M. J. (1999), *Hercynian, Pan-African, Proterozoic and Archean Ion-microprobe Zircon Ages for a Betic-Rif Core Complex, Alpine Belt, W Mediterranean — Consequences for its P-T-t Path*. Contributions to Mineralogy and Petrology *134*, 134–149.

(Received May 7, 2002, revised September 12, 2002, accepted September 25, 2002)

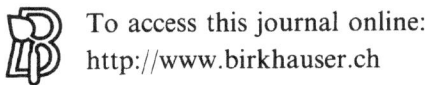

Pure appl. geophys. 161 (2004) 489–519
0033–4553/04/030489–31
DOI 10.1007/s00024-003-2460-7

Pure and Applied Geophysics

The Alpine Rif Belt (Morocco): A Case of Mountain Building in a Subduction-Subduction-Transform Fault Triple Junction

A. Chalouan[1] and A. Michard[2]

Abstract — The Rif belt forms with the Betic Cordilleras an asymmetric arcuate mountain belt (Gibraltar Arc) around the Alboran Sea, at the western tip of the Alpine orogen. The Gibraltar Arc consists of an exotic terrane (Alboran Terrane) thrust over the African and Iberian margins. The Alboran Terrane itself includes stacked nappes which originate from an easterly, Alboran-Kabylias-Peloritani-Calabria (Alkapeca) continental domain, and displays Variscan low-grade and high-grade schists (Ghomarides-Malaguides and Sebtides-Alpujarrides, respectively), shallow water Mesozoic sediments (mainly in the Dorsale Calcaire passive margin units), and infracontinental peridotite slices (Beni Bousera, Ronda). During the Late Cretaceous?-Eocene, the Alboran Terrane was likely located south of a SE-dipping Alpine-Betic subduction (cf. Nevado-Filabride HP-LT metamorphism of central-eastern Betics). An incipient collision against Iberia triggered back-thrust tectonics south of the deformed terrane during the Late Eocene-Oligocene, and the onset of the NW-dipping Apenninic-Maghrebian subduction. The early, HP-LT phase of the Sebtide-Alpujarride metamorphism could be hypothetically referred to the Alpine-Betic subduction, or alternatively to the Apenninic-Maghrebian subduction, depending on the interpretation of the geochronologic data set. Both subduction zones merged during the Early Miocene west of the Alboran Terrane and formed a triple junction with the Azores-Gibraltar transform fault. A westward roll back of the N-trending subduction segment was responsible for the Neogene rifting of the internal Alboran Terrane, and for its coeval, oblique docking onto the African and Iberian margins. Seismic evidence of active E-dipping subduction, and opposite paleomagnetic rotations in the Rif and Betic limbs of the Gibraltar Arc support this structurally-based scenario.

Key words: Alpine belt, tectonics, orogenic arc, terrane, collage, subduction rollback, Western Mediterranean, Morocco, Betic Cordilleras.

Introduction

The Moroccan Rif belt is the westernmost segment of the 2000-km-long Maghrebide belt which fringes North Africa and Sicily south of the Mediterranean Sea (Fig. 1). The Rif belt is also the southern limb of the Betic-Rif arcuate mountain belt which seems to seal Africa to Iberia and almost closes the Alboran Sea to the

[1] Département de Géologie, Faculté des Sciences, BP 1014, Rabat 10000, Maroc.
E-mail: chalouan@fsr.ac.ma
[2] Ecole Normale Supérieure, 24 rue Lhomond, 75231 Paris Cedex 05, France.
E-mail: michard@geologie.ens.fr

Figure 1
The West Mediterranean domain: Sketch map with location of the Betic-Rif orogen. Empty arrows: Present-day horizontal compression direction after BALLING and BANDA (1992) and BUFORN et al. (1995). Azores-Gibraltar fault zone in the Gulf of Cadiz after MALDONADO et al. (1999). Ed: Edough massif; GB: Gorringe Bank.

west. Whereas the Maghrebide belt continues into the Apennines through the Calabrian Arc, the Betic Cordilleras continuation towards Corsica and the Alps is conjectural, due to the Oligocene-Miocene opening of the Algerian-Provençal basin. Hence, the Rif belt belongs to a puzzling orogenic bend which seems to close the Mediterranean Alpine belt upon itself, and to conceal the eastern projection of the Iberia-Africa plate boundary, i.e., the Azores-Gibraltar fault zone. Contradictory models have been proposed for the Betic-Rif (Maghrebide) orogen, which was depicted either as, i) a collisional orogen unrelated to any subduction zone (KORNPROBST and VIELZEUF, 1984; PLATT and VISSERS, 1989; partially VISSERS et al., 1995; TURNER et al., 1999; MONTEL et al., 2000); or, ii) a subduction-related belt, with a NW-dipping subduction zone extending from Gibraltar to the Apennines (BUFORN et al., 1995; ZECK, 1996; LONERGAN and WHITE, 1997; CALVERT et al., 2000; CABY et al., 2001); or else, iii) an orogen involving two subduction zones, i.e.,, the SE-dipping Alpine-Betic subduction and the NW-dipping Apenninic-Maghrebian subduction, with a microcontinent in between (Andrieux et al., 1971; DURAND-DELGA and FONTBOTÉ, 1980; REHAULT et al., 1984; TORRES-ROLDAN et al., 1986; GUERRERA et al., 1993; DOGLIONI et al., 1998, 1999; FRIZON de LAMOTTE et al., 2000; CHALOUAN et al., 2001; MICHARD et al., 2002). Likewise, MALDONADO et al. (1999)

retain the occurrence of two opposite subductions beneath the Alboran domain, but assume that the southern subduction (*sensu lato*) mainly concerns the delaminated, lower African lithosphere. Indeed, the models ii) and iii) clearly consider the correlations between the Alboran-Kabylias-Peloritani-Calabria metamorphic units (Alkapeca; BOUILLIN *et al.*, 1986), but only the two-subduction models completely take into account the critical correlations with the Western Alps (MICHARD *et al.*, 1991, 2002; DOGLIONI *et al.*, 1998, 1999).

In this study, we first summarize the structure and evolution of the Rif domain, with emphasis on deep structure, metamorphism and geochronology. Then we infer a tectonic history involving two successive subduction zones, i.e., the Alpine-Betic and Apenninic-Maghrebian subductions zones. However, contrasting with our previous publications (CHALOUAN *et al.*, 2001; MICHARD *et al.*, 2002) we infer that the Rif-Betic metamorphism can be described through two alternative scenarios instead of a single one, as long as more convincing dates are not obtained for the early, high-pressure, low-temperature (HP-LT) parts of this metamorphism.

Geology and Geophysics

Shallow Structure and Stratigraphy

The Rif belt displays three major stratigraphic/structural domains (Fig. 2). The *External Zones* (equivalent to the Algerian Tell units) consists of a fold-thrust belt detached from the attenuated African crust along Upper Triassic evaporitic redbeds (Fig. 3A). From SW to NE and bottom to top, three groups of stacked units are shown, i.e., the Prerif, Mesorif and Intrarif units (SUTER, 1980). South of Al-Hoceima (Alhucemas), the Upper Jurassic-Lower Cretaceous Intrarif metasediments overlay the serpentinized lherzolites of the Beni Malek massif, interpreted as an obducted serpentinite ridge from the tip of the African paleomargin (MICHARD *et al.*, 1992; ELAZZAB *et al.*, 1997). Pre-orogenic sediments characterize the External Zones up to the Lower-Middle Eocene, whereas varied turbiditic deposits develop there from the Middle-Late Eocene onward (CHALOUAN *et al.*, 2001, their Fig. 4). Note that the External Rif sediments are about three times thicker than the corresponding Sub-Betic sediments, which belong to another continental margin.

The *Maghrebian Flysch Nappes* overthrust the External Zones, except some inliers that overlay the Internal Zones through Early Miocene olistostromes (BOUILLIN *et al.*, 1973; DURAND-DELGA, 1980). It is currently accepted that these unrooted nappes originate from an oceanic/transitional crust-floored "Maghrebian Trough" extending between Africa and the Internal Zone domain during the Jurassic-Early Miocene times (WILDI, 1983; HOYEZ, 1989; DURAND-DELGA *et al.*, 2000). The Maghrebian Flyschs are known from Sicily up to the Gibraltar area, but hardly extend in Spain east of Granada.

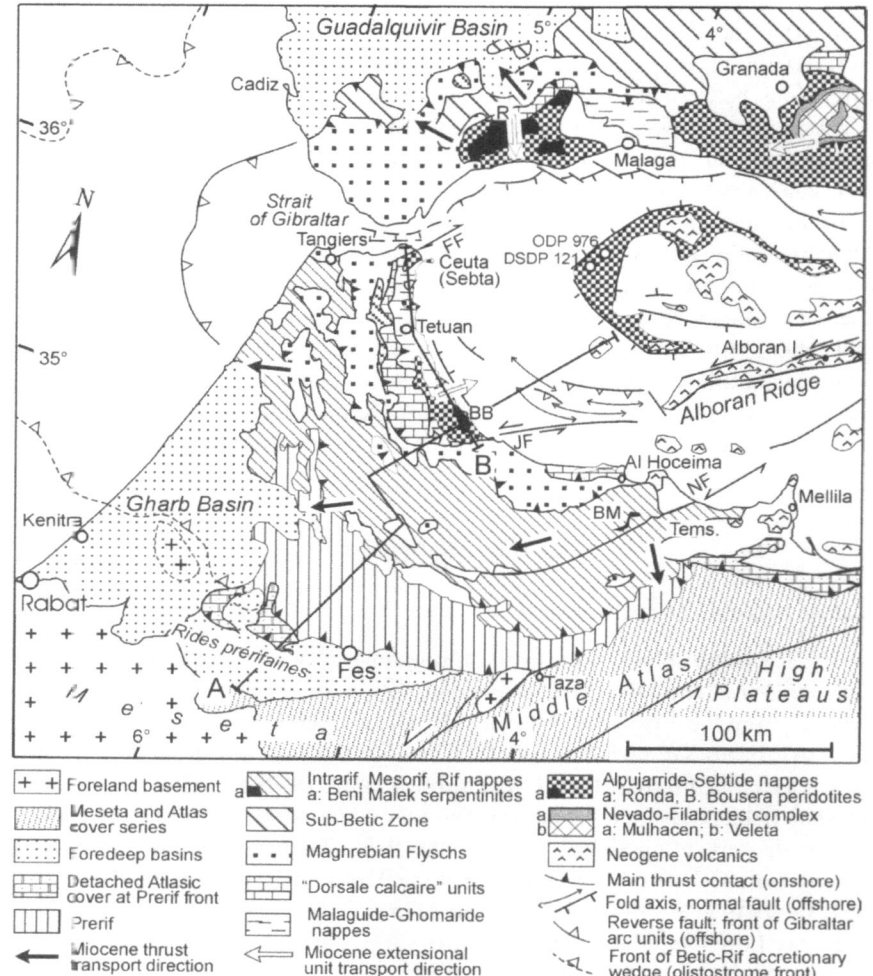

Figure 2

Structural map of the Rif belt, west-central Betic Cordilleras, Alboran basin and Gulf of Cadiz after
CHALOUAN *et al.* (2001), MARTÍNEZ-MARTÍNEZ and AZAÑON (1997), COMAS *et al* (1999), and MALDONADO
et al. (1999), respectively. See Figure 1 for location. Tortonian synclines (External Rif) and Plio-
Quaternary basins in white. Shallow granitic basement in southwest Gharb basin (dashed area with
crosses) after RIMI *et al.* (1998). Kinematic directions after FRIZON DE LAMOTTE *et al.* (1991), and
MARTÍNEZ-MARTÍNEZ and AZAÑON (1997). A, B: Traces of cross-sections Figure 3; BB: Beni Bousera;
BM: Ben Malek; FF: Jebel Fahies fault; FF: Jebha fault; NF: Nekor fault; R: Ronda; Tems.: Temsamane.

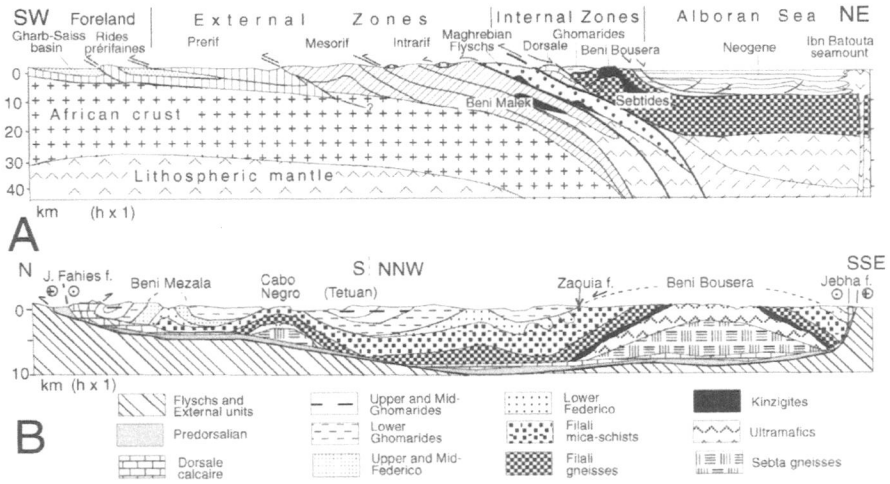

Figure 3

Structural profiles across the Rif orogen (location: Fig. 2). A: Transverse cross-section of the Rif belt and SW Alboran basin; thickness of external sedimentary prism and underlying crust after FAVRE (1995); Alboran basin after CHALOUAN *et al.* (1997) and TORNÉ *et al.* (2000); deep geometry of the African–Alboran transition inferred from geology and intermediate-depth seismicity (SEBER *et al.*, 1996). B: Longitudinal crosssection of the Internal Zones. The occurrence of crystalline rocks benath the Beni Bousera peridotites is hypothetic, and inferred from that of the Sebta or Monte Hacho gneisses at Ceuta (Sebta) and of the Ojen crystalline rocks at the bottom of the Ronda ultramafics in Spain.

The *Internal Zones*, which the Rif belt shares with the Betic Cordilleras, consist of three nappe complexes of continental origin. These nappes include, from SW to NE and structurally from top to bottom (Figs. 2, 3), i) the stacked carbonate slabs of the Dorsale Calcaire, with Upper Triassic-Liassic platform to passive margin deposits (pelagic deposits as early as the Sinemurian in the external part of the platform), followed upward by Jurassic-Late Cretaceous pelagic layers, and Priabonian-Aquitanian clastics and olistostromes (WILDI *et al.*, 1977; BENYAICH *et al.*, 1986; EL HATIMI *et al.*, 1991; EL KADIRI *et al.*, 1992); ii) four stacked Ghomaride = Malaguide nappes, each one consisting of low-grade Paleozoic metasediments unconformably overlain by Triassic redbeds, Liassic limestones, Paleocene-Eocene calcarenites (CHALOUAN and MICHARD, 1990; MAATÉ, 1996), and eventually by post-nappe late Oligocene-Miocene clastic deposits (FEINBERG *et al.*, 1990) ; and iii) the Sebtide = Alpujarride nappes, affected by a low- to high-grade Alpine metamor-phism (see below), and deformed by the late-metamorphic Beni Mezala and Beni Bousera antiforms. The Sebtide nappes include from top to bottom the Federico units (Permian-Triassic phyllites and quartzites, and Middle Triassic meta-carbon-ates), the Filali unit (high-grade schists and gneisses), the Beni Bousera unit (kinzigite/granulite slivers overlying 2-km-thick ultramafics), and lastly the Sebta

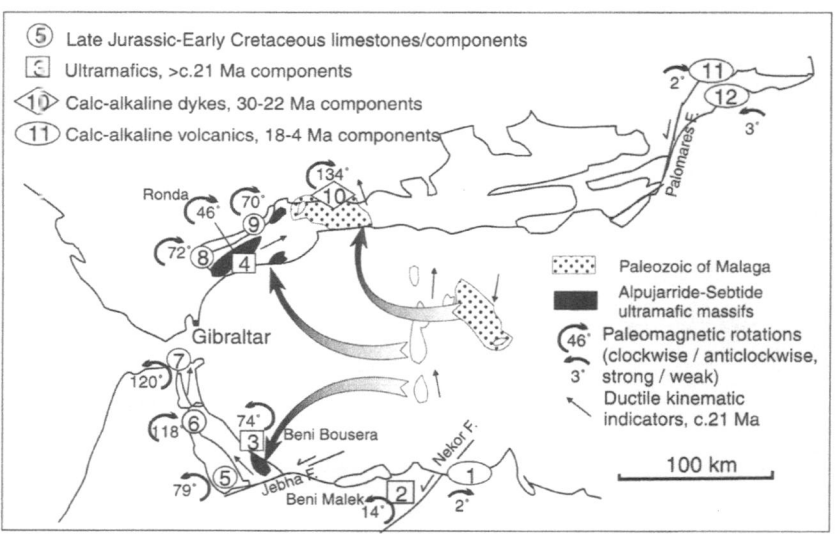

Figure 4
Paleomagnetic rotations (mean site values) in the Betic-Rif arcuate mountain belt, after FEINBERG *et al.*
(1996), modified. The inferred approximate N-S strike of the pre-Early Miocene orogen is schematically
shown in the mid-Alboran area. This also restores the main ductile shear directions of the west Gibraltar
Arc in a rectilinear to smoothly arcuate N-S trend. References: 1: NAJID *et al.* (1981); 2: ELAZZAB and
FEINBERG (1994); 3: SADDIQI *et al.* (1995); 4: FEINBERG *et al.* (1996); 5–9: PLATZMAN (1992), PLATZMAN
et al. (1993), ALLERTON *et al.* (1994); 10: PLATZMAN *et al.* (2000), CALVO *et al.* (2001); 11, 12: CALVO *et al.*
(1997).

(Ceuta) orthogneiss unit (KORNPROBST, 1974; MICHARD *et al.*, 1997). The Sebtides
units strictly compare with the west Alpujarride nappes (Benarraba imbrications,
Jubrique schists, Ronda kinzigites and peridotites, and Ojen gneiss, respectively;
BALANYÁ *et al.*, 1997). Similar rocks have been cored from the Alboran Sea basement
(COMAS *et al.*, 1999).

The Rif Internal units, their Betic equivalents, and the intervening Alboran
basement constitute the *Alboran Terrane*, in the sense of CHALOUAN *et al.* (2001) and
MICHARD *et al.* (2002), thrust onto the African margin units (Rif External Zones), and
roughly symmetrically onto the Sub-Betic Iberian units of western Betics. In the central
and eastern Betics, however, large antiformal windows allow another complex, i.e., the
Nevado-Filabride nappes, to crop out beneath the Alpujarrides. The upper Nevado-
Filabride units (Mulhacén Complex) include both continental rocks (Paleozoic
orthogneisses and schists, Permian and Triassic metasediments; MARTÍNEZ-MARTÍNEZ
and AZAÑÓN, 1997) and serpentinites, gabbros and basalts of arguable oceanic (Puga
et al., 1999, 2002a,b) or transitional crust origin (GOMEZ-PUGNAIRE *et al.*, 2000) partly
dated from the Late Jurassic (PUGA *et al.*, 2002a,b). The Mulhacén Complex was
affected by an early Alpine HP-LT metamorphism. The underlying HT-LP Veleta

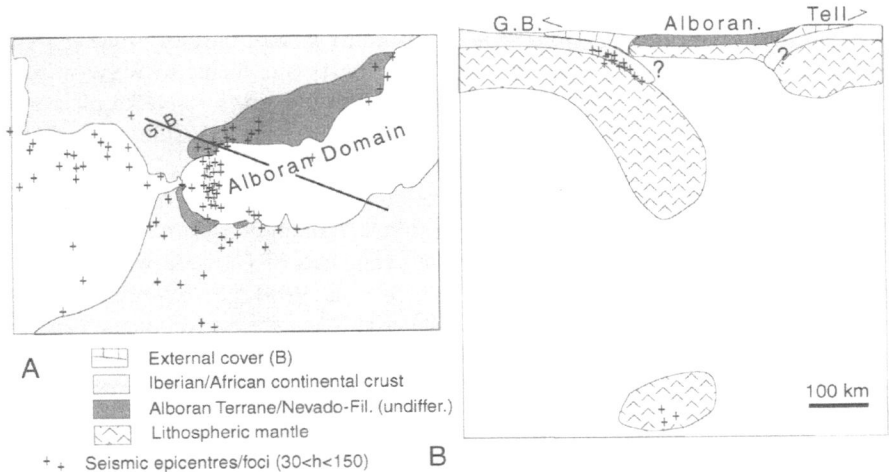

Figure 5

Intermediate-depth seismicity (A, B) and interpretation of seismic tomography data (B) in the Gibraltar Arc region. (A) after BUFORN *et al.* (1995, 1997), (B) after CALVERT *et al.* (2000), modified for the interpretation of the Tell margin.

Complex only consists of polymetamorphic Paleozoic schists probably originating from distal parts of the Iberian margin (ANDRIEUX *et al.*, 1989).

Kinematics

Active tectonics is characterized in the Alboran area by NNW-trending horizontal compression direction (Fig. 1), as recorded by shallow seismicity and fault plane solutions (BALLING and BANDA, 1992; BUFORN *et al.*, 1995; MEGHRAOUI *et al.*, 1996). The effect of the Iberia-Africa convergence is also recorded in Miocene-Pleistocene major strike-slip faults onland and in the Alboran basin (LEBLANC and OLIVIER, 1984; CAMPOS *et al.*, 1992; MALDONADO *et al.*, 1992; MAUFFRET *et al.*, 1992) such as the Nekor (Fig. 2) and Palomares faults (Fig. 4). The SW-trending sinistral fault system seems to continue southwestward in the Middle-Atlas shear zone (JACOBSHAGEN, 1992; BIERMANN, 1995; ELAZZAB and EL WARTITI, 1998; BERNINI *et al.*, 2000; ANDEWEG and CLOETINGH, 2001). The structure of the Rif foreland was deeply marked by the compressional reactivation of ancient NE- and E-trending crustal faults (PIQUÉ *et al.*, 1987, 1998; ZOUHRI *et al.*, 2001). However, tectonic displacements in the Betic-Rif belt have been considerably more complicated than expected if only caused by the Iberia-Africa convergence.

Paleogene to Early Miocene displacements are seldom recorded in the metamorphic units of the Gibraltar Arc (see Metamorphic evolution section). In contrast, the Miocene transport directions were clearly established on structural criteria (Fig. 2).

The nappe transport direction changes from SSE-ward, east of the NE-trending Nekor-Temsamane sinistral shear zone, to SW-ward, west of the latter zone (FRIZON DE LAMOTTE, 1987; FRIZON de LAMOTTE *et al.*, 1991), and finally to WSW-ward in the westernmost External Rif (Morley, 1987). By contrast, top-to-NE displacement of extensional units characterizes the Internal Zones (e.g., Zaouia fault, Fig. 6 hereafter), as well as the southwest Alboran basin (Chalouan *et al.*, 1997). In the Betic External Zones, WNW to NW thrust transport directions prevail from the Gibraltar area to the easternmost part of the belt (FRIZON de LAMOTTE *et al.*, 1991), except along the External-Internal boundary zone east of Granada where conspicuous backthrust structures occur (ALLERTON *et al.*, 1994). In central Betics, the Alpujarride-Malaguide nappe stack was diplaced SW-ward on top of the Nevado-Filabrides in ductile to brittle extensional conditions (MARTÍNEZ-MARTÍNEZ and AZAÑÓN, 1997), whereas in western Betics, the thrust contact at the bottom of the Alpujarride-Malaguide was inverted as a SE-dipping low-angle extensional fault (GARCÍA-DUEÑAS *et al.*, 1992; CRESPO-BLANC and CAMPOS, 2001).

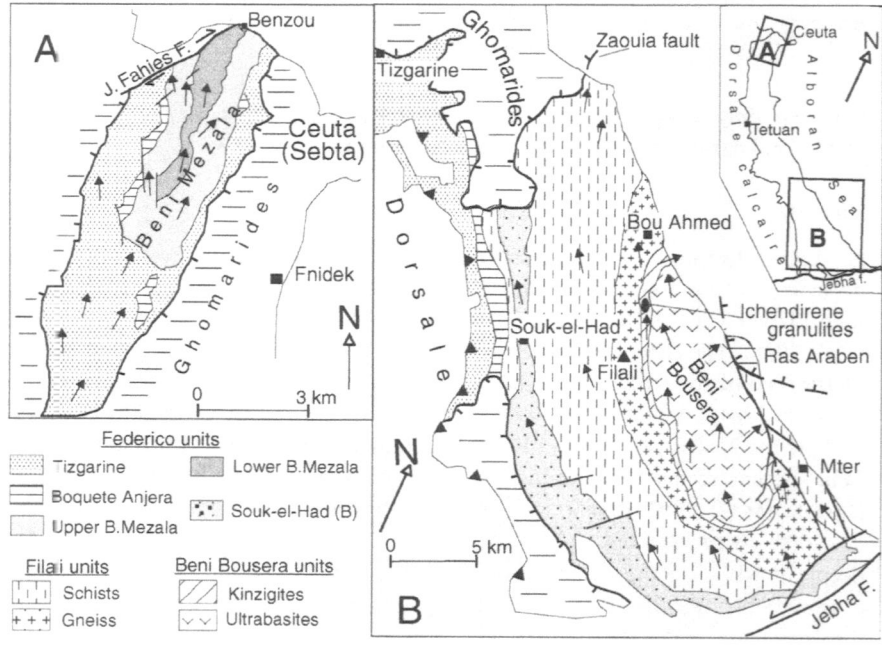

Figure 6

Metamorphic structure of the Sebtide nappes, after BOUYBAOUENE (1993), SADDIQI (1995) and MICHARD *et al.* (1997). Arrows: HT-LP stretching lineations. The Federico units mainly consist of Permian metapelites, Triassic quartzites and carbonates.

Paleomagnetic studies also yielded kinematic information (Fig. 4). Rotations of the Late Jurassic and Late Cretaceous limestones are anticlockwise (ACW) in the Rif Dorsale units, except in the Tetouan area (PLATZMAN et al., 1993), and clockwise in the Betic Dorsale and Penibetic (Internal Sub-Betic) Zones (PLATZMAN, 1992; ALLERTON et al., 1994; VILLALAÍN et al., 1994). Similar opposite rotations are observed in the Beni Bousera (SADDIQI et al., 1995) and Ronda peridotites and associated granite dykes, allowing FEINBERG et al. (1996) to date the rotations from the c. 20 Ma-old cooling of these rocks. In the Beni Malek ultramafics and overlying metasediments of eastern Intrarif, ELAZZAB and FEINBERG (1994) observed a high-temperature component with a 14° ACW rotation, and an unrotated lower-T component. These HT and LT components can be interpreted as related to the superimposed regional metamorphic events, i.e., a greenschist-facies and very-low grade metamorphic events, dated at 28 Ma and 7–8 Ma, respectively (MONIÉ et al., 1984). If so, the Beni Malek data suggest a moderate ACW rotation of eastern Intrarif between 28–7 Ma. Indeed, rotations were over by the time (c.18–4 Ma) when the Trans-Alboran volcanism (HERNANDEZ et al., 1987) occurred, at least in the central and eastern parts of the orogen (NAJID et al., 1981; CALVO et al., 1997). The Early Miocene rotation pattern, complemented by the clockwise rotations observed in the Malaga extensional allochthon (PLATZMAN et al., 2000; CALVO et al., 2001), suggests that prior to c. 21 Ma, the future Gibraltar Arc was a roughly N-trending orogenic segment (PLATZMAN, 1992). This major conclusion is strongly supported by the change in orientation of the ductile kinematic indicators on top of the Beni Bousera and Ronda ultramafics from top-to-the NW to top-to-the NE, respectively (Fig. 4, and Fig. 6 hereafter).

Deep Structure

Gravimetric modeling suggests that, from the Rif foreland (Moroccan Meseta and Atlas domain) to the Internal Zones (Fig. 3), the Moho depth weakly decreases from c. 30 km to c. 27 km (flexural rebond), then increases northward up to c. 35 km (FAVRE, 1995). Seismic refraction data and gravity modeling show that the Alboran Sea is underlain by a thinned continental crust, about 15–20 km thick in the central basin (HATZFELD et al., 1978; BANDA et al., 1993; WATTS et al., 1993; GALINDO-ZALDIVAR et al., 1998). The crust thickens again northward beneath the Betics, up to c. 38 km under the Sierra Nevada, before stabilizing around 35 km in the Betic foreland (BANDA et al., 1993; GALINDO-ZALDIVAR et al., 1997). The Moho deepens to 30–32 km beneath the Gibraltar Strait (TORNÉ et al., 2000). The lithospheric mantle thickness also shows strong lateral variations in the Alboran domain. Tomography studies imaged a pronounced low-velocity anomaly beneath the Alboran Sea (BLANCO and SPAKMAN, 1993; CALVERT et al., 2000). According to TORNÉ et al. (2000), 3-D gravity modeling suggests that the base of the lithosphere shallows from 140–120 km depth beneath the mountainous arc and the Gibraltar Strait, to c. 45 km depth in the Alboran Sea, which is consistent with heat flow data (POLYACK et al., 1996; FERNÀNDEZ et al., 1998). A travel-time tomography profile from western Betics to eastern Tell (Fig. 5) imaged a high-velocity body dipping SE-ward from lithospheric depths to depths of c. 350 km beneath the Alboran low-velocity region, and a detached body at c. 600 km near a cluster of deep earthquakes (CALVERT et al., 2000). The latter cluster was repeatedly interpreted as related to a relic of subducted slab (UDÍAS et al., 1976; BLANCO and SPAKMAN, 1993; ZECK, 1996; MALDONADO et al., 1999). The N-S line of intermediate-depth earthquakes extending from the west Betics to the Rif coast (BUFORN et al., 1995, 1997) coincides with the W-E transition from high- to low-velocities. Gravity modeling and seismic attenuation allowed MORALES et al. (1999) to display the continental nature of the seismic low-velocity zone, therefore interpreted as the result of active subduction of the Iberian crust. Likewise, ANDEWEG and CLOETINGH (2001) interpret the low heat flow values (40 mWm^{-2}) east of the Gibraltar Strait, and the strong heat flow gradient between the latter area and the western Alboran basin (110 mWm^{-2}), as the dynamic effects of active underthrusting of the Iberian crust under the Alboran crust.

Petrology and Geochronology

Metamorphic Evolution

The African margin units only display low-grade greenschist facies recrystallization which affected exclusively the eastern and most internal parts of the External Zones (FRIZON DE LAMOTTE, 1985). This event was dated at 28 Ma (MONIÉ et al., 1984), consistent with the Upper Oligocene unconformity observed in the area (FAVRE, 1992). A younger (7–8 Ma), very low-grade event also occurred in the eastern Rif, prior to the Tortonian post-nappe sedimentation. In contrast, the Alboran Terrane units display widespread, and often high-grade Alpine metamorphic imprints. These imprints are virtually lacking in the Dorsale Calcaire and Ghomaride (Malaguide) nappes, except at the very bottom of the latter nappes, where greenschist-facies recrystallizations, dated at c. 25 Ma (MONTIGNY et al., 2004), overprinted the Hercynian greenschist assemblages (CHALOUAN and MICHARD, 1990).

In fact, Alpine metamorphism essentially characterizes the Sebtides (Alpujarrides) units. In the Permian-Triassic metapelites of the Beni-Mezala antiform (Fig. 6A), the peak P-T conditions increase downward from greenschist (Tizgarine unit) to HP-greenschist (Boquete Anjera), to blueschist (upper Beni Mezala), and eventually to eclogite-facies conditions in the lower Beni Mezala unit (BOUYBAOU-ENE, 1993; GOFFÉ et al., 1996; MICHARD et al., 1997; VIDAL et al., 1999). In the latter unit, the retrograde P-T path (Fig. 7A) reveals a roughly isothermal unloading from 1.5–1.8 GPa down to c. 0.6 GPa, followed by unloading under decreasing T (AGARD et al., 1999). The southern Federico units of the Beni Bousera antiform (Fig. 6B) do

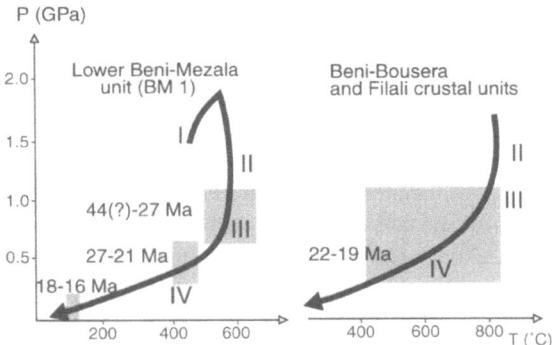

Figure 7

P-T path evolution of the main Sebtide units, estimated by the calculation of phase equilibria in the observed mineralogical assemblages, after BOUYBAOUENE (1993), GOFFÉ *et al.* (1996), VIDAL *et al.* (1999), AGARD *et al.* (1999), with tentative dating of the tectonic-metamorphic phases after MONTIGNY *et al.* (2004). I: burial stage; II: isothermal unloading; III: HT-LP contraction; IV: syn-rift exhumation tectonics.

not show HP-LT relic assemblages. The lowermost, Souk-el-Had unit seemingly suffered a metamorphic climax under lower P/T gradient than its northern equivalent (BM1 unit), with peak conditions close to 600 °C, 1.2 GPa, and late cordierite and andalusite growth under lower P. In the underlying Filali schists and gneisses, T calculations give conditions ranging from 530 °C at the top to 780 °C at the bottom, at a maximum pressure of c. 0.7 GPa before the late stage of andalusite growth (KORNPROBST, 1974; EL MAZ and GUIRAUD, 2001). In the Beni Bousera unit, the kinzigite slivers (garnet-sillimanite ± kyanite-graphite metapelites) on top of the ultramafics show peak P-T conditions in the range 800–850 °C, 0.9–1.3 GPa, locally 760–820 °C, > 1.6 GPa (Ichendirene HP-granulite; BOUYBAOUENE *et al.*, 1998).

In the Beni Mezala units, white micaceous assemblages from retrogressed HP-LT Permian metapelites yielded K/Ar ages comprised between 19.4 ± 1.2 Ma and 27.4 ± 0.6 Ma, and a green amphibole yielded an age of 44.7 ± 1.6 Ma (MONTIGNY *et al.*, 2004). $^{39}Ar/^{40}Ar$ measurement on the micas with the older age yielded high-T degassing steps between 38–40 Ma. The latter date could either record the occurrence of excess Ar, or represent a minimum age for the blueschist/eclogite-facies metamorphism which preceded the isothermal unloading stage. In contrast, the onset of cooling seems confidently dated at c. 27 Ma (Fig. 7A). In the Beni Bousera antiform, biotites from both the Filali schists and kinzigites reveal a cluster of K/Ar ages around 22 ± 2 Ma (MONTIGNY *et al.*, 2004), in agreement with the multimethod age determinations previously performed on various metamorphics, pyroxenites and anatectic leucogranites from the Beni-Bousera antiform and Spanish equivalents (e.g., REISBERG *et al.*, 1989; MONIÉ *et al.*, 1991, 1994; ZECK *et al.*, 1992; KUMAR *et al.*, 1996; BLICHERT-TOFT *et al.*, 1999; PLATT and WHITEHOUSE, 1999; SÁNCHEZ-RODRIGUEZ and GEBAUER, 2000). Although the presence of Hercynian high-grade crustal rocks in the Sebtide-Alpujarride nappes can be contended (MICHARD et al., 1997; BOUYBAO-UENE et al., 1998; SÁNCHEZ-RODRIGUEZ and GEBAUER, 2000; MONTEL et al., 2000), the multimethod age cluster around 22 ± 2 Ma can be viewed as reflecting a rapid cooling of the deep Sebtide-Alpujarride units following an Alpine HT-LP climax of metamorphism close to 750–800 °C, 1.0 GPa (Fig. 7B).

In both the pre-Permian and Mesozoic rocks, the main foliation and the NNE- to NW-trending stretching lineation (Fig. 6) developed contemporaneously with the retromorphic minerals, and are associated with dominant top-to-the-N kinematic indicators. These structures are overprinted by crenulation folds trending parallel to the local antiform axis, i.e., roughly parallel to the stretching lineation (SADDIQI et al., 1988; ZAGHLOUL, 1994; MICHARD et al., 1997). These structurally recorded tectonic-metamorphic phases are labelled II and III in Figure 7, after BALANYÁ et al. (1997). Phase I corresponds to the subduction event responsible for the HP-LT conditions observed in the Beni Mezala and in various Alpujarride units (GOFFÉ et al., 1989; TUBÍA and GIL IBARGUCHI, 1991; BOUYBAOUENE, 1993; AZAÑÓN, 1994). Phase IV refers to the extensional tectonics which initiated under HT-LP conditions (emplacement of anatectic dykes), and then operated under decreasing T in brittle conditions (e.g., Zaouia low-angle normal fault, Fig. 6B).

Late-orogenic Magmatism

Late-orogenic, calc-alkaline magmatism is widespread in the entire Alboran region ("Trans-Alboran province", partly shown in Fig. 2; HERNANDEZ et al., 1987). The earliest occurrences are concentrated in the Malaga area, where basaltic-andesitic dyke swarm with arc-tholeite affinities yielded K/Ar ages of 23–20 Ma (TORRES-ROLDAN et al., 1986), and $^{40}Ar/^{39}Ar$ ages of 30–18 Ma (TURNER et al., 1999). The Alboran Island andesites yielded K/Ar ages of 18–7 Ma (APARICIO et al., 1991), but most of the Trans-Alboran calc-alkaline volcanism (andesites, shoshonitic trachy-andesites, rhyolites of eastern Betics, Alboran basin and eastern Rif) took place during the Middle-Late Miocene (HERNANDEZ et al., 1987; MAURY et al., 2000). According to TURNER et al. (1999), these volcanics could derive from decompression melting of asthenosphere, although alternative interpretations, involving interaction between subducting lithosphere, asthenospheric mantle and upper plate continental crust are advocated by LONERGAN and WHITE (1997), CARMINATI et al. (1998), ZECK and WHITEHOUSE (1999), and MAURY et al. (2000).

Volcanic outpours still occurred in the eastern Rif and neighbouring Atlas foreland from the end of Messinian to Early Quaternary. This post-orogenic volcanism mainly consists of undersaturated alkalic basalts. According to PIQUÉ et al. (1998), transition from calc-alkaline to alkaline magmatism might be related to separation of the "Central Maghreb indenter" from the Moroccan Meseta through NE-trending sinistral shear zones (e.g., Middle Atlas shear zone; ELAZZAB and EL WARTITI, 1998; BERNINI et al., 2000).

Mountain-building Process

Pre-orogenic Setting

During the Late Jurassic-Early Cretaceous, i.e., during the climax of the Neo-Tethys opening (SAVOSTIN *et al.*, 1986), the Alboran Terrane is the western part of an Alkapeka block (Fig. 8A), bounded to be located hundreds of kilometres east of its present-day position as, i) the Triassic-Liassic sequences of the Ghomaride-Malaguide and Dorsale nappes strongly contrast with that of Morocco and southern

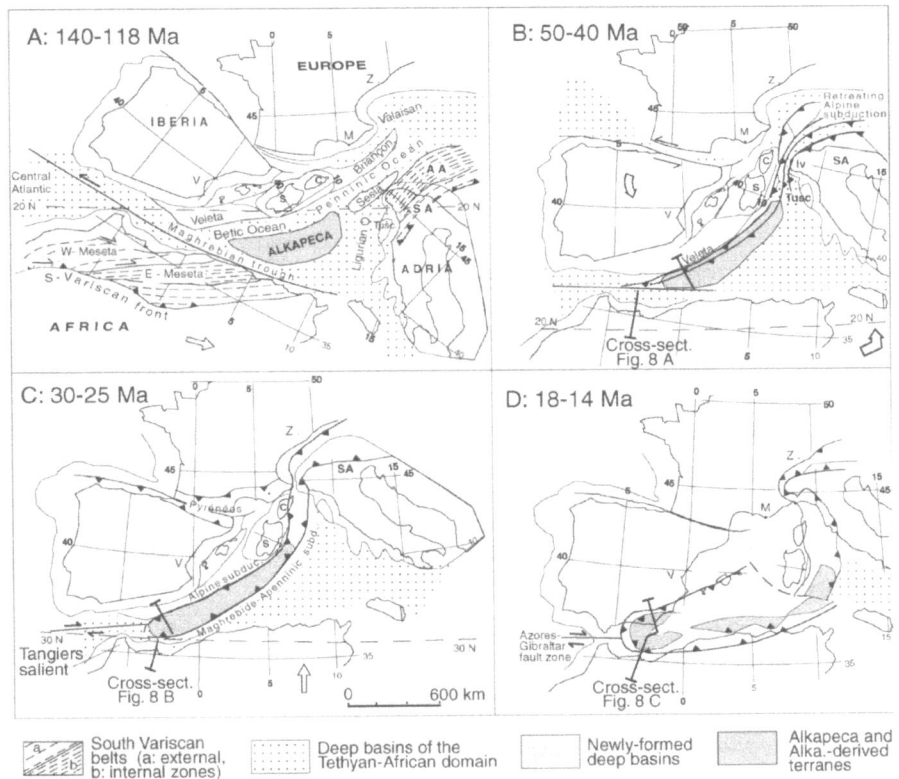

Figure 8

The Betic-Rif mountain building : broad paleogeographic and tectonic evolution in map view. Positions and displacements (empty arrows) of Africa, Iberia and Adria versus Europe according to SAVOSTIN *et al.* (1986), ROEST and SRIVASTAMA (1991), DERCOURT *et al.* (1993). Western Alps after FROITZHEIM *et al.* (1996) and STAMPFLI *et al.* (1998). Miocene setting (D) after FACCENNA *et al.* (1997) and BOUILLIN *et al.* (1998). AA: Austroalpine; C: Corsica; Iv.: Ivrea zone; M: Marseilles; SA: South-Alpine; S-M: Sesia-Margna; Tusc: Tuscany; V: Valencia; Z: Zürich.

Iberia, and compare with that of southern Apulia, Tuscany and Austroalpine areas (BOUILLIN *et al.*, 1986; MICHARD *et al.*, 1991), and ii) the Alkapeca Variscan evolution is similar to that of Eastern Morocco and Tuscany, rather than of southern Sardinia (CHALOUAN and MICHARD, 1990; MICHARD *et al.*, 2002). In line with ANDRIEUX *et al.* (1989) and GUERRERA *et al.* (1993), we contend that the Alboran/ Alkapeca block was surrounded by two Neo-Tethyan arms, namely the Maghrebian-Ligurian ocean to the SE, and the Betic-Alpine one to the NW. Then, Alkapeca should be a southern equivalent of the Sesia block, drifted away from the Apulian lower plate of the Tethyan rift (FROITZHEIM *et al.*, 1996; MANATSCHAL and BERNOULLI, 1999). Such an origin is consistent with the occurrence of Variscan granulites and peridotites in the Sesia/Ivrea zones and Alkapeca blocks (Beni Bousera, Ronda and their Kabylian equivalents; REUBER *et al.*, 1982; BOUYBAOUENE *et al.*, 1998), while they are lacking in the Briançonnais and related Alpine units (MICHARD *et al.*, 1993). U-Pb high-resolution datings of zircon at 183–131 Ma in the Ronda ultramafics point to their early exhumation being related to the Pangea break-up (SÁNCHEZ RODRIGUEZ and GEBAUER, 2000), coeval with the Ivrea ultramafic uplift (HANDY *et al.*, 1999).

At that time, the future Ghomaride-Malaguide and Dorsale nappes can be restored at the center and southern margin of the Alboran/Alkapeca block, respectively—a setting which is preserved up to the Eocene (Fig. 9A, left). This is supported by stratigraphic evidence of continuity from the Maghrebian Flysch Trough to the Dorsale Calcaire passive margin (cf. occurrence of "Predorsalian" transitional units; DURAND-DELGA and OLIVIER, 1988), to the Ghomaride-Mala-guide platform (EL KADIRI *et al.*, 1992; MAATÉ, 1996; OLIVIER *et al.*, 1996). Accordingly, the Sebtide-Alpujarride units would have been located at the northern margin of the continental block (DURAND-DELGA and OLIVIER, 1998; MICHARD *et al.*, 1992, 2002; CHALOUAN *et al.*, 2001). However, following WILDI *et al.* (1977), one can assume as an alternative hypothesis, consistent with the stratigraphic data (see above), that the Sebtide-Alpujarride rocks formed the basement of the unrooted Dorsale nappes (Fig. 9A, right).

Alpine-Betic Subduction

From the onset of the Africa-Europe convergence (80 Ma) to the Middle Eocene, Iberia and Africa are brought together by c. 300 km (Fig. 8B). The North-African margin, Maghrebian Flysch Trough, and southern-central Alboran Terrane (Dor-sale, Ghomarides-Malaguides) remain virtually undisturbed (Fig. 9A), except along some (sinistral) strike-slip faults related to the terrane drift along North Africa, and evidenced in the Dorsale and African margin units (RAOULT, 1974; WILDI *et al.*, 1977; KUHNT and OBERT, 1991). A branch of this transform system could have been located between the southern (Massylian) and northern (Mauretanian) subbasins which display deep water Infra-Numidian clays and shallow water *Microcodium*

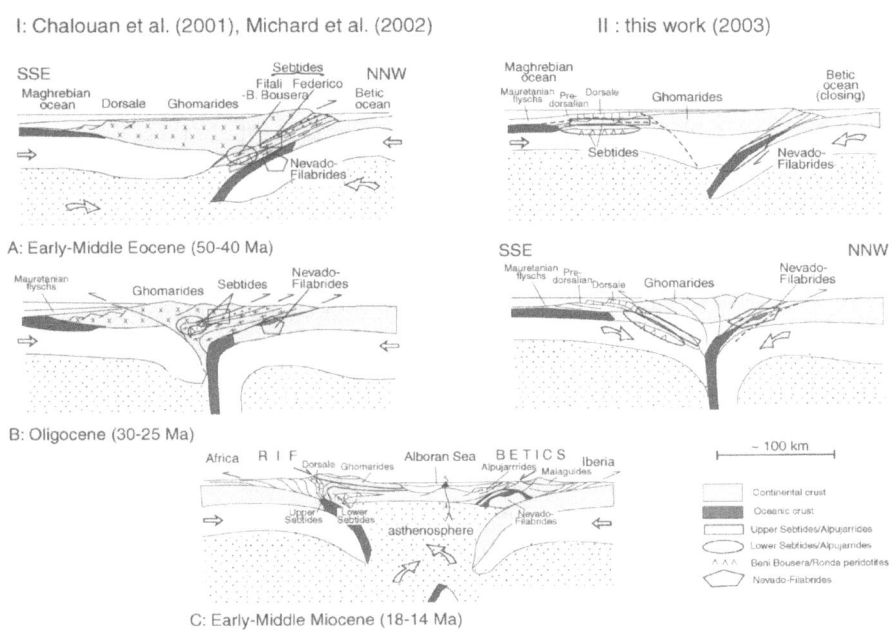

Figure 9

The Betic-Rif mountain building: Two alternative scenarios, depending on the age (still poorly constrained) of the HP-LT metamorphic event in the Sebtides-Alpujarrides. Location of the sketch cross-sections: Figure 8.

limestones and Nummulitic calcarenites during the Paleogene, respectively. In contrast, the Mulhacén oceanic/transitional crust domain is seemingly involved at that time in a SE-dipping subduction zone. This hypothesis relies on, i) the occurrence of Alpine blueschist- and eclogite-facies rocks in this nappe complex; ii) the paleontologic constraints which bound this HP-LT metamorphism to be younger than the early Late Cretaceous post-ophiolitic calcschists (TENDERO et al., 1993), older than the poorly dated (Eocene?) Soportújar Formation which reworks the high-grade Mulhacén rocks before being affected by a lower-grade tectonic-metamorphic event (PUGA et al., 1996), and significantly older than the Late Oligocene-Early Miocene deposits that covered the Alpujarrides nappes after their emplacement onto the Nevado-Filabrides (DURAND-DELGA et al., 1993; LONERGAN and MANGE-RADETZKY, 1994); iii) $^{39}Ar/^{40}Ar$ dates clustering between 48–36 Ma in the Mulhacén Complex (MONIÉ et al., 1991; DE JONG et al., 1992; NIETO et al., 1997; PUGA et al., 2002a), with additional Sm/Nd ages of 91 ± 21 Ma and 62 ± 11 Ma (NIETO et al., 1997), and a 51.6 ± 1.3 Ma phengite $^{39}Ar/^{40}Ar$ age from the underlying Veleta Complex (PUGA et al., 2002a). SHRIMP zircon dates at about 15 Ma from Mulhacén pyroxenites were interpreted as closely reflecting the age of the HP-LT event (LÓPEZ SÁNCHEZ-VIZCAÍINO *et al.*, 2001), but alternatively can reflect the dramatic HT extensional event which affected the Alboran area during the Miocene (see below). Then we contend that the SE-dipping, Late Cretaceous-Paleogene subduction of Western Alps (SCHMID *et al.*, 1996; MICHARD *et al.*, 1996; STAMPFLI *et al.*, 1998) and Corsica (MALAVIEILLE *et al.*, 1998; BRUNET *et al.*, 2000) should have extended southwestward along the northern branch of the Alpine ocean up to the Betic transect. Note that Early-Late Cretaceous synmetamorphic shear zones are described in Great Kabylia (SAADALLAH and CABY, 1996; CHEILLETZ *et al.*, 1999).

The fate of the Sebtide-Alpujarride rocks during the same time interval depends on their initial location. If they should have been located north of the Alboran Terrane, they might be involved in the Alpine-Betic subduction zone (Fig. 9A, left). Alternatively, provided they were located south of the Alboran Terrane, they would escape any deformation and metamorphism (Fig. 9A, right). Indeed, the age of the Sebtide-Alpujarride HP-LT event is quite poorly known yet, due to the strong HT-LP overprint at c. 22 Ma (see Metamorphic evolution). Although scarce K/Ar and $^{39}Ar/^{40}Ar$ dates at 38-44 Ma were obtained in the Beni Mezala, the older, and most convincing dates are close to 27 Ma in the Sebtides (MONTIGNY *et al.*, 2004), and 25 Ma in the Alpujarrides (MONIÉ *et al.*, 1991).

Apenninic-Maghrebian Subduction

During the Late Eocene-Oligocene, most of the Alboran Terrane (Ghomaride-Malaguide domain) is emerged, with rare conglomeratic formations (Fig. 9B). Clastic sediments invade the Dorsale margin and neighbouring Mauretanian basin (Beni Ider Flysch), i.e., the northern part of the Maghrebian Trough, while the infra-Numidian clays characterizes the southern part of the same trough. In the African margin itself, transpressional synmetamorphic deformation occurs in the eastern Intrarif (FRIZON DE LAMOTTE, 1985), south-verging thrusts affect the eastern Mesorif (FAVRE, 1992), and sandy sediments accumulate in Prerif depocenters (BENYAICH, 1991; MORLEY, 1992; CHALOUAN *et al.*, 2001). This seemingly testifies to the onset of backthrust tectonics in response to the Betic subduction slow-down as more and more Iberian crust is involved in the subduction zone. A similar, and coeval back-thrust event occurs in the Western Alps (PIANA and POLINO, 1995; SCHMID *et al.*, 1996). Along strike in the Apennine transect (Fig. 8C), the Eocene-Oligocene backthrust zone changes to a west-dipping subduction zone, due to the occurrence of oceanic crust at the leading edge of the Adria plate (DOGLIONI *et al.*, 1998). As a similar tectonic setting prevails all along the African plate leading edge, we might admit with DOGLIONI *et al.* (1998) and others (GUERRERA *et al.*, 1993; ZECK, 1996; LONERGAN and WHITE, 1997; FRIZON DE LAMOTTE *et al.*, 2000) that the Apenninic subduction continues southwestward into a Maghrebian subduction up to the Alboran area.

In the latter area, the role of the Maghrebian subduction can be envisioned in two alternative ways, according to the age which is assumed for the Sebtide-Alpujarride HP-LT event. Should we retain an Eocene age, then the signature of the Oligocene subduction is essentially the closure of the Maghrebian Trough and building of a SE-vergent orogenic wedge (Fig. 9B, left). Conversely, if we retain an Oligocene age for the Sebtide-Alpujarride metamorphism, then we can assume that it occurred within the Maghrebian subduction zone itself (Fig. 9B, right). Exhumation of the HP-LT rocks began shortly after during the Paleogene in synorogenic conditions as in many other Mediterranean Alpine belts (AVIGAD et al., 1997; BALANYÁ et al., 1997; AZAÑÓN and CRESPO BLANC, 2000; CHALOUAN et al., 2001).

During the late Oligocene-Miocene, the SE-ward retreat (rollback mechanism) and progressive curvature of the subduction zone (Fig. 8D) triggered the boudinage of the backarc domain, the drift of Corsica, Sardinia and Balearic Islands away from Iberia, and the fragmentation of the Alkapeca orogenic prism, with the Alboran Terrane drifting towards eastern Morocco (GUEGUEN et al., 1997; LONERGAN and White, 1997; MALDONADO et al., 1999; VERGÉS and SÀBAT, 1999; FRIZON DE LAMOTTE et al., 2000). Coarse Aquitanian turbidites (Numidian flysch; Hoyez, 1989) record this stage in the Maghrebian Trough before it closes at c. 19 Ma. Next, from the Early Miocene to the mid-Tortonian, docking of the Alboran Terrane onto the African margin developed, with a growing accretionary prism at the bottom of the exotic units, formed by unrooted Dorsale slivers, detached Maghrebian flyschs, and more and more sediments and low-grade meta-sediments from the continental margin (Fig. 9C). This formed a "Proto-Rif belt", the foredeep of which was the still undeformed Prerif domain. Flexural bending of the African lithosphere in front of the Proto-Rif prism resulted in the Langhian-Serravallian onlap over the Mesozoic cover of the Atlas foreland (Rides Prérifaines, Figs. 2, 3; ZIZI, 1996). The Maghrebian subduction progressively adopt a N-directed trend at its western end (Figs. 8C,D), being pinned there to the tip of the Eocene-Oligocene orogen and to the Tangiers crustal salient, a major structure of the African margin clearly shown on gravimetric maps, and related to the late Hercynian-Atlasian fault network (MENVIELLE and LE MOUËL, 1985; FAVRE, 1995). Then, the Maghrebian subduction zone was likely able to merge with the Alpine-Betic one, giving birth to the strong curvature of the Betic-Rif orocline. The occurrence of thinned continental crust (Iberian and African margins) west of the subduction zone favoured the westward thrusting of the upper plate units, as documented in the Gulf of Cadiz by MALDONADO et al. (1999). The oblique convergence of the Alboran Terrane with respect to Africa is recorded by the sinistral Nekor and Jebha faults in the SE Rif, and dextral Jebel Fahies fault N of Ceuta, which operated as lateral ramps of the SW-moving tectonic prism (LEBLANC and OLIVIER, 1984). The Jebha fault is coincident with, and was likely guided by a transition zone from normal to thinned crust in the Tangiers salient of the African margin (OLIVIER, 1981–1982; MORLEY, 1987).

Alboran Rifting and Coeval Contraction

In the backarc domain, the internal parts of the Alboran orogen collapsed through a dramatic rifting process, dated from the late Oligocene-Middle Miocene (Figs. 3D, 9C; GARCÍA-DUEÑAS *et al.*, 1992; MALDONADO *et al.*, 1992; WATTS *et al.*, 1993; CHALOUAN *et al.*, 1997, 2001; COMAS *et al.*, 1999). The oldest unconformable "post-nappe" deposits over the Ghomaride units are late Oligocene-Aquitanian in age (FEINBERG *et al.*, 1996; CHALOUAN *et al.*, 1995), whereas those over the Alpujarrides, or including HP-LT mineral clasts are dated from the latest Aquitanian-Burdigalian (DURAND-DELGA *et al.*, 1993; LONERGAN and MANGE-RAJETSKY, 1994; EL KADIRI *et al.*, 2001; PUGLISI *et al.*, 2001). These stratigraphic data are consistent with the cooling ages of the Sebtide-Alpujarride rocks being concentrated around 22 ± 2 Ma (MONTIGNY *et al.*, 2004). The low-angle normal fault on top of the Filali-Beni Bousera units (Zaouia fault, Fig. 6B) is also dated at c. 25–20 Ma. At the onset of the Alboran Sea extensional event, anatectic melts form at depth (SÁNCHEZ-GÓMEZ *et al.*, 1995; SÁNCHEZ-RODRIGUEZ and GEBAUER, 2000), then the Trans-Alboran calc-alkaline volcanism develops, indicative of interaction between subducting lithosphere, asthenospheric mantle and upper plate continental crust (LONERGAN and WHITE, 1997; CARMINATI *et al.*, 1998; ZECK and WHITEHOUSE, 1999; MAURY *et al.*, 2000). During the Middle Miocene, the Numidian units at the crest of the accretionary prism are submitted to erosion (DIDON and FEINBERG, 1979), and allowed to detach and slide inward on the Alboran basin slope (Jbel Zemzen klippe, Fig. 2). Such Maghrebian Flysch inliers are widespread in Algeria, associated with olistostromes (BOUILLIN *et al.*, 1973; DURAND-DELGA, 1980).

The late Miocene-Pliocene events poorly change the Rif-Betic structure. During the early-middle Tortonian, contraction goes on in the Rif orogenic prism, and causes the External Rif nappe stacking and folding (locally associated with very-low-grade metamorphism of the Early-Middle Miocene sediments at c. 7–8 Ma; MONIÉ *et al.*, 1984), the detachment of the Prerif cover in the Triassic evaporite level, and the synsedimentary thrusting of the detached Prerif on top of the Gharb foredeep series (SEPTFONTAINE, 1983; WERNLI, 1987). The so-called "olistostrome front" of the Prerif nappe, i.e., the front of the Miocene-Pliocene imbrications (LITTO *et al.*, 2001) can be connected with that of the Sub-Betic - Guadalquivir basin area (FLINCH *et al.*, 1996; FERNÀNDEZ *et al.*, 1998) through the submarine structures of the Gulf of Cadiz (Fig. 2) described by MALDONADO *et al.* (1999). Strike-slip and reverse fault activity and folding occur in the Alboran basin (Alboran Ridge, Fig. 2; MALDONADO *et al.*, 1992; CHALOUAN *et al.*, 1997). In contrast, during the late Tortonian-early Pliocene interval, extensional faulting characterizes the External Rif (South-Rifian Seaway), likely due to the superimposed effects of the accretionary prism collapse, and African crust flexural bending (MOREL, 1989; FLINCH, 1996; SAMAKA *et al.*, 1997).

Finally, during the late Pliocene-Pleistocene, the External Rif is again submitted to contractional tectonics, resulting in the Central Rif "post-nappe synclines," and reverse faults at the Prerif-foredeep boundary (e.g., Rides Prérifaines) (FAUGÈRES, 1978; AHMAMOU and CHALOUAN, 1988; ZIZI, 1996; MEDINA, 1995; LITTO et al., 2001). This contraction event is shown all around the Gibraltar Arc (WEIJERMARS et al., 1985; MOREL, 1989) with a roughly radiating compressional axis, which discards the hypothesis that it could only result from the ongoing Africa-Iberia convergence, and supports the role of arcuation of a retreating slab in the Alboran area.

Discussion and Conclusion

Our "two-subduction scenario," whether in its early version (CHALOUAN et al., 2001; Fig. 9, left) or in our new working hypothesis (Fig. 9, right) conflicts on critical points with some previous models.

Peridotite Emplacement

Based on petrologic and (mostly) geochronological arguments, the Beni Bousera-Ronda ultramafic massifs have been, and still are interpreted as hot asthenospheric diapirs emplaced during the Neogene and responsible for the Alpujarride-Sebtide metamorphism (LOOMIS, 1975; PLATT and VISSERS, 1989; BLICHERT-TOFT et al., 1999; MONTEL et al., 2000). As admitted by VISSERS et al. (1995), this is inadequate to account for the occurrence of Alpine HP-LT assemblages in varied Sebtide-Alpujarride rocks (GOFFÉ et al., 1989; TUBÍA and GIL IBARGUCHI, 1991; BOUYBAO-UENE, 1993; AZAÑÓN, 1994; GOFFÉ et al., 1996; VIDAL et al., 1999), and for their dating at > 25–27 Ma (MONIÉ et al., 1991; MONTIGNY et al., 2004). In our view, the Beni Bousera-Ronda lherzolites emplaced as tectonic slivers during the Paleogene subduction/collision process, at the expense of infracontinental lithospheric mantle (KORNPROBST, 1974) uplifted at upper crustal levels during the Tethyan rifting (REUBER et al., 1982; MICHARD et al., 1991). Indeed, the Ojen eclogites and Ronda pyroxenites yielded U-Pb SHRIMP zircon ages between 183 ± 3 and 131 ± 3 Ma which seemingly record the Tethyan extensional event (SÁNCHEZ-RODRIGUEZ and GEBAUER, 2000). A Sm/Nd age of 235.1 ± 1.7 Ma from Alpujarride garnets may also record the Triassic breakup of Pangea (ARGLES et al., 1999).

Collision vs. Subduction-collision Orogen. I: Eocene Subduction

The authors who assign a major tectonic-metamorphic role to a Neogene asthenospheric diapir tend to depict the Betic-Rif orogen as a collisional orogen, unrelated to any true subduction process, and devoid of any exotic terrane (PLATT and VISSERS, 1989; VISSERS et al., 1995; MONTEL et al., 2000). VISSERS et al. (1995) differ from PLATT and VISSERS (1989) as they admit that the Africa-Iberia convergence was underway by the late Eocene, with stacking of continental units in the hanging-wall of a subduction zone, dipping presumably to the NW (cf.

VAN DER WAL and VISSERS, 1993). Likewise, ZECK (1996) hypothesises an African subduction beneath Iberia at the onset of the Betic-Rif orogeny during the Eocene-Oligocene, and makes no reference to an exotic Alboran domain. A similar view is expressed by CABY et al. (2001) for the east Algerian transect (Edough massif). In contrast, we insist on the occurrence of lost oceanic/transitional crust floored areas (recorded on the one hand by the Maghrebian Flyschs and associated ophiolitic slivers, and on the other hand by the ultramafics and metabasites of the Mulhacén Complex) on both sides of a widely displaced Alboran/Alkapeca continental block. If one considers, i) this paleogeographic restoration (based on stratigraphical/structural evidences); ii) the strong, Eocene HP-LT metamorphic imprint in the upper Nevado-Filabrides (Mulhacén Complex); and iii) the obvious continuity of the Alpine belt at the plate tectonic scale before the Mediterranean basin opening (ALVAREZ, 1976), then one can hardly escape the conclusion that the Late Cretaceous-Eocene, SE-dipping subduction of the Western Alps extended up to the Betic transect. Note that the SE dip of the Eocene subduction in Corsica, although questioned by PRINCIPI and TREVES (1984), DURAND-DELGA et al. (2000), and PADOA and DURAND-DELGA (2001), was re-assessed, based on geological and/or seismic data (CROP profiles) by MALAVIEILLE et al. (1998), DOGLIONI et al. (1998, 1999), MARRONI et al. (2001) and FINETTI et al. (2001).

Collision vs. Subduction-collision Orogen. II: Oligocene-Neogene Subduction

VISSERS et al. (1995) and ZECK (1996) argue that the Oligocene-Miocene evolution of the Rif-Betic orogen essentially depends on some form of detachment or convective removal of lithospheric mantle, shortly before 22 Ma. According to VISSERS et al. (1995), "there should have been an immediate increase of surface elevation, possibly reflected in the increased clastic sedimentation during the Oligocene-Early Miocene in the flysch basins on the African margin and in the Gibraltar area." However, i) the Beni Ider Flysch, actually nourished from the Internal zones, accumulated during the Oligocene-Aquitanian interval, c. 33–20 Ma; and ii) the Numidian Flysch, the sedimentation of which is actually concentrated in the latest Oligocene-Aquitanian interval, c. 24–20 Ma, has been nourished from Saharan sources (HOYEZ, 1989). In fact, there is a large chronologic decoupling between, on the one hand, the surface uplift in the Internal Zones, stratigraphically bracketed between the Paleocene-Eocene (shallow water limestones, c. 55–40 Ma) and the mid-upper Oligocene (c. 30–25 Ma; CHALOUAN et al., 2001, with references therein), and on the other hand, the onset of the Alboran Sea extensional event and associated rapid cooling, dated at 22 ± 2 Ma.

A Miocene, collision-induced "subduction" which would have only concerned the delaminated lower lithosphere of the African margin was proposed by DOCHERTY and BANDA (1992) and MALDONADO et al. (1999). Contrary to these authors and to VISSERS et al. (1995), but in line with LONERGAN and WHITE (1997) and FRIZON DE

LAMOTTE *et al.* (2000), we contend that the late Oligocene-Neogene Betic-Rif tectonics is better explained by considering the evolution of a NW-dipping Apenninic-Maghrebian subduction zone. The birth of this young subduction zone as a result of incipient blocking of the older, Alpine subduction zone was repeatedly discussed by DOGLIONI *et al.* (1998, 1999). The structural, magmatic, and seismic/ tomographic signature of this subduction is clearly documented on the Northern Apennine-Corsica transect (SERRI *et al.*, 1993; BOCCALETTI *et al.*, 1997; VAN DER MEULEN *et al.*, 1999; BRUNET *et al.*, 2000). This subduction was responsible for the Mediterranean backarc basin opening, and for the dispersal of the deformed Alkapeka fragments in Alboran, Kabylia, Sicily (Peloritani) and Calabria (REHAULT *et al.*, 1984; MALINVERNO and RYAN, 1986; ROYDEN, 1993; GUEGUEN *et al.*, 1997; LONERGAN and WHITE, 1997; JOLIVET and FACCENNA, 2000). The Apenninic subduction was also responsible for the Oligocene HP-LT metamorphism of Tuscany (BRUNET *et al.*, 2000) and Calabria (ROSSETTI *et al.*, 2001). In the Alboran transect, the Appeninic-Maghrebian subduction is evidenced by the Early Miocene closure of the Maghrebian Trough and coeval building of an orogenic wedge thrust onto the African margin, but the potential, Oligocene metamorphic record of this subduction is controversial. If we attach a geologic meaning to the K/Ar and ^{40}Ar/^{39}Ar ages at 44 Ma and 40–38 Ma respectively, obtained in the upper Sebtides, the Sebtide-Alpujarride HP-LT metamorphism could be ascribed to the Betic subduction (Fig. 9, left). If we alternatively retain a younger age, based on the significant dates at 25 Ma (central Alpujarrides) and 27 Ma (Sebtides) reported by MONIÉ (1991) and MONTIGNY *et al.* (2004), then the Sebtide-Alpujarride metamorphism must be ascribed to the Oligocene-Miocene subduction zone (Fig. 9, right). Whatever the final choice may be between these alternative hypotheses, the Alpine-Betic and Apenninic subduction zones had to meet west of the Alboran Terrane, i.e., close to the Azores-Gibraltar transform fault. The latter plate boundary was characterized by dextral transpression from the Late Eocene onward (ROEST and SRIVASTAVA, 1991). Then, from the Early Miocene onward, the Gibraltar Arc developed as a compressional Subduction-Subduction-Transform fault triple junction.

Acknowledgments

This work would not have been possible without the cooperation of numerous Moroccan colleagues and students, among whom we will only cite late Prof. Ahmed Benyaich. Field and laboratory works were supported by the French-Moroccan Inter-Universities Cooperation Program, Actions Intégrées 93/629 and 98/161, and by the Moroccan Programme d'Appui à la Recherche Scientifique, Sci. Univers, n° 26. In Spain, we were nicely guided by Victor García-Dueñas, Paco González-Lodeiro and their colleagues, and supported by the Institut National des Sciences de l'Univers, Programme DBT, thème 5. We are indebted to Carlo Doglioni, Claudio

Facenⁱa, and Dominique Frizon de Lamotte for constructive criticism of our recent papers on the area, and to Michel Durand-Delga, Raymond Montigny and an anonymous reviewer for careful review of an early draft of this paper.

REFERENCES

AGARD, P., JULLIEN, M., GOFFÉ, B., BARONNET, A., and BOUYBAOUENE, M. (1999), The Evidence for High-temperature (300 °C) Smectite in Multistage Clay-mineral Pseudomorphs in Pelitic Rocks (Rif, Moroeco), Eur J. Mineral. 11, 655–668.

AHMAMⁱU, M. and CHALOUAN, A. (1988), Distension synsédimentaire plio-quaternaire et rotation anti-horaire des contraintes au Quaternaire ancien sur la bordure nord du bassin du Saïs (Maroc), Bull. Inst. Sci. Rabat 12, 19–26.

ALLERTON, S., REICHERTER, K., and PLATT, J.P. (1994), A Structural and Paleomagnetic study of a Section through the Eastern Subbetic, Southern Spain, J. Geol. Soc. London 151, 659–668.

ALVAREZ, W. (1976), The Former Continuation of the Alps, Geol. Soc. Am. Bull. 87, 891–896.

ANDEWEG, B. and CLOETINGH, S. (2001), Evidence for an Active Sinistral Shear Zone in the Western Alboran Region, Terra Nova 13, 44–50.

ANDRIEUX, J., FONTBOTE, J.M., and MATTAUER, M. (1971), Sur un modèle explicatif de l'arc de Gibraltar, Earth Planet. Sci. Lett. 12, 191–198.

ANDRIEUX, J., FRIZON DE LAMOTTE, D., and BRAUD, J. (1989), A Structural Scheme for the Western Mediterranean Area in Jurassic and Early Cretaceous Times, Geodinamica Acta 3, 5–15.

APARICIⁱ, A., MITJAVILA, J. M., ARANA, V., and VILLA, I. M. (1991), La edad del volcanismo de las islas Columbrete Grande y Alboran (Mediterraneo occidental), Boll. Geol. Minero 102, 562–570.

ARGLES, T.W., PRINCE, C.I., FOSTER, G.L., and VANCE, D. (1999), New Garnets for Olds? Cautionary Tales from Young Mountain Belts, Earth Planet. Sci. Lett. 172, 301–309.

AVIGAD D., JOLIVET, L., and AZAÑÓN, J.M. (1997), Backarc Extension and Denudation of Mediterranean Eclogites, Tectonics 16, 924–941.

AZAÑÓN, J.M. (1994), Metamorphism de alta presión/baja temperatura, baja presión/alta temperatura y tectónica del Complejo Alpujárride (Cordilleras Bético-Rifeñas), Thesis Doct. Univ. Granada (Spain)-Ecole Normale Supérieure (Paris), 300 pp.

AZAÑÓN, J.M. and CRESPO-BLANC, A. (2000), Exhumation during a Continental Collision Inferred from the Tectonometamorphic Evolution of the Alpujarride Complex in the Central Betics (Alboran Domain, SE Spain, Tectonics 19, 549–565.

BALANYÁ, J.C., GARCÍA-DUEÑAS, V., AZAÑÓN, J.M., and SÁNCHEZ-GÓMEZ, M. (1997), Alternating Contractional and Extensional Events in the Alpujarride Nappes of the Alboran Domain (Betics, Gibraltar Arc), Tectonics 16, 226–238.

BALLINC, N. and BANDA, E. (1992), Europe's lithosphere, recent activity. In (Blundel D., Freeman R., and Muelⁱr S., eds.), A Continent Revealed: The European Geotraverse (Cambridge Univ. Press), pp. 111–135.

BANDA, E., GALLART, J., GARCÍA-DUEÑAS, V., DAÑOBEITIA, J., and MAKRIS, J. (1993), Lateral Variation of the Crust in the Iberia Peninsula: New Evidence for the Betic Cordillera, Tectonophysics 221, 53–66.

BENYAICH, A. (1991), Evolution tectono-sédimentaire du Rif externe centro-occidental (régions de M'sila et Ouezzane, Maroc). La marge africaine du Jurassique au Crétacé inférieur: les bassins néogènes d'avant-fosse, Thèse Doctorat Etat, Univ. Pau, 308 pp.

BENYAICH, A., DURAND-DELGA, M., FEINBERG, H., MAATÉ, A., and MAGNÉ, J. (1986), Implications de niveaux du Miocène inférieur dans les rétrocharriages de la Dorsale rifaine (Maroc): signification à l'échelle de l'arc de Gibraltar, C. R. Acad. Sci. 302, 587–592.

BERNINI, M., BOCCALETTI, M., MORATTI, G., and PAPANI, G. (2000), Structural Development of the Taza-Guercⁱf Basin as a Constraint for the Middle Atlas Shear Zone Tectonic Evolution, Marine Petrol. Geol. 17, 391–408.

BIERMANN, C. (1995), The Betic Cordilleras (SE Spain) Anatomy of a Dualistic Collision Type Orogenic Belt, Geol. Mijnb. 41, 167–182.

BLANCO, M.J. and SPAKMAN, W. (1993), *The P-wave Velocity Structure of the Mantle below the Iberian Peninsula. Evidence for a Subducted Lithosphere below Southern Spain*, Tectonophys. *221*, 13–14.

BLICHERT-TOFT, J., ALBARÈDE, F., and KORNPROBST, J. (1999), *Lu-Hf Isotope Systematics of Garnet Pyroxenites from Beni Bousera, Morocco: Implications for Basalt Origin*, Science *283*, 1303–1306.

BOCCALETTI, M., GIANELLI, G., and SANI, F. (1997), *Tectonic Regime, Granite Emplacement and Crustal Structure in the Inner Zone of the Northern Apennines (Tuscany, Italy): A New Hypothesis*, Tectonophys. *270*, 127–143.

BOUILLIN, J.P., DURAND-DELGA, M., GÉLARD, J.P., LEIKINE, M., RAOULT, J.F., RAYMOND, D., TEFIANI, M., and VILA, J.M. (1973), *Les olistostromes d'âge miocène inférieur liés aux flyschs allochtones kabyles de l'orogène alpin d'Algérie*, Bull. Soc. géol. Fr. (7) *15*, 340–344.

BOUILLIN, J. P., DURAND-DELGA, M., and OLIVIER P., (1986), *Betic-Rifian and Tyrrhenian Arcs: Distinctive features, genesis, and developpment stages*. In (Wezel F. C., ed.) *The Origin of Arcs* (Elsevier Sci. Publ., Amsterdam), pp. 281–304.

BOUILLIN, J.-P., POUPEAU, G., TRICART, P., BIGOT-CORMIER, F., MASCLE, G., and TORELLI, L. (1998), *Premières données thermo-chronologiques sur les socles sarde et kabylo-péloritain submergés dans le canal de Sardaigne (Méditerranée occidentale)*, C. R. Acad. Sci. *326*, 561–566.

BOUYBAOUENE, M.L. (1993), *Etude pétrologique des métapélites des Sebtides supérieures, Rif interne, Maroc*, Thèse Doct. Etat, Univ. Mohamed V, Rabat, 160 pp.

BOUYBAOUENE, M. L., GOFFÉ, B., and MICHARD, A. (1998), *High-pressure Granulites on Top of the Beni Bousera Peridotites, Rif Belt, Morocco: A Record of an Ancient Thickened Crust in the Alboran Domain*, Bull. Soc. géol. Fr. *169*, 153–162.

BRUNET, C., MONIÉ, P., JOLIVET, L., and CADET, J.P. (2000), *Migration of Compression and Extension in the Tyrrhenian Sea, Insights from $^{40}Ar/^{39}Ar$ Ages on Micas along a Transect from Corsica to Tuscany*, Tectonophys. *321*, 127–155.

BUFORN, E, SANZ DE GALDEANO, C., and UDÍAS, A. (1995), *Seismo-tectonics of the Ibero-Maghrebian Region*, Tectonophys. *248*, 247–261.

BUFORN, E, COCA, P., UDÍAS, A., and LASA, C. (1997), *Source Mechanism of Intermediate and Deep Earthquakes in Southern Spain*, J. Seismol. *1*, 113–130.

CABY, R., HAMMOR, D., and DELOR, C. (2001), *Metamorphic Evolution, Partial Melting and Miocene Exhumation of Lower Crust in the Edough Metamorphic Core Complex, West Mediterranean Orogen, Eastern Algeria*, Tectonophys. *342*, 239–273.

CALVERT, A., SANDVOL, E., SEBER, D., BARAZANGI, M., ROECKER, S., MOURABIT, T., VIDAL, F., ALGUACIL, G., and JABOUR, N. (2000), *Geodynamic Evolution of the Lithosphere and Upper Mantle beneath the Alboran Region of the Western Mediterranean: Constraints from Travel-time Tomography*, J. Geophys. Res. *105*, B5, 10,871–10,898.

CALVO, M., VEGAS, R., and OSETE, M.-L. (1997), *Paleomagnetic Results from Upper Miocene and Pliocene Rocks from the Internal Zone of the Eastern Betic Cordilleras (Southern Spain)*, Tectonophys. *277*, 271–283.

CALVO, M., CUEVAS, J., and TUBÍA, J.M. (2001), *Preliminary paleomagnetic results on Oligocene-early Miocene mafic dykes from southern Spain*, Tectonophys. *332*, 333–345.

CAMPOS, J., MALDONADO, A., and CAMPILLO, A.C. (1992), *Post-Messinian Evolutional Patterns of the Central Alboran Sea*, Geo-Marine Lett. *12*, 173–178.

CARMINATI, E., WORTEL, M.J.R., SPAKMAN, W., and SABADINI, R. (1998), *The Role of Slab Detachment Processes in the Opening of the Western-central Mediterranean Basins: Some Geological and Geophysical Evidence*, Earth Planet. Sci. Lett. *160*, 651–665.

CHALOUAN, A. and MICHARD, A., (1990), *The Ghomarides Nappes, Rif Coastal Range, Morocco: A Variscan Chip in the Alpine Belt*, Tectonics 9, 1565–1583.

CHALOUAN, A., OUAZANI-TOUHAMI, A., MOUHIR, L., SAJI, R., and BENMAKHLOUF, M. (1995), *Les failles normales à faible pendage du Rif interne (Maroc) et leur effet sur l'amincissement crustal du domaine d'Alboran*, Geogaceta *17*, 107–109.

CHALOUAN, A., SAJI, R., MICHARD, A., and BALLY, A.W. (1997), *Neogene Tectonic Evolution of the Southwestern Alboran Basin as Inferred from Seismic Data off Morocco*, AAPG Bull. *81*, 1161–1184.

CHALOUAN, A., MICHARD, A., FEINBERG, H., MONTIGNY, R., and SADDIQI, O. (2001), *The Rif Mountain Building (Morocco): A New Tectonic Scenario*, Bull. Soc. géol. Fr. *172*, 5, 603–616.

CHEILLETZ, A., RUFFET, G., MARIGNAC, C., KOLLI, O., GASQUET, D., FÉRAUD, G., and BOUILLIN, J.P. (1999), $^{39}Ar/^{40}Ar$ Dating of Shear Zones in the Variscan Basement of Greater Kabylia (Algeria). Evidence of an Eo-Alpine Event at 128 Ma (Hauterivian-Barremian Boundary): Geodynamic Consequences, Tectonophys. 306, 97–116.

COMAS, M.C., PLATT, J.P., SOTO, J.I., and WATTS, A.B. (1999), The origin and tectonic history of the Alboren basin: Insights from ODP leg 161 results, In Proc. Ocean Drill. Program Sci. Results 161 (eds Zahn R., Comas M.C., and Klaus A.), pp. 555–580.

CRESPO-BLANC, A. and CAMPOS, J. (2001), Structure and Kinematics of the South Iberian Paleomargin and its Relationship with the Flysch Trough Units: Extensional Tectonics within the Gibraltar Arc Fold-and-thrust Belt (Western Betics), J. Struct. Geol. 23, 1615–1630.

DE JONG, K., WIJBRANS, J.R., and FÉRAUD, G. (1992), Repeated Thermal Resetting of Phengites in the Mulhacén Complex (Betic Zone, Southeastern Spain) Shown by $^{40}Ar/^{39}Ar$ Step Heating and Single Grain Laser Probe Dating, Earth Planet. Sci. Lett. 110, 173–191.

DERCOURT, J., RICOU, L. E., and VRIELINCK, B., Eds. (1993), Atlas of Tethys Environmental Maps, Paris, Gauthier-Villars, 307 pp.

DIDON, J. and FEINBERG, H. (1979), Données nouvelles sur l'âge et la signification géodynamique du Miocène des Béni-Issef: Importance de la tectonique burdigalienne dans le Rif septentrional (Maroc), C. R. somm. Soc. géol. Fr. (7) 21, 183–187.

DOCHERTY, J.I.C. and BANDA, E. (1992), A Note on the Subsidence History of the Northern Margin of the Alboran Sea Basin., Geo-Marine Lett. 12, 82–87.

DOGLIONI, C., MONGELLI, F., and PIALLI, G. (1998), Boudinage of the Alpine Belt in the Apenninic Backarc, Mem. Soc. Geol. It. 52, 457–468.

DOGLIONI, C., FERNANDEZ, M., GUEGUEN, E., and SABAT, F. (1999), On the Interference between the Early Apennines-Maghrebides Backarc Extension and the Alps-Betics Orogen in the Neogene Geodynamics of the Western Mediterranean, Bull. Soc. geol. It. 118, 75–89.

DURAND-DELGA, M. (1980), Le Méditerranée occidentale: étapes de sa genèse et problèmes structuraux liés à celle-ci, Mém. h.-s. Soc. géol. Fr. 10, 203–224.

DURAND-DELGA, M. and FONTBOTÉ, J.M. (1980), Le cadre structural de la Méditerranée occidentale, 26th Int. Geol. Congr., Paris, Coll. C5, 67–85.

DURAND-DELGA, M. and OLIVIER, P. (1988), Evolution of the Alboran block margin from Early Mesozoic to Early Miocene time. In The Atlas System of Morocco (ed. Jacobshagen, V.H.) Lecture Notes Earth Sci. 15, 465–480.

DURAND-DELGA, M., FEINBERG, H., MAGNÉ, J., OLIVIER, P., and ANGLADA, R. (1993), Les formation oligo-miocènes discordantes sur les Malaguides et les Alpujarrides et leurs implications dans l'évolution géodynamique des Cordillères bétiques (Espagne) et de la Méditerranée d'Alboran, C. R. Acad. Sci. 317, 679–687.

DURAND-DELGA, M, ROSSI, P., OLIVIER, P., and PUGLISI, D. (2000), Situation structurale et nature ophiolitique de roches basiques jurassiques associées aux flyschs maghrébins du Rif (Maroc) et de Sicile (Italie), C. R. Acad. Sci. 331, 29–38.

ELAZZAB D. and FEINBERG, H. (1994), Paléomagnétisme des roches ultrabasiques du Rif externe (Maroc), C. R. Acad. Sci. 318, 351–357.

ELAZZAB D., GALDEANO, A., FEINBERG, H., and MICHARD, A. (1997), Prolongement en profondeur d'une écaille ultrabasique allochtone: traitement des données aéromagnétiques et modélisation 3D des péridotites des Beni Malek (Rif, Maroc), Bull. Soc. géol. Fr. 168, 667–683.

ELAZZAB D. and EL WARTITI, M. (1998), Paléomagnétisme des laves du Moyen-Atlas (Maroc): rotations récentes, C. R. Acad. Sci. 327, 509–512.

EL HATIMI, N., DUÉE, G., and HERVOUET, Y. (1991), La Dorsale calcaire du Haouz: ancienne marge continentale passive téthysienne (Rif, Maroc), Bull. Soc. géol. Fr. 162, 79–90.

EL KADIRI, K., LINARES, A., and OLORIZ, F. (1992), La Dorsale calcaire rifaine (Maroc septentrional): Evolution stratigraphique et géodynamique durant le Jurassique-Crétacé, Notes Mém. Serv. Géol. Maroc 336, 217–265.

EL KADIRI, K., CHALOUAN, A., EL MRHINI, A., HLILA, R., LÓPEZ-GARIDO, A., SANZ DE GALDEANO, C., SERAND, F., and KERZAZI, K. (2001), Les formations sédimentaires de l'Oligocène supérieur-Miocène *inférieur dans l'unité ghomaride des Béni-Hozmar (secteur de Talembote, Rif septentrional, Maroc)*, Eclogae geol. Helv. *94*, 313–320.

EL MAZ, A. and GUIRAUD, M. (2001), *Paragenèse à faible variance dans les métapélites de la série de Filali (Rif interne marocain): description, interprétation et conséquence géodynamique*, Bull. Soc. géol. Fr. *172*, 469–485.

FACCENNA, C., MATTEI, M., FUNICIELLO, R., and JOLIVET, L. (1997), *Styles of Backarc Extension in the Central Mediterranean*, Terra Nova 9, 126–130.

FAUGÈRES, J. C. (1978), *Les Rides sud-rifaines: Evolution sédimentaire et structurale d'un bassin atlantico-mésogéen de la marge africaine*, Thèse Doct. Etat, Univ. Bordeaux I, 480 pp.

FAVRE, P. (1992), *Géologie des massifs calcaires situés au front S de l'unité de Kétama (Rif, Maroc)*, Publ. Dépt. Géol. Paléonto. Univ. Genève 11, 138 pp. and 178 pp.

FAVRE, P. (1995), *Analyse quantitative du rifting et de la relaxation thérmique de la partie occidentale de la marge transformante nord-africaine : le Rif externe (Maroc). Comparaison avec la structure actuelle de la chaîne*, Geodinamica Acta *8*, 59–81.

FEINBERG, H., MAATÉ, A., BOUHDADI, S., DURAND-DELGA, M., MAATÉ, M., MAGNÉ, J., and OLIVIER, Ph. (1990), *Significations des dépôts de l'Oligocène supérieur-Miocène inférieur du Rif interne (Maroc) dans l'évolution géodynamique de l'arc de Gibraltar*, C. R. Acad. Sci. *310*, 1487–1495.

FEINBERG, H., SADDIQI, O., and MICHARD, A. (1996), *New constraints on the bending of the Gibraltar Arc from paleomagnetism of the Ronda peridotite (Betic Cordilleras, Spain)*, In Paleomagnetism and Tectonics of the Mediterranean Region (eds. Morris, A. and Tarling, D. H.), Geol. Soc. London Spec. Publ. *105*, 43–52.

FERNÁNDEZ, M., BERÁSTEGUI, X., PUIG, C., GARCÍA-CASTELLANOS, D., JURADO, M.J., TORNÉ, M., and BANKS, C. (1998), *Geophysical and geological constraints on the evolution of the Guadalqivir foreland basin, Spain*. In Cenozoic foreland basins of Western Europe (eds. Mascle, A., Puigdefàbregas, C., Luterbacher H.P. and Fernàndez, M.), Geol. Soc. London Spec. Publ. *134*, 29–48.

FINETTI, I.R., BOCCALETTI, M., BONINI, M., DEL BEN, A., GELETTI, R., PIPAN, M., and SANI, F. (2001), *Crustal Section Based on CROP Seismic Data across the North Tyrrhenian-Northern Apennines-Adriatic Sea*, Tectonophys. *343*, 135–163.

FLINCH, F.J. (1996), *Accretion and extensional collapse of the external Western Rif (Northern Morocco)*. In Peri-Tethys Memoir 2: Structure and Prospects of Alpine Basins and Forelands (eds. Ziegler P.A. and Horvath F.), Mém. Mus. nation.. Hist. natur. *170*, 61–85.

FRIZON DE LAMOTTE, D. (1985), *La structure du Rif oriental (Maroc). Rôle de la tectonique longitudinale et importance des fluides*, Mém. Sci. Terre Univ. P. et M. Curie, 85–03

FRIZON DE LAMOTTE, D. (1987), *Un exemple de collage synmétamorphe: La déformation miocène des Temsamane (Rif externe, maroc)*, Bull. Soc. géol. Fr. *3*, 337–344.

FRIZON DE LAMOTTE, D., ANDRIEUX, J., and GUEZOU, J.-C. (1991), *Cinématique des chevauchements néogènes dans l'Arc bético-rifain: discussion sur les modèles géodynamiques*, Bull. Soc. géol. Fr. *169*, 611–626.

FRIZON DE LAMOTTE, D., SAINT BEZAR, B., BRACÈNE, R., and MERCIER, E. (2000), *The Two Main Steps of the Atlas Building and Geodynamics of the Western Mediterranean*, Tectonics *19*, 740–761.

FROITZHEIM, N., SCHMID, S.M., and FREY, M. (1996), *Mesozoic Paleogeography and the Timing of Eclogite-facies Metamorphism in the Alps: A Working Hypothesis*, Eclogae geol. Helv. *89*, 81–110.

GALINDO-ZALDÍVAR, J., JABALOY, A., GONZÁLEZ-LODEIRO, F., and ALDAYA, F. (1997), *Crustal Structure of the Central Sector of the Betic Cordilleras (SE Spain)*, Tectonics *16*, 18–37.

GALINDO-ZALDÍVAR, J., GONZÁLEZ-LODEIRO, F., JABALOY, A., MALDONADO, A., and SCHREIDER, A.A. (1998), *Models of Magnetic and Bouguer Gravity Anomalies for the Deep Structure of the Central Alboran Sea basin*, Geo-Marine Lett. *18*, 10–18.

GARCÍA-DUEÑAS, V., BALANYÁ, J.C., and MARTÍNEZ-MARTÍNEZ, J. M. (1992), *Miocene Extensional Detachment in the Outcropping Basement of the Northern Alboran Basin (Betics) and their Tectonic Implications*, Geo-Marine Lett. *12*, 88–95.

GOFFÉ, B., MICHARD, A., GARCÍA-DUEÑAS, V., GONZÁLEZ-LODEIRO, F., MONIÉ, P., CAMPOS, J., GALINDO-ZALDIVAR, J., JABALOY, A., MARTINEZ, J. M., and SIMANCAS, J. F. (1989), *First Evidence of High-pressure, Low-temperature Metamorphism in the Alpujarride Nappes, Betic Cordilleras (SE Spain)*, Eur. J. Mineral. *1*, 139–142.

GOFFÉ, B., AZAÑÓN, J.M., BOUYBAOUENE, M.L., and JULLIEN, M. (1996), *Metamorphic Cookeite in Alpine Metapelites from Rif (Northern Morocco) and Betic Chains (Southern Spain)*. Eur. J. Mineral. *6*, 897–911.

GÓMEZ-PUGNAIRE, M.T., ULMER, P., and LÓPEZ-SÁNCHEZ VIZCAÍNO, V. (2000), *Petrogenesis of the Mafic Igneous Rocks of the Betic Cordilleras: A Field, Petrological and Geochemical Study*, Contrib. Mineral. Petrol. *139*, 436–457.

GUEGUEN, E., DOGLIONI, C., and FERNÀNDEZ, M. (1997), *Lithospheric Boudinage in the Western Mediterranean Backarc Basin*, Terra Nova *9*, 184–187.

GUERRERA, F., MARTIN-ALGARRA, A., and PERRONE, V. (1993), *Late Oligocene-Miocene syn-/-late-orogeric Successions in Western and Central Mediterranean Chains from the Betic Cordillera to the Southern Apennines*, Terra Nova *5*, 525–544.

HANDY, M.R., FRANZ, L., HELLER, F., JANOTT, B., and ZURBIGGEN, R. (1999), *Multistage Accretion and Exhumation of the Continental Crust (Ivrea Crustal Section, Italy and Switzerland)*, Tectonics *18*, 1154–1177.

HATZFELD, D. and WORKING GROUP FOR DEEP SEISMIC SOUNDING (1978), *Crustal Seismic Profiles in the Alboran Sea. Preliminary Results*, Pure Appl. Geophys. *116*, 167–180.

HERNANDEZ, J., DE LAROUZIÈRE, F. D., BOLZE, J., and BORDET, P. (1987), *Le magmatisme néogène bético-rifain et le couloir de décrochement trans-Alboran*, Bull. Soc. géol. Fr. (*8*) 3, 257–267.

HOYEZ, B. (1989), *Le Numidien et les flyschs oligo-miocènes de la bordure sud de la Méditerranée occidentale*, Thèse Doct. Etat, Univ. Lille, 464 pp.

JACOBSHAGEN, V. (1992), *Major Fracture Zones of Morocco: The South Atlas and the Transalboran Fault Systems*, Geol. Rundsch. *81*, 185–197.

JOLIVET, L. and FACCENNA, C. (2000), *Mediterranean Extension and the Africa-Eurasia Collision*, Tectonics *19*, 1095–1106.

KORNPROBST, J. (1974), *Contribution à l'étude pétrographique et structurale de la zone interne du Rif (Maroc septentrional)*, Notes Mém. Serv. géol. Maroc *251*, 256 pp.

KORNPROBST, J. and VIELZEUF, D. (1984), *Transcurrent crustal thinning: a mechanism for the uplift of deep continental crust/upper mantle associations*. In *Kimberlites and related rocks* (ed. Kornprobst J.), Developments in Petrology Series *11*B, 347–359.

KUHNT, W. and OBERT, D. (1991), *Evolution crétacée de la marge tellienne*, Bull. Soc. géol. Fr. *162*, 515–522.

KUMAR, N., REISBERG, L., and ZINDLER, A. (1996), *A Major and Trace Elements and Strontium, Neodymium, and Osmium Isotopic Study of a Thick Pyroxenite Layer from the Beni Bousera Ultramafic Complex of Northern Morocco*, Geochim. Cosmochim. Acta *60*, 1429–1444.

LEBLANC, D. and OLIVIER, P. (1984), *Role of Strike-slip Faults in the Betic-Rifian Orogeny*, Tectonophys. *101*, 345–355.

LITTO, W., JAAIDI, E., DAKKI, M., and MEDINA, F. (2001), *Etude sismo-structurale de la marge nord du bassin du Gharb (avant-pays rifain, Maroc) : mise en évidence d'une distension d'âge miocène supérieur*, Eclogæ geol. Helv. *94*, 63–73.

LONERGAN, L. and MANGE-RAJETSKY, M. (1994), *Evidence for Internal Zone Unroofing from Foreland Basin Sediments, Betic Cordillera, SE Spain*, J. geol. Soc. London *151*, 515–529.

LONERGAN, L. and WHITE, N. (1997), *Origin of the Betic-Rif Mountain Belt*, Tectonics *16*, 504–522.

LOOMIS, T.P. (1975), *Tertiary Mantle Diapirism, Orogeny and Plate Tectonics East of the Strait of Gibraltar*, Am. J. Sci. *275*, 1–30.

LÓPEZ SÁNCHEZ-VIZCAÍNO, V., RUBATTO, D., GÓMEZ-PUGNAIRE, M.T., TROMMSDORFF, V., and MÜNTENER, O. (2001), *Middle Miocene High-pressure Metamorphism and Fast Exhumation of the Nevado-Filábride Complex, SE Spain*, Terra Nova *13*, 327–332.

MAATÉ, A. (1996), *Estratigrafia y evolucion paleogeografica alpina del dominio Ghomaride (Rif interno, Marruecos)*, Thesis Univ. Granada, 397 pp.

MALAVIEILLE, J., CHEMENDA, A., and LARROQUE, C. (1998), *Evolutionary Model for Alpine Corsica: Mechanism for Ophiolite Emplacement and Exhumation of High-pressure Rocks*, Terra Nova *10*, 317–322.

MALDONADO, A., CAMPILLO, A.C., MAUFFRET, A., ALONSO, B., WOODSIDE, J., and CAMPOS, J. (1992), *Alboran Sea Late Cenozoic Tectonic and Stratigraphic Evolution*, Geo-Marine Lett. *12*, 179–186.

MALDONADO, A., SOMOZA, L., and PALLARÉS, L. (1999), *The Betic Orogen and the Iberian-African Boundary in the Gulf of Cadiz: Geological Evolution (Central North Atlantic)*, Marine Geol. *155*, 9–43.

MALINVERNO, A. and RYAN, W.B.F. (1986), *Extension in the Tyrrhenian Sea and Shortening in the Apennines as a Result of Arc Migration Driven by Sinking of the Lithosphere*, Tectonics 5, 227–245.

MANATSCHAL, G. and BERNOULLI, D. (1999), *Architecture and Tectonic Evolution of Nonvolcanic Margins: Present-day Galicia and Ancient Adria*, Tectonics 18, 1099–1119.

MARRONI, M., MOLLI, G., OTTRIA, G., and PANDOLFI, L. (2001), *Tectono-sedimentary Evolution of the External Liguride Units (Northern Apennines, Italy): Insights in the Pre-collisional History of a Fossil Ocean-continent Transition Zone*, Geodinamica Acta 14, 307–320.

MARTÍNEZ-MARTÍNEZ, J.M. and AZAÑÓN, J.M. (1997), *Mode of Extensional Tectonics in the Southeastern Betics (SE Spain): Implications for the Tectonic Evolution of the Peri-Alboran Orogenic System*, Tectonics 16, 205–225.

MAUFFRET, A., MALDONADO, A. and CAMPILLO, A.C. (1992), *Tectonic Framework of the Eastern Alboran and Western Algerian Basins, Western Mediterranean*, Geo-Marine Lett. 12, 104–110.

MAURY, R.C., FOURCADE, S., COULON, C., EL AZZOUZI, M., BELLON, H., COUTELLE, A., OUABADI, A., SEMROUD B., MEGARTSI M., COTTON, J., BELANTEUR, O., LOUNI-HACINI A., PIQUÉ A., CAPDEVILA, R., HERNANDEZ, J., and REHAULT, J.-P. (2000), *Post-collisional Neogene Magmatism of the Mediterranean Maghreb Margin: A Consequence of Slab Breakoff*, C. R. Acad. Sci. 331, 159–173.

MEDINA, F. (1995), *Present-day State of Stress in Northern Morocco from Focal Mechanism Analysis*, J. Struct. Geol. 17, 1035–1046.

MEGHRAOUI, M., MOREL, J. L., ANDRIEUX, J., and DAHMANI, M. (1996), *Tectonique plio-quaternaire de la chaîne tello-rifaine et de la mer d'Alboran. Une zone complexe de convergence continent-continent*, Bull. Soc. géol. Fr. 167, 141–158.

MENVIELLE, M. and LE MOUËL, J.-L. (1985), *Existence d'une anomalie de conductivité dans le Haut-Atlas marocain et concentration des courants telluriques à l'échelle régionale*, Bull. Soc. Géol. Fr. (8) 1, 553-558.

MICHARD, A., GOFFÉ, B., CHALOUAN, A., and SADDIQI, O. (1991), *Les corrélations entre les Chaînes bético-rifaines et les Alpes et leurs conséquences*, Bull. Soc. Géol. Fr. 162, 1151–1160.

MICHARD, A., FEINBERG, H., ELAZZAB, D., BOUYBAOUENE, M., and SADDIQI, O., (1992), *A Serpentinite Ridge in a Collisional Paleomargin Setting: The Beni Malek Massif, External Rif, Morocco*, Earth Planet. Sci. Lett. 113, 435–442.

MICHARD, A., CHOPIN, C., and HENRY, C. (1993), *Compression versus Extension in the Exhumation of the Dora-Maira Coesite-bearing Unit, Western Alps, Italy*., Tectonophys. 221, 173–193.

MICHARD, A., GOFFÉ, B., CHOPIN, C., and HENRY, C. (1996), *Did the Western Alps Develop Through an Oman-type Stage? The Geotectonic Setting of High-pressure Metamorphism in Two Contrasting Tethyan Transects*, Eclogae geol. Helv. 89, 43–80.

MICHARD, A., GOFFÉ, B., BOUYBAOUENE, M., and SADDIQI, O. (1997), *Late Hercynian-Mesozoic Thinning in the Alboran Domain : Metamorphic Data from the Northern Rif, Morocco*, Terra Nova 9, 1–8.

MICHARD, A., CHALOUAN, A., FEINBERG, H., GOFFÉ, B., and MONTIGNY, R. (2002), *How Does the Alpine Belt End between Spain and Morocco?* Bull. Soc. géol. Fr. 173, 3–15.

MONIÉ, P., FRIZON DE LAMOTTE, D., and LEIKINE, M. (1984), *Etude géologique préliminaire par la méthode $^{39}Ar/^{40}Ar$ du métamorphisme alpin dans le Rif externe (Maroc). Précisions sur le calendrier tectonique tertiaire*, Rev. Géol. Dynam. Géogr. Phys. 25, 307–317.

MONIÉ, P., GALINDO ZALDIVAR, J., GONZALEZ LODEIRO, F., GOFFÉ, B., and JABOLOY, A., (1991), *$^{40}Ar/^{39}Ar$ Geochronology of Alpine Tectonism in the Betic Cordilleras (Southern Spain)*, J. Geol. Soc. London 148, 288–297.

MONIÉ, P., TORRES-ROLDAN, R.L., and GARCÍA-CASCO, A. (1994), *Cooling and Exhumation of the Western Betic Cordilleras, $^{40}Ar/^{39}Ar$ Thermochronological Constraints on a Collapsed Terrane*, Tectonophysics 238, 353–379.

MONTEL, J.-M., KORNPROBST, J., and VIELZEUF, D. (2000), *Preservation of Old U-Th-Pb Ages in Shielded Monazites: Example from the Beni-Bousera Hercynian Kinzigites (Morocco)*, J. metam. Geol. 18, 335–342.

MONTIGNY, R., MICHARD, A., CHALOUAN, A., SADDIQI, O., BOUYBAOUENE, M., GOFFÉ, B., and FEINBERG, H. (2004), *K-Ar and $^{40}Ar/^{39}Ar$ Study of Metamorphic Rocks from the Internal Rif (Morocco); Geochronological Constraints on the Peridotite Emplacement and Exhumation Tectonics in the Gibraltar Arc, Revisited*, Bull. Soc. géol. Fr. 174, in press.

MORALES, J., SERRANO, I., JABALOY, A., GALINDO-ZALDIVAR, J., ZHAO, D., TORCAL, F., VIDAL, F., and GONZÁLEZ-LODEIRO, F. (1999), *Active Continental Subduction Beneath the Betic Cordillera and the Alboran Sea*, Geology *27*, 735–738.

MOREL, J.C. (1989), *Etats de contrainte et cinématique de la chaîne rifaine (Maroc) du Tortonien à l'actuel*, Geodinamica Acta *3*, 283–294.

MORLEY, C.K. (1987), *Origin of a Major Cross-element Zone: Moroccan Rif*, Geology *15*, 761–764.

MORLEY, C.K. (1992), *Notes on Neogene Basin History of the Western Alboran Sea and its Implications for the Tectonic Evolution of the Rif-Betic Orogenic Belt*, J. Afr. Earth Sci. *14*, 57–65.

NAJID, D., WESTPHAL, M., and HERNANDEZ, J. (1981), *Paleomagnetism of Quaternary and Miocene Lavas from North-East and Central Morocco*, J. Geophysics *49*, 149–152.

NIETO, J.M., PUGA, E., MONIÉ, P., DÍAZ DE FEDERICO, A., and JAGOUTZ, E. (1997), *High Pressure Metamorphism in Metagranites and Orthogneisses from the Mulhacén Complex (Betic Cordilleras, Spain)*, Terra Nova *1* (abstract suppl.), 22–23.

OLIVIER, Ph. (1981-1982), *L'accident de Jebha-Chrafate (Rif, Maroc)*, Rev. Géol. dyn. Géogr. Phys. *22*, 201–212.

OLIVIER, Ph., DURAND-DELGA, M., MANIVIT, H., FEINBERG, H., and PEYBERNES, B. (1996), *Le substratum jurassique des flyschs maurétaniens de l'ouest des Maghrébides: l'unité de Ouareg (région de Targuist, Rif, Maroc)*, Bull. Soc. géol. France *167*, 609–616.

PADOA, E. and DURAND-DELGA, M. (2001), *L'unité ophiolitique du Rio Magno en Corse alpine, élément des Ligurides de l'Apennin septentrional*, C. R. Acad. Sci. *333*, 285–293.

PIANA, F. and POLINO, R. (1995), *Tertiary Structural Relationships Between Alps and Apennines: The Critical Torino Hill and Montferrato Area, Northwestern Italy*, Terra Nova *7*, 138–143.

PIQUÉ, A., DAHMANI, M., JEANNETTE, D., and BAHI, L. (1987), *Permanence of Structural Lines in Morocco since Precambrian to Present*, African J. Earth Sci. *6*, 247–256.

PIQUÉ, A., A BRAHIM, L., EL AZZOUZI, M., MAURY, R., BELLON, H., SEMROUD, B., and LAVILLE, E. (1998), *Le poinçon maghrébin: contraintes structurales et géochimiques*, C. R. Acad. Sci. *326*, 575–581.

PLATT, J.P. and VISSERS, R.L.M. (1989), *Extensional Collapse of Thickened Continental Lithosphere: A Working Hypothesis for the Alboran Sea and Gibraltar Arc*, Geology *17*, 540–543.

PLATT, J.P. and WHITEHOUSE, M.J. (1999), *Early Miocene High-temperature Metamorphism and Rapid Exhumation in the Betic Cordillera (Spain): Evidence from U-Pb Zircon Ages*, Earth Planet. Sci. Lett. *171*, 591–605.

PLATZMANN, E.S. (1992), *Paleomagnetic Rotations and Kinematics of the Gibraltar Arc*, Geology *20*, 311–314.

PLATZMANN, E.S., PLATT, J.P., and OLIVIER, P. (1993), *Paleomagnetic Rotations and Fault Kinematics in the Rif Arc of Morocco*, J. Geol. Soc. London *150*, 707–718.

PLATZMANN, E.S., PLATT, J.P., and KELLEY, S.P. (2000), *Large Clockwise Rotations in an Extensional Allochthon, Alboran Domain (Southern Spain)*, J. Geol. Soc. London *17*, 1187–1197.

POLYAK, B.G., FERNÁNDEZ, M., KHUTORSKOY, M.D., SOTO, J.I., BASOV, I.A., COMAS, M.C., KHAIN, V.Y., AGAPOVA, G.V., MAZUROVA, I.S., NEGREDO, A., TOCHITSKY, V.O., DE LA LINDE, J., BOGDANOV, N.A., and BANDA, E. (1996), *Heat Flow in the Alboran Sea, Western Mediterranean*, Tectonophys. *263*, 191–218.

PRINCIPI G. and TREVES, B. (1984), *Il sistema Corso-Appenninico come prisma d'accrezione. Riflessi sul problema generale del limite Alpi-Appennini*, Mem. Soc. geol. It. *28*, 549–576.

PUGA, E., NIETO, J.M., DÍAZ DE FEDERICO, A., PORTUGAL, M., and REYES, M. (1996), *The Intra-Orogenic Soportújar Formation of the Mulhacén Complex: Evidence for the Polycyclic Character of the Alpine Orogeny in the Betic Cordilleras*, Eclogae geol. Helv. *89*, 129–162.

PUGA, E, NIETO, J.M., DÍAZ DE FEDERICO, A., BODINIER, J.L., and MORTEN, L. (1999), *Petrology and Metamorphic Evolution of Ultramafic Rocks and Dolerite Dykes of the Betic Ophiolitic Association (Mulhacén Complex, SE Spain): Evidence of eo-Alpine Subduction following an Ocean-floor Metasomatic Process*, Lithos *49*, 23–56.

PUGA, E., DÍAZ DE FEDERICO, A., and NIETO, J.M. (2002a), *Tectonostratigraphic Subdivision and Petrological Characterisation of the Deepest Complexes of the Betic Zone: A Review*, Geodinamica Acta *15*, 23–43.

PUGA, E., RUIZ CRUZ, M.D., and DÍAZ DE FEDERICO, A. (2002b), *Polymetamorphic Amphibole Veins in Metabasalts from the Betic Ophiolitic Association at Cóbdar, Southern Spain: Relics of Ocean-floor Metamorphism Preserved through the Alpine Orogeny*, Canad. Mineral. *40*, 67–83.

PUGLISI, D., ZAGHLOUL, M.N. and MAATÉ, A. (2001), *Evidence of Sedimentary Supply from Plutonic Sources in the Oligocene-Miocene Flyschs of the Rifian Chain (Morocco): Provenance and Paleogeographic Implications*, Bull. Soc. Geol. It. *120*, 55–68.

RAOULT, J. F. (1974), *Géologie du centre de la Chaîne numidique (Nord du Constantinois, Algérie)*, Mém. Soc. géol. Fr. *53*–121, 163 pp

REISBERG, L., ZINDLER, A., and JAGOUTZ, E. (1989), *Further Sr and Nd Isotopic Results from Peridotites of the Ronda Ultramafic Complex*, Earth Planet. Sci. Lett. *96*, 161–180.

REHAULT, J.-P., BOILLOT, G., and MAUFFRET, A. (1984), *The Western Mediterranean Basin Geological Evolution*, Marine Geol. *55*, 447–477.

REUBER, I., MICHARD, A., CHALOUAN, A., JUTEAU, T., and JERMOUMI, B. (1982), *Structure End Emplacement of the Alpine-type Peridotite from Beni Bousera, Rif, Morocco: A Polyphase Tectonic Interpretation*, Tectonophys. *82*, 231–251.

RIMI, A., CHALOUAN, A. and BAHI, L. (1998), *Heat Flow in the Westernmost Part of the Alpine Mediterranean System (The Rif, Morocco)*, Tectonophys. *285*, 135–146.

ROEST, W.R. and SRIVASTAVA, S.P. (1991), *Kinematics of the Plate Boundaries between Eurasia, Iberia, and Africa in the North Atlantic from the Late Cretaceous to the Present*, Geology *19*, 613–616.

ROSSETTI, F., FACCENNA, C., GOFFÉ, B., MONIÉ, P., ARGENTIERI, A., FUNICIELLO, R., and MATTEI M. (2001), *Alpine Structural and Metamorphic Signature of the Sila Piccola Massif Nappe Stack (Calabria, Italy): Insights for the Tectonic Evolution of the Calabrian Arc*, Tectonics *20*, 112–133.

ROYDEN, L.H. (1993), *Evolution of Retreating Subduction Boundaries Formed during Continental Collision*, Tectonics *12*, 629–638.

SAADALLAH, A. and CABY, R. (1996), *Alpine Extensional Detachment Tectonics in the Grande Kabylie Metamorphic Core Complex of the Maghrebides (Northern Algeria)*, Tectonophys. *267*, 257–273.

SADDIQI, O., REUBER, I., and MICHARD, A. (1988), *Sur la tectonique de dénudation du manteau infracontinental dans les Beni Bousera, Rif septentrional, Maroc*, C. Acad. Sci. *307*, 657–662.

SADDIQI, O., FEINBERG, H., ELAZZAB, D., and MICHARD, A. (1995), *Paléomagnétisme des Péridotites des Beni Bousera (Rif interne, Maroc): Conséquences pour l'évolution Miocene de l'Arc de Gibraltar*, C. R. Acad. Sci. *321*, 361–368.

SAMAKA, F., BENYAICH, A., DAKKI, M., HÇAINE, M., and BALLY, A.W. (1997), *Origine et inversion des bassins miocènes supra-nappes du Rif Central (Maroc). Etude de surface et de subsurface. Exemple des bassins de Taounate et de Tafrannt*, Geodinamica Acta *10*, 30–40.

SÁNCHEZ-GOMEZ, M., GARCÍA-DUEÑAS, V., and MUÑOZ, M. (1995), *Relations structurales entre les péridotites de la Sierra Bermeja et les unités alpujarrides sous-jacentes (Benahavis, Ronda, Espagne)*, C. R. Acad. Sci. *321*, 885–892.

SANCHEZ-RODRIGUEZ, L. and GEBAUER, D. (2000), *Mesozoic Formation of Pyroxenites and Gabbros in the Ronda Area (Southern Spain), Followed by Early Miocene Subduction Metamorphism and Emplacement into the Middle Crust: U-Pb Sensitive High-resolution Ion Microprobe Dating of Zircon*, Tectonophys. *316*, 19–44.

SAVOSTIN, L.A., SIBUET, J.-C., ZONENSHAIN, L.P., LE PICHON, X., and ROULET, M.-J. (1986), *Kinematic Evolution of the Tethys Belt from the Atlantic Ocean to the Pamirs since the Triassic*, Tectonophys. *123*, 1–35.

SCHMID, S.M., PFIFFNER, O.A., FROITZHEIM, N., SCHÖNBORN, G., and KISSLING, E. (1996), *Geophysical-Geological Transect and Tectonic Evolution of the Swiss-Italian Alps*, Tectonics *15*, 1036–1064.

SEBER, D., BARAZANGI, M., IBENBRAHIM, A., and DEMNATI, A. (1996), *Geophysical Evidence for Lithospheric Delamination beneath the Alboran Sea and the Rif-Betic Mountains*, Nature *379*, 785–790.

SEPTFONTAINE, M. (1983), *La formation du Jbel Binet (Rif externe oriental, Maroc), un dépôt "anté-nappes" d'âge miocène supérieur. Implications paléotectoniques*, Eclogae geol. Helv. *76*, 581–609.

SERRI, G., INNOCENTI, F., and MANETTI, P. (1993), *Geochemical and Petrological Evidence of the Subduction of Delaminated Adriatic Continental Lithosphere in the Genesis of the Neogene-Quaternary Magmatism of Central Italy*, Tectonophys. *223*, 117–147.

STAMPFLI, G.M., MOSAR, J., MARQUER, D., MARCHANT, R., BAUDIN, T., and BOREL, G. (1998), *Subduction and Obduction Processes in the Swiss Alps*, Tectonophys. *296*, 159–204.

SUTER, G. (1980), *Carte structurale de la Chaîne rifaine au 1:500.000*, Notes Mém. Serv. Géol. Maroc *245*b.

TENDERO, J.A., MARTIN-ALGARRA, A., PUGA, E., and DIAZ DE FEDERICO, A. (1993), *Lithostratigraphie des métasédiments de l'association ophiolitique Névado-Filabride (SE Espagne) et mise en évidence d'objets ankéritiques évoquant des foraminifères planctoniques du Crétacé: conséquences paléogéographiques*, C. R. Acad. Sci. *316*, 1115–1122.

TORNÉ, M., FERNÀNDEZ, M., COMAS, M.C., and SOTO, J.I. (2000), *Lithospheric Structure Beneath the Alboran Basin: Results from 3D Gravity Modeling and Tectonic Relevance*, J. Geophys. Res. *105*, B2, 3209–3228.

TORRES-ROLDAN, R. L., POLI, G., and PESSERILLO, A. (1986), *An Early Miocene Arc-tholeitic Magmatic Dike Event from the Alboran Sea: Evidence for Precollisional Subduction and Backarc Crustal Extension in the Westernmost Mediterranean*, Geol. Rundsch. *75*, 219–234.

TUBÍA, M.J. and GIL IBARGUCHI, J.I. (1991), *Eclogites of the Ojen nappe: a Record of Subduction in the Alpujarride complex (Betic Cordlleras, southern Spain)*, J. Geol Soc. London *148*, 801–804.

TURNER, S.P., PLATT, J.P., GEORGE, R.M.M., KELLEY, S.P., PEARSON, D.G., and NOWELL, G.M. (1999), *Magmatism Associated with Orogenic Collapse of the Betic-Alboran Domain, SE Spain*, J. Petrol. *40*, 1011–1036.

UDÍAS, A., LÓPEZ ARROYO, A., and MÉZCUA, J. (1976), *Seismotectonics of the Azores-Alboran Region*, Tectonophys. *31*, 259–289.

VAN DER MEULEN, M.J., KOUWENHOVEN, T.J., VAN DER ZWAN, G.J., MEULENKAMP, J.E., and WORTEL, M.J.M (1999), *Late Miocene Uplift in the Romagnan Apennines and the Detachment of Subducted Lithosphere*, Tectonophys. *315*, 319–335.

VAN DER WAL, D. and VISSERS, R.L.M. (1993), *Uplift and Emplacement of Upper Mantle Rocks in the Western Mediterranean*, Geology *21*, 1119–1122.

VÉRGES, J. and SÀBAT, F. (1999), *Constraints on the Neogene Mediterranean kinematic evolution along a 1000 km transect from Iberia to Africa*. In *The Mediterranean basins: Tertiary Extension within the Alpine Orogen (eds. Durand B., Jolivet L., Horváth F. and Séranne M.)*, Geol. Soc. London Spec. Publ. *156*, 63–80.

VIDAL, O., GOFFÉ, B., BOUSQUET, R., and PARRA, T. (1999), *Calibration and Testing of an Empirical Chloritoid-Chlorite Mg-Fe Exchange Thermometer and Thermodynamic Data for Daphnite*, J. Metamorphic Geol. *17*, 25–39.

VILLALA N, J.J., OSETE M.L., VEGAS, R., GARCÍA-DUEÑAS V., and HELLER, F. (1994), *Widespread Neogene Remagnetization in Jurassic Limestones of the South-Iberian Paleomargin (Western Betics, Gibraltar Arc)*, Phys. Earth Planet. Int. *85*, 15–33.

VISSERS, R.L.M., PLATT, J.P., and VAN DER WAL, D. (1995), *Late Orogenic Extension of the Betic Cordillera and the Alboran Domain: A Lithospheric View*, Tectonics *14*, 786–803.

WATTS, A.B., PLATT, J.P., and BUHL, P. (1993), *Tectonic Evolution of the Alboran Sea Basin*, Basin Research *5*, 153–177.

WEIJERMARS, R., ROEP, T.H., VAN DEN EECKHOUT, B., POSTMA, R., and KLEVERLAAN, K. (1985), *Uplift History of a Betic Fold Nappe Inferred from Neogene-quaternary Sedimentation and Tectonics (in the Sierra Alhamilla and Almeria, Sorbas and Tabernas Basins of the Betic Cordilleras, SE Spain)*, Geol. Mijnbouw *64*, 397–411.

WERNLI, R. (1987), *Micropaléontologie du Néogène post-nappes du Maroc septentrional et description systématique des foraminifères planctoniques*, Notes Mém. Serv. geol. Maroc *331*, 1–266.

WILDI, W. (1983), *La chaîne tello-rifaine (Algérie, Maroc, Tunisie): Structure, stratigraphie et évolution du Trias au Miocène*, Rev. Géol. Dyn. Géogr. Phys. *24*, 201–297.

WILDI, W., NOLD, M., and UTTINGER, J. (1977), *La Dorsale Calcaire entre Tetouan et Assifane (Rif interne, Maroc)*, Eclogae geol. Helv. *70*, 371–415.

ZAGHLOUL, M.N. (1994), *Les unités Federico septentrionales (Rif interne, Maroc): inventaire des déformations et contexte géodynamique*, Thèse Doct. Univ. Mohamed V, Rabat

ZECK, H.P. (1996), *Betic-Rif Orogeny: Subduction of Mesozoic Tethys Lithosphere under Eastward Drifting Iberia, Slab Detachment Shortly before 22 Ma, and Subsequent Uplift and Extensional Tectonics*, Tectonophysics *254*, 1–16.

ZECK, H.P., MONIÉ, P., VILLA, I.M., and HANSEN, B.T. (1992), *Very High Rates of Cooling and Uplift in the Alpine Belt of the Betic Cordillerras*, Southern Spain, Geology *20*, 79–82.

ZECK, H.P. and WHITEHOUSE, M.J. (1999), *Hercynian, Pan-African, Proterozoic and Archean Ion-microprobe Zircon Ages from a Betic Core Complex, Alpine Belt, W Mediterranean – Consequences for its P-T-T path*, Contrib. Mineral. Petrol. *134*, 134–149.

ZIZI, M. (1996), *Triassic-Jurassic extension and Alpine inversion in Northern Morocco*. In *Peri-Tethys Memoir 2: Structure and Prospects of Alpine Basins and Forelands* (eds. Ziegler, P. And Horvath, F.), Mém. Mus. nation Hist. natur. *170*, 87–101.

ZOUHRI, L., LAMOUROUX, C. and BURET C. (2001), *La Mamora, charnière entre la Meseta et le Rif: son importance dans l'évolution géodynamique post-paléozo du Maroc*, Geodinamica Acta *14*, 361–372.

(Received January 30, 2002, accepted November 5, 2002)

 To access this journal online:
http://www.birkhauser.ch

Pure appl. geophys. 161 (2004) 521–540
0033–4553/04/030521–20
DOI 10.1007/s00024-003-2461-6

❙ Pure and Applied Geophysics

Recent Tectonic Deformations and Stresses in the Frontal Part of the Rif Cordillera and the Saïss Basin (Fes and Rabat Regions, Morocco)

K. Bargach[1], P. Ruano[2], A. Chabli[1], J. Galindo-Zaldívar[2],
A. Chalouan[1], A. Jabaloy[2], M. Akil[1], M. Ahmamou[1],
C. Sanz de Galdeano[3], and M. Benmakhlouf[4]

Abstract — Prerif Ridges are located at the frontal part of the Rif Cordillera, which develops at the Eurasian-African plate boundary. The ridges are formed by recent tectonic structures that also deform foreland basins (Saïss and Gharb basins) and the foreland (Moroccan Meseta). The position of the ridges is the consequence of inversion tectonics undergone in the area. The ENE-WSW trend of the northern edge of the Neogene Saïss basin is determined by the location of Mesozoic basins. Although Prerif ridges probably started to develop since the Early Miocene, the most active deformation phase affecting Pliocene rocks consisted of N-S to NW-SE oriented compression. Striated pebbles show that this compression has prolate stress ellipsoids. The deformation produces southwards vergent folds and NNW-SSE striae on reverse faults at the base of the ridges. The flexure of the Paleozoic basement by the emplacement of the Ridges produced extensional deformation and the development of the Saïss foreland basin. The extension in this basin is oblate and features a well determined NNE-SSW trend near the Ridges, whereas it becomes prolate and pluridirectional near the foreland edge represented by the Rabat region. This part of the Moroccan Meseta, commonly considered to be stable, is deformed by sets of orthogonal joints and faults with short slip that affect up to Quaternary sediments. Southwestward, the Meseta rocks are also deformed by transcurrent faults, which indicate NW-SE and N-S trends of compression. The NW-SE approximation of Eurasia and Africa determines a regional stress field with the same trend of compression. Regional stresses are notably disturbed by the development of the active structures in the Rif, which exhibit alternating trends of compression and extension. The clearest evidence of the relationship between the local deformation and the general plate motion is found at the deformation front of the Cordillera, that is, the Prerif Ridges.

Key words: Prerif ridges, Saïss Basin, Moroccan Meseta, recent and active tectonics, paleostresses.

Introduction

The Betic and Rif Cordilleras constitute a tectonic arc that has developed as a consequence of the relative motion between the African and European plates and the

[1] Département des Sciences de la Terre, Faculté des Sciences, Université Mohammed V, Rabat, Morocco.
[2] Departamento de Geodinámica, Universidad de Granada, 18071-Granada, Spain.
E-mail: jgalindo@ugr.es
[3] Instituto Andaluz de Ciencias de la Tierra, C.S.I.C.-Universidad de Granada, 18071-Granada, Spain.
[4] Département de Géologie, Faculté des Sciences de Tétouan, Morocco.

westward displacement of their Internal Zones. The current plate motion determined for this area indicates an oblique convergence of 4 mm/yr with NW-SE trend (DEMETS *et al.*, 1990). The present-day stresses in this sector of the plate boundary are also characterized by a NW-SE maximum compression (ZOBACK and BURKE, 1993). However, the deformation inside these cordilleras is very heterogeneous, as highlighted by the studies of Plio-Quaternary tectonics in the Rif (MEGHRAOUI *et al.*, 1996). Most of the Internal Zones in the Betic Cordillera (PLATT and VISSERS, 1989; GONZÁLEZ-LODEIRO *et al.*, 1996) and in the northern part of the Rif (CHALOUAN *et al.*, 1995; OUAZANI-TOUHAMI and CHALOUAN, 1995) underwent extensional deformations from the Early Miocene, alternating with compressive episodes. Moreover, large basins have developed since the Early Miocene, such as the Alboran Sea, located on the Internal Zones of these cordilleras, likewise affected by compressional and extensional deformations (CHALOUAN *et al.*, 1997).

The Rif (Fig. 1) is made up of the Internal Rif (or Internal Domain), which derives from the Alboran microplate (domain or terrain), the Flysch Domain of probable oceanic origin (DURAND-DELGA *et al.*, 2000) entirely detached from its substratum and the External Rif (External Domain). The latter represents the deposits on the African margin and mainly consists of the Intrarif, Mesorif and Prerif.

The location of the major alpine structures of ENE-WSW orientation was determined by the trend of transcurrent and transtensional fault sets related to the opening of the Atlantic Ocean. One example is the fracture lineation located between the Jebha and Arbaoua (MORLEY, 1987). The alpine deformations are more intense and earlier in the Internal Rif (Oligocene to Early Miocene) than in the External Rif (ranging from Middle Miocene to the present). In the External Rif there is a lateral diachronism in the age of the deformations: the emplacement of the Prerif thrust sheets and the deformation in the mountain chain front occurred during Middle-Late Miocene in the eastern sectors (eastward of Fes), and since the Latest Miocene in the western sectors (Prerif Ridges, Saïss and Gharb basins, Atlantic margin; FEINBERG, 1986).

Paleostress and recent stress studies in the Rif indicate that since the Tortonian, there has been an anticlockwise rotation of compression trends (AHMAMOU and CHALOUAN, 1988; AÏT BRAHIM and CHOTIN, 1989; MOREL, 1989). The trend of compression was NE-SW during Early Tortonian and up to Messinian when the Prerif thrust sheets and the SW-ward thrusting of the Ketama-Tanger units on the Mesorif were emplaced (LESPINASSE, 1975; FRIZON DE LAMOTTE, 1985, 1991; Morley, 1987). AHMAMOU and CHALOUAN (1988) determine that the more recent rotation of the compressive stress trend, from NNE-SSW up to NNW-SSE, took place during the Early Quaternary at the front of the mountain chain. Earthquake focal mechanisms of the northeastern Rif (MEDINA and CHERKAOUI, 1992; MEDINA, 1995) indicate that the present-day stress field is complex, with a regional field characterized by NE-SW oriented extension that is associated in some areas with NW-SE

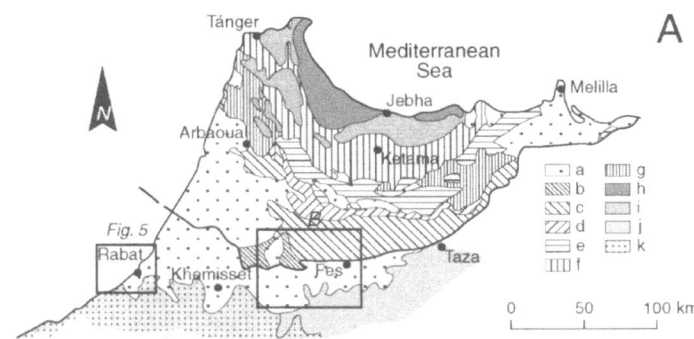

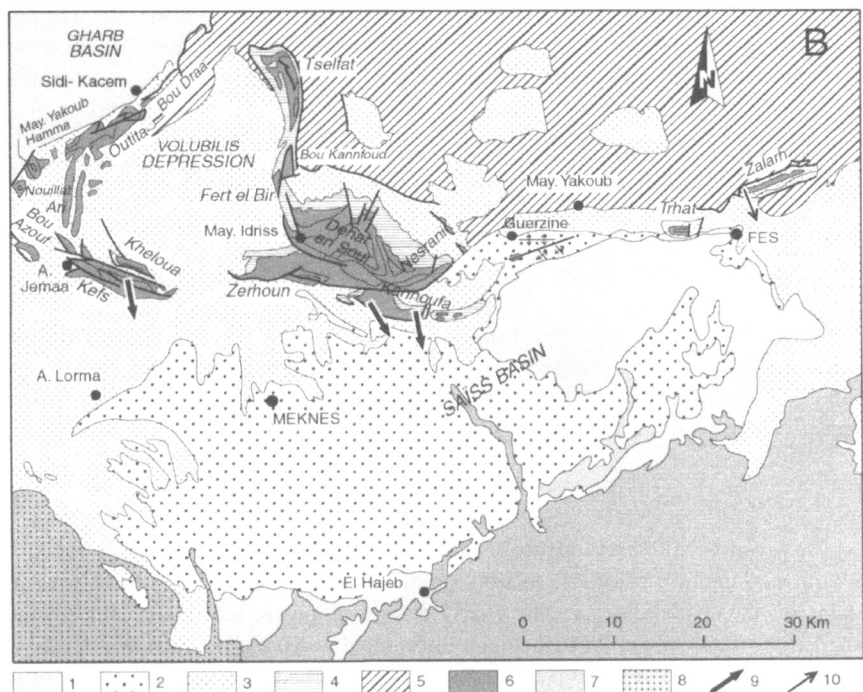

Figure 1

Geological setting (A) and detailed map of the Prerif ridges (B). a. Neogene and Quaternary rocks. b. Prerif ridges. c. External Prerif. d. Internal Prerif. e. Mesorif. f. Intrarif (Ketama-Tánger units). g. thrust nappes of Intrarif origin. h. Internal Zones. i. Flysch. j. Atlas. k. Meseta. Sedimentary rocks of Saïss and Gharb basins (1. Quaternary; 2. Pliocene; 3. Upper Miocene; 4. Middle Miocene); 5. Olistostromes of the External Prerif; 6. Prerif ridges; 7. Atlas; 8. Meseta; 9. displacement trend of the hanging walls on the southern borders of the Prerif ridges; 10. displacement trend of the hanging walls in low-angle faults.

oriented compression. In addition, MEDINA (1995) determines the alternation of recent stresses, sometimes compressive with a NE-SW trend. In the frontal part of the Cordillera, that is, the Prerif Ridges, MEDINA and CHERKAOUI (1992) determine a transcurrent earthquake focal mechanism (34.0 °S, 5.2 °W) with WNW-ESE trending compression.

Taking into account this geological framework, it is interesting to establish the main characteristics of the fault kinematics and the evolution of the recent tectonic stresses in the frontal part of the Rif Cordillera and in its foreland, where alpine deformations related to the recent development of the Cordillera are easy to establish because there are fewer superimposed tectonic phases. The objective of this research is to provide an outline of the Plio-Quaternary tectonics of the frontal zone of the External Rif, at the southern border of the Prerif Ridges; and of the propagation of the deformations to the Saïss foreland basin. In addition, the area near Rabat may prove to be representative of the deformation that affects the contact between the foreland basin (Gharb basin) and the foreland (Moroccan Meseta and Atlas) in areas far from the mountain front. These new data help to establish the relationship between fault kinematics, recent stresses and present-day plate motion.

Structure of the Southern Front of the Rif Cordillera and its Foreland

The area studied in this contribution is located at the front of the Rif Cordillera and its foreland. In this wide area we selected two sectors located near Fes and near Rabat taking into account the optimal exposure of outcrops and their locations, respectively near and far from the Rif mountain front.

Prerif Ridges

The Prerif thrust sheets form the frontal part of the Rif Cordillera. They are a tectonic-sedimentary complex, thrust over the South-Rif corridor (former foreland basin) or the Middle Atlas. The Prerif Ridges comprise elongated hills, formed mainly by Jurassic rocks. The ridges are interpreted as part of the Meseta-Atlas cover of the foreland, involved in thrusts of the External Rif—due to the contraction of the African margin—during compression from Late Miocene extending to Middle Pliocene (FAUGUÈRES, 1978; ZIZI, 1996a,b).

The Prerif ridges (Fig. 1) are grouped into two great ensembles separated by the Volubilis Depression. The western group is arched in a horseshoe shape, convex toward the west. This ensemble is formed by the djebels (ridges) of Bou Draa, Outita, Moulay Yakoub Hamma, Nouillat, Ari, Koudiat Bou Azouf, Kheloua and Kefs. The western boundary of the Prerif ridges is probably a great fault with a NE-SW trend, especially visible in the seismic sections. It separated the ridges in the east, undergoing uplifting since Late Miocene, from the Gharb basin to the west, downthrown during the same period (CIRAC, 1985; LITTO et al., 2001). The ridges of Tselfat, Bou Kannfoud, Fert-el-Bir, Dehar-en-Sour, Zerhoun, Nesrani and Kannoufa make up the eastern group, forming a more open arc than the western group. To the east, the Jbel Trhat and Jbel Zalarh are two tiny insulated Ridges outcropping on both sides of Fes. The Prerif ridges are generally interpreted as tectonic elements compressed and transported southwestward between two fault zones with NE-SW orientation (VIDAL and FAUGUÈRES, 1975).

The Prerif ridges are formed by a sedimentary sequence starting with Triassic rocks (evaporites, red clays and basalts) that crop out very locally, yet are well represented in the seismic profiles; a thick Jurassic sequence (reaching 2000 m), with dolostones and limestones at its base (Lowermost Jurassic), marly limestones (Lower Jurassic) and sand and marls (Bajocian); and locally, a marly Cretaceous series in the Eastern Ridges. Unconformable Lower and Middle Miocene marls as well as sandstones of Middle-Upper Miocene cover the Mesozoic series of the ridges. The Mesozoic sedimentary sequence of the ridges is separated from the basement rocks of the Meseta at the level of the Triassic rocks. Geophysical studies (ZIZI, 1996a; b) based mainly on seismic reflection profiles complement the surface observations, and indicate that the Triassic rocks are thicker under the area of the ridges than below the Saïss basin.

Field data show that the ridges generally correspond to anticlinal hinges of SW and S vergence located in the mountain front and associated with thrusts of the Prerif nappe sheets, deforming up to the Neogene basins of Saïss and Gharb. They are complex folds, with a kilometric length and periclinal ends (Fig. 1). In the study area there are vergent folds such as Jbel Zalarh and Jbel Trhat (Fig. 2), and fault related folds, including Jbel Kannoufa (Fig. 3), Jbel Zerhoun and Jbel Kefs.

Jbel Trhat is a fold with an E-W oriented axis, practically undeformed by faults (Fig. 2). The fold is tighter toward the west, where layers of the southern side become overturned and a periclinal end can be observed. The fold asymmetry clearly indicates southward tectonic transport. This fold developed after Pliocene because conglomerates of this age are also overturned (Fig. 2).

Jbel Kannoufa (Fig. 3) is a vergent fold clearly associated with reverse faults. The fold has an E-W elongation and its cross section shows that it is very asymmetrical, with a long normal limb and a short vertical or overturned limb showing southward vergence. The overturned limb is cut by reverse faults.

Although most of the ridges were investigated in the field, it was difficult to find kinematic structures because of poor access, and the main faults associated with the ridges are generally covered by rock-fall. Kinematic criteria were observed clearly only at the ridges located at the northern edge of the Saïss basin. Most of these structures are striae on low-angle reverse faults and indicate a top-to-the S-SE transport (Fig. 1). The striae were recognized on the thrust on the Plio-Quaternary sediments of the Saïss basin at the ridges formed by fold-related faults (Kefs and

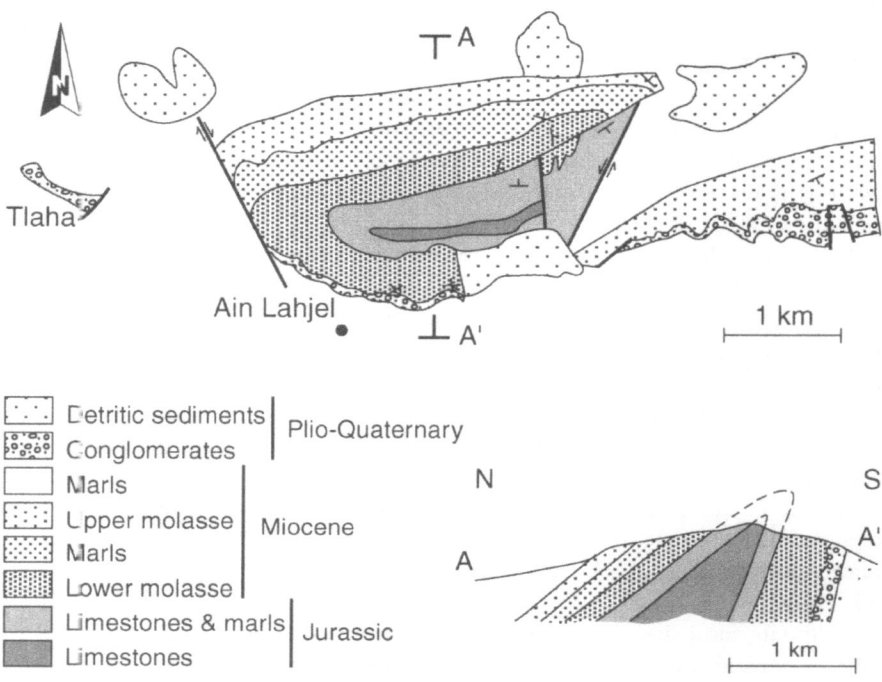

Figure 2
Geological sketch map and detailed cross section of Jbel Trhat. Location in Figure 1.

Kanncufa) and on small faults with low dip on Jurassic limestones of the southern edge of Jbel Zalarh.

Saïss-Gharb Basins

The Saïss-Gharb basins conform a depression that extends eastward from the Atlantic to the Taza Strait. This basin opened in Late Miocene after the collapse of the northern edges of the Western Meseta and the Middle Atlas. The Saïss basin is located to the east and contains the towns of Fes and Meknes, while the Gharb basin is located to the west.

Seismic profiles and boreholes carried out in the Gharb basin show the nature of its basement: Paleozoic granites are followed by a Triassic-Jurassic lacustrine sequence, a Miocene sequence similar to that of Saïss basin, and a very thick Pliocene marine sequence. The Prerif thrust sheets are interfingered with Upper Miocene rocks in the eastern part of the basin, and with Pliocene rocks in its western part.

The Saïss basin behaved as a subsiding marine basin during the Late Miocene, then was lacustrine during Late Pliocene and Quaternary (TALTASSE, 1953;

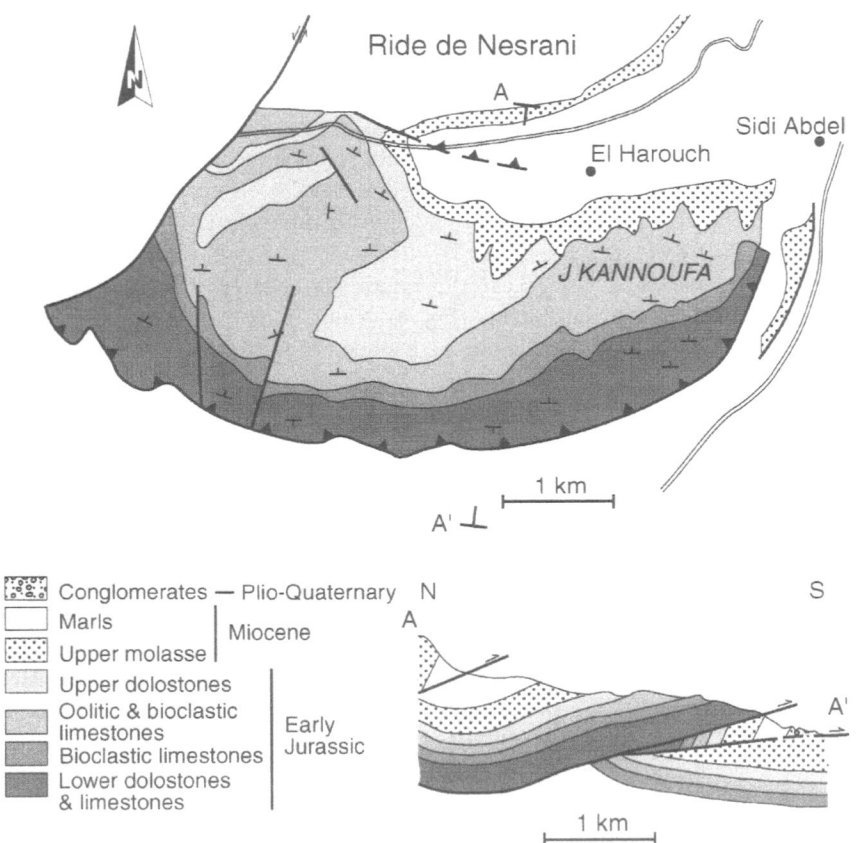

Figure 3
Geological sketch map and detailed cross section of Jbel Kannoufa. Modified from HADDAOUI (2000).
Location in Figure 1.

FEINBERG, 1986). The Gharb basin underwent rapid subsidence, especially during the Pliocene, at the same time as it was thrusted by the Prerif thrust sheet (FEINBERG, 1986; FLINCH, 1996; LITTO et al., 2001). The northern edge of Saïss basin was subjected to significant compressive deformations and the thrusting of the External Rif and the Prerif ridges. Boreholes and seismic profiles across the Saïss basin show that it has a wedge-shaped infill thickening northward and a Paleozoic substratum of the type seen in the Meseta. The sedimentary cover sequence starts with Triassic and Jurassic series that respectively correspond to clays and evaporites (extending to 1000 m thickness) and to carbonates (reaching 2000 m thickness) affected by normal faults of the Triassic-Jurassic rifting. Over these rocks is a marly sequence of Late Miocene age with a thickness between 500 and 1000 m, a few tens of meters of marine sand ("Sable fauve") of Early-Middle Pliocene age, and a lacustrine sequence of Late Pliocene-Quaternary age made up of conglomerates, marls and limestones.

The Pliocene and Quaternary infill of the basin exhibits the recent evolution of the area. Pliocene limestones have undergone compressional and extensional deformations. In the northern part of the basin, in the area of Guerzine (Fig. 1), these rocks are deformed by normal NNE-SSW syn-sedimentary faults, and later by folds *en echelon* with axes of N120°E to N150°E that are related with N120°E to E-W sinistral faults, and finally by reverse faults of NE-SW orientation (AHMAMOU and CHALOUAN, 1988).

South of Jbel Kefs, very near the ridges, the Pliocene "Sables fauves" are deformed by normal fault systems (Fig. 4). Faults have variable trends between N70°E and N100°E, and they feature high-angle northward and southward dips.

Meseta-Atlas Foreland

The Meseta-Atlas foreland is mostly hidden under the Neogene formations of the Saïss and Gharb basins, and locally (*Col of Taza*) it is directly thrusted by the External Rif. This tectonic foreland can be subdivided into two parts: the western segment is represented by the hercynian Meseta, and the eastern part is represented by the Middle Atlas. The Atlas foreland, mainly formed by carbonatic Mesozoic series, behaved differently along its northern edge during the alpine shortenings. In its western part, it had a relatively stable tectonic behavior; it was thus deformed by NE-SW faults with normal or reverse slip. Mesozoic sedimentary sequences preserved their nearly horizontal position (*Causse moyen-atlasique* or Tabular Middle Atlas, COLO, 1961; CHARRIÈRE, 1990). In its eastern part, the Middle Atlas evidences significant alpine deformations and generally NE-SW oriented fractures (folds and reverse faults of the folded Middle Atlas).

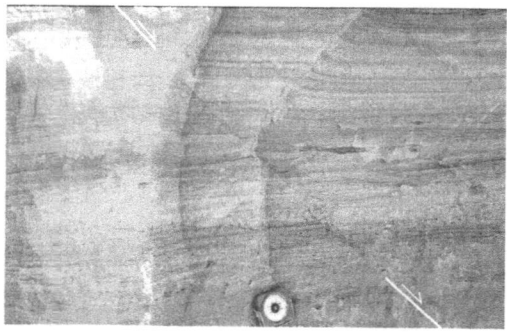 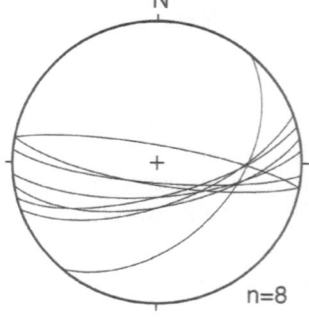

Figure 4
Normal faults at Ain Jemaa (western end of Jbel Kefs). A. field view; B. stereographic projection. Lower hemisphere. n. number of fault planes.

The Meseta foreland is made up principally of Paleozoic sequences (Cambrian to Permian) that underwent a penetrative hercynian deformation (and locally Caledonian in the Sehoul block; HASSANI, 1990). Later, Triassic rifting produced horsts and grabens related to NE-SW normal faults; the graben of Khémisset (approximately 20×50 km) constitutes a good example, still visible at the surface. Near Rabat, the basement is covered by Miocene subhorizontal yellow marls (FEINBERG and LORENZ, 1970), eolic and marine Pliocene and Quaternary sediments (CHOUBERT and AMBROGGI, 1953; WERLNI, 1978) and Quaternary basalts. Three Quaternary marine levels (Maarifian, Anfatian and Rabatian) have been identified (BOURCART et al., 1949).

During the alpine deformation the Meseta behaved like a rigid block. It did not undergo notable deformations, other than a weak reactivation of the Triassic normal faults and a reactivation of the depression of the Khémisset graben, eventually invaded by the Miocene Sea. The area near Rabat (Fig. 5) may be considered representative of the deformations assigned to the foreland of the Rif Cordillera. In this area, the sedimentary Neogene and Quaternary rocks are affected by brittle deformations—mainly tension joints and normal and transcurrent faults—in the outcrops of Motorway, Douar Doum, Rabat and Temara (Fig. 5). The first three outcrops are located north of Temara, which is situated near the Palaeozoic outcrops.

The outcrop of Motorway (Fig. 5) shows a complete sequence from Late Miocene until the Quaternary. It is deformed by subvertical fractures with NE-SW and NNW-SSE preferent trends (Fig. 5), which cut most of the older levels. Some fractures show displacement with the southward or westward downthrown block. Moreover, one of the NNW-SSE joints was reactivated as a dextral fault.

In the locality of Douar Doum (Fig. 5) Pliocene calcarenite levels (Mograbian; CHOUBERT, 1965) overlie Miocene marls. There are several subvertical fractures with ENE, SE and SSE orientations, which also show subvertical striations and decimeter slips. Some joints were reactivated as faults with vertical displacements.

The outcrops of Tour Hassan and Oudayas, situated at Rabat (Fig. 5) are located on Quaternary marine rocks of Anfatian, Maarifian and Rabatian origin (BOUCART et al., 1949; AKIL, 1990). There are two well developed sets of tensional joints with NNE-SSW and ESE-WNW orthogonal orientations. The joints are sometimes open and some are filled with more recent sediments (Fig. 6). Moreover, syn-sedimentary E-W normal faults of decameter slip determine the position of channels between the Maarifian and Anfatian levels (Fig. 6).

At the Temara quarry (Fig. 5) superposed marine levels are affected by sets of joints with highly variable orientation (NNE, ENE, SW and SSW) and some faults. The representations of the joints found in each level (Fig. 5) indicate that the Pliocene rocks are deformed mainly by two sets of orthogonal tension joints with ENE and SSE orientations. The Messaoudian and Maarifian rocks show more complex orientations with four sets of joints, the most predominant ones having E-W

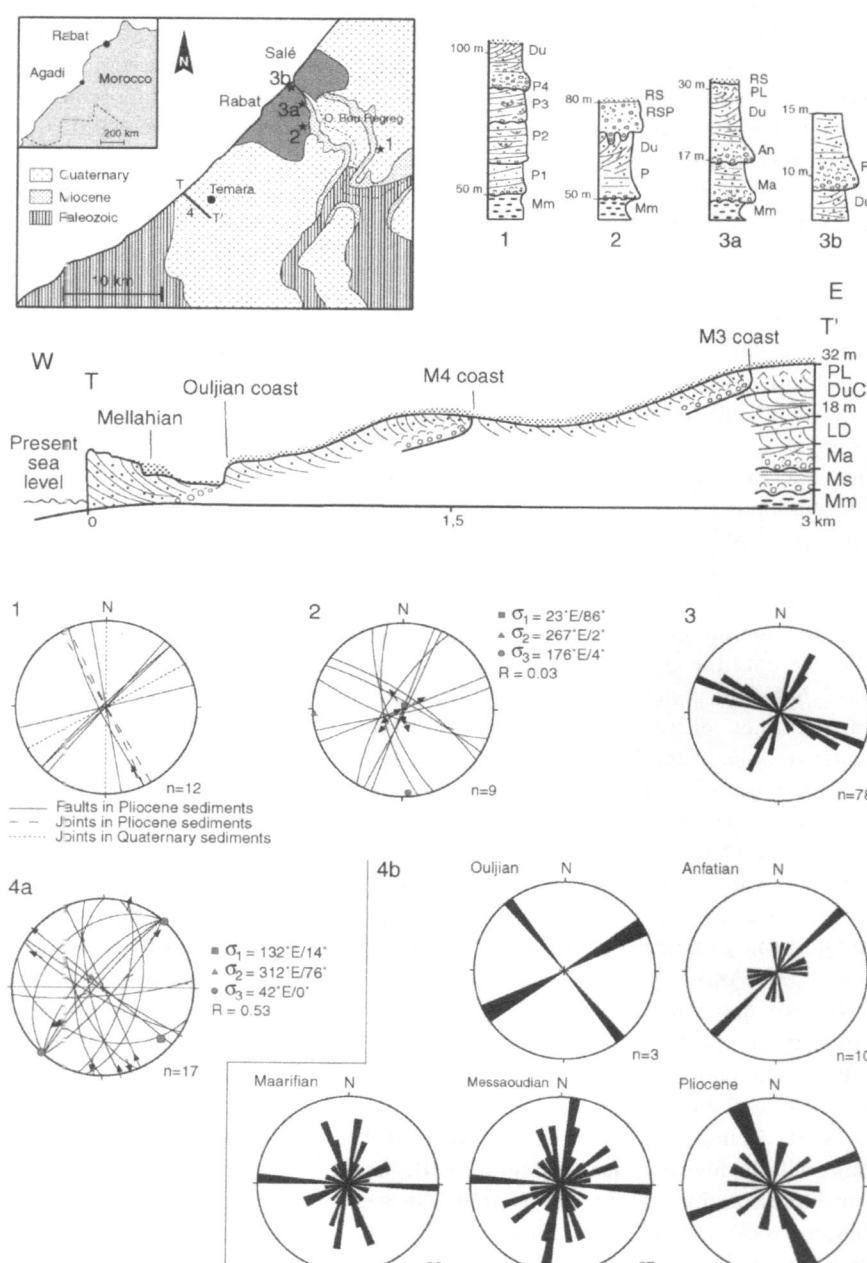

orientation. On the Anfatian rocks a NE-SW preferent orientation can be seen, which indicates a NW-SE trend of extension. The Ouljien sediments, though less deformed, show the same preferent orientations as the Pliocene rocks with two sets of orthogonal tensional joints. The faults are transcurrent, transtensional and normal; and they present a complex model of faulting, with widely varying orientations (Fig. 5).

Paleostresses

In most of the Cordilleras, the study of the brittle structures provides a wealth of information about the stresses and deformation that continued in an area—most notably in the External Zones—free of penetrative structures. In the Rif Cordillera, several of the studied outcrops are deformed by faults and joints that yield clues as to the recent stress field of the mountain front (Prerif ridges) and the foreland basin (Saïss and Gharb basins), and as to the propagation of stresses to the foreland. The determination of stresses (orientation of main axes and axial ratio of stress ellipsoid) from fault striae orientations on outcrops and on pebbles using the research grid method (GALINDO-ZALDÍVAR and GONZALEZ-LODEIRO, 1988) has produced good results in different sectors of the Betic cordillera (GALINDO-ZALDÍVAR et al., 1993).

In Jbel Trhat (Figs. 1 and 2), located on the topographic front of the Prerif ridges, deformed Pliocene conglomerates lie near the frontal ridge fold. Most of the conglomerate pebbles manifest solution pits and striations, thus indicating subvertical compression. Analysis of the striations on pebbles (Fig. 7) reveals the main deformation to be prolate, with a subvertical axis of compression. The development of the pebble striations can be related to the same deformation that produced the fold with an E-W subhorizontal axis. Therefore, if the fold is restored, these data suggest that after Pliocene the area was subjected to prolate N-S directed compression, which produced the fold with southward vergence.

In the Saïss basin, located on the foreland, stresses show a complex setting. The Pliocene *sables fauves* at the northern edge indicate that the area underwent a recent N-S oriented extension (Fig. 4). In Ain Lorma, then, the microfaults that developed in lacustrine limestones of the central part of the basin indicate two distinct stages of

◄

Figure 5

Geological setting of the Rabat region and location of the studied outcrops. 1. Motorway; 2. Douar Doum; 3a. Rabat, Hassan Tower; 3b. Rabat, Oudayas; 4. Temara (T-T' section in geological sketch). Sedimentary sequences of each outcrop are indicated. Mm, Miocene; P, P1, P2 and P3, marine Pliocene; P4, Pliocene limestones; Ms, Messaoudian; Ma, Maarifian; An, Anfatian; Du, aeolian dune; R, Rabatian; RS, red sands; RSP, red sand with pebble; LD, lacustrine deposits; DuC, dune complex; PL, powdered limestones. Brittle structures and paleostress determinations are represented in stereographic projection, lower hemisphere. Axial ratio $R = (\sigma_2 - \sigma_3)/(\sigma_1 - \sigma_3)$; 4a. Temara faults ; 4b. Temara joints grouped by age of affected rocks.

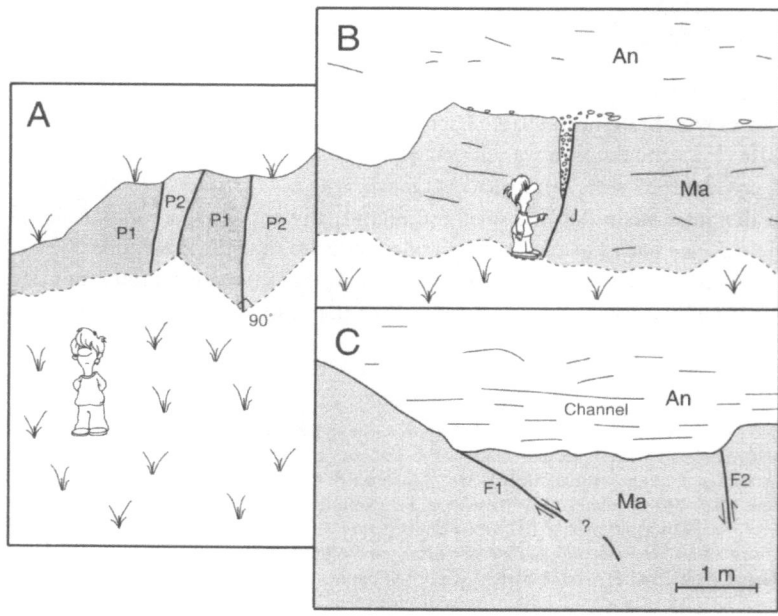

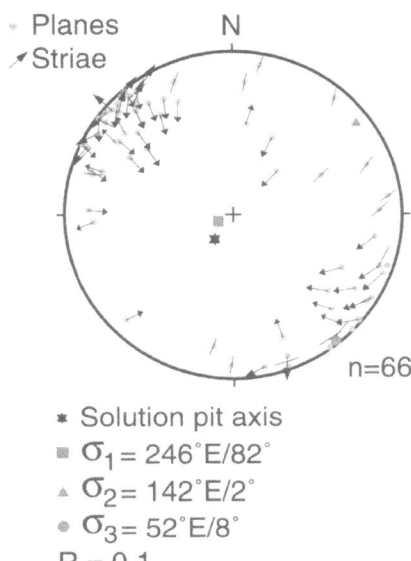

Figure 7

Striated pebbles at the southern part of Jbel Trhat. Field view and stereographic projection, lower hemisphere, of striations showing the results of the determinations of paleostress. Axial ratio
$$R = (\sigma_2 - \sigma_3)/(\sigma_1 - \sigma_3).$$

deformation characterized by extension with a NNE-SSW trend and an oblate stress ellipsoid, and triaxial deformation with NE-SW compression and orthogonal extension (Fig. 8). The study of the western part of the Saïss basin (AHMAMOU and CHALOUAN, 1988) indicates the superposition since Pliocene of WNW-ESE directed extension, NE-SW oriented compression and, finally, NW-SE compression.

The foreland displays weak though complex deformation. The Pliocene and Quaternary rocks of the area near Rabat (Figs. 4 and 5) are generally affected by sets of orthogonal joints developed in an extensional stress field. The orientation of the sets varies in different outcrops, however NE-SW and SE-NW orientations are frequent. The joints were reactivated as normal faults in Rabat and Douar Doum, the latter outcrop (Fig. 5, Table 1) showing nearly radial extension. Thus, the stress

◀

Figure 6

Brittle deformations in Hassan Tower and Oudayas, field examples and interpretative sketches. A. Two orthogonal sets of tensional joints, P1 and P2 joint planes. B. Subvertical N45°E tensional joint (J) in Maarifian marine rocks (Ma) filled and sealed by Anfatian rocks (An). C. Normal faults with strikes comprised between N80°E (F1) and N120°E (F2) that deform Maarifian rocks (Ma) and determine the position of a channel located at the bottom of the Anfatian rocks (An).

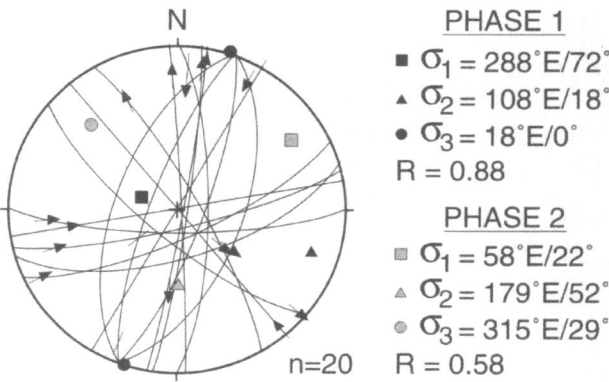

Figure 8
Paleostress determinations from microfaults of Pliocene limestones at Ain Lorma. Axial ratio
$R = (\sigma_2 - \sigma_3)/(\sigma_1 - \sigma_3)$.

field must be generally extensional, the mean and minor stresses having close values; this would facilitate the permutation of the stress axes and the development of two sets of orthogonal joints.

The Motorway and Temara outcrops are deformed by transcurrent faults (Fig. 5). The paleostress determination from the faults of the Temara quarry indicates the existence of transcurrent stresses, with triaxial ellipsoids characterized by a NW-SE subhorizontal maximum compression axis and a NE-SW tension axis. The presence of some faults that are not associated with these stresses (dextral subvertical faults of N170°E orientation, similar to the transcurrent dextral fault of the Motorway outcrop), indicate that a NNE-SSW compression was also active. In the Motorway outcrop, the fault corresponds to a reactivated joint, indicating very recent deformation.

Discussion

The Prerif ridges are elongated hills associated with folds and reverse faults developed by the southward propagation of alpine deformation related to the Rif Cordillera. The development of this type of structure requires the presence of a detachment level which, in the Prerif thrust sheets, is generally located in the Triassic-Paleozoic basement boundary. The location of the ridges was also determined by the existence of Triassic rocks at depth, found in the Triassic basins. Seismic reflection profiles (ZIZI, 1996a,b) show that Mesozoic normal faults affect the basement below the compressive alpine structures represented by the Prerif ridges. This area underwent inversion tectonics (Fig. 9C) that determine the preferent NE-SW and

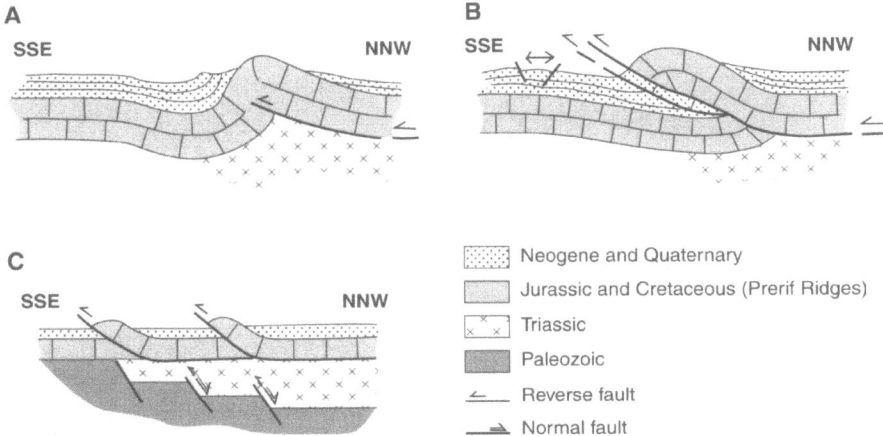

Figure 9

Model of development of the Prerif ridges. A. Several ridges (as Jbel Trhat and Jbel Zalarh) are probably formed by folds related to blind faults. Folds deform up to Plio-Quaternary rocks that become locally vertical. B. Most of the ridges are formed by S-SSE vergent folds-related-faults. Extensional deformation in footwall may be a consequence of the flexure of the basement during the ridge emplacement. C. The ridges are located along the border of a Triassic basin that has undergone tectonic inversion.

ENE-WSW orientation of the ridges, probably related to the same orientation of the normal faults of the Mesozoic basin boundaries, where there was an accumulation of Triassic marls and gypsum. This tectonic setting resembles that observed to the north, at the Jebha-Arbaoua lineation, with similar orientation as that at the front of the Prerif ridges, which is interpreted as an alpine tectonic inversion of Mesozoic faults (MORLEY, 1987). At any rate, inversion was not so intense as to erase the normal character of the faults at the basement. The southward thinning of Triassic rocks probably stopped the propagation of recent compressive deformation towards the Saïss basin. The folds observed in the Guerzine area (Fig. 1) (AHMAMOU and CHALOUAN, 1988) may well be related to blind thrusts affecting the sedimentary infill of the basin (Fig. 10).

Alpine tectonics produced the inversion of the normal basement faults which resulted in new folds and faults that are well represented in the Jurassic and Cretaceous limestones. Ridge development probably took place in two stages. The first was associated with the propagation of the NE-SW orientated compression of Internal Zones of the Cordillera and produced a southwestward tectonic transport from Early Tortonian up to the Messinian (MOREL, 1989). This tectonic transport trend is well documented in the frontal thrust sheets of the Ketama-Tánger units (LESPINASSE, 1975; FRIZON DE LAMOTTE, 1985; MORLEY 1987; FRIZON DE LAMOTTE et al., 1991) located in the internal sector of the Cordillera. Later, there was an anticlockwise rotation of the trend of compression, to almost N-S during the

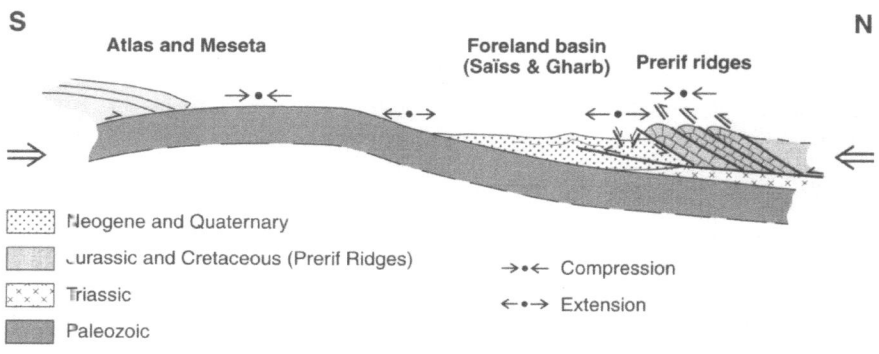

Figure 10
General simplified sketch of the frontal part of the Rif Cordillera, showing the areas in compression from Plio-Quaternary age to the present.

Quaternary. This rotation was well documented by AÏT BRAHIM and CHOTIN (1989) in the boundary between the Subrif Strait and the Prerif and by general studies of paleostresses and deformations in the Rif overall (MOREL, 1989; MEGHRAOUI *et al.*, 1996). Our field observations on striated pebbles and on fault kinematics indicate that the ridges located on the northern edge of the Saïss basin, which represents the front of the Rif mountains, developed during a N-S to NNW-SSE prolate stage of compression (Fig. 2). This deformation, which occurred at the most external zones of the Cordillera, is compatible with the NW-SE trend of present-day convergence between the Eurasian and African plates (DEMETS *et al.*, 1990) and the regional stresses (ZOBACK and BURKE, 1993). Prerif ridges can constitute active structures and, moreover, are associated with very frequent thermal springs and seismicity (the destructive Meknes earthquake of 1755). The earthquake focal mechanism that indicates WNW-ESE compression (MEDINA and CHERKAOUI, 1992) may be related to reactivated faults at the basement, and it is not enough to determine the present-day stress field.

Although the ridges are associated with asymmetrical folds with southward vergence, there are ridges with well-developed reverse faults at their base (such as Jbel Kefs and Jbel Kannoufa) and others where no remarkable reverse fault crops out at surface (Jbel Trhat and Jbel Zalarh) but which deform the same rocks (Figs. 9A-B). One possible explanation of this setting may be that these active structures have different degrees of development. The structures start to develop as fold related faults with a blind thrust at the fold core, and finally cut the overturned fold limb.

At the northern edge of the Saïss basin there is a very abrupt change between the compressive stresses that deform the ridges and the mainly extensional stresses that deform the sedimentary infill of the foreland basin. In this area, the trend of maximum compressive stresses of the prolate ellipsoid of the Prerif ridges became approximately the trend of minimum stress of the oblate ellipsoid determined in the

Saïss basin (Fig. 8). This extension (Fig. 10) must be related to the flexure of Paleozoic basement as a response to the emplacement of the ridges at the frontal part of the Rif Cordillera. The extensional deformations, intense and well directed near the ridges, become weaker and complex far from the mountain front (Fig. 10). The stress field is very complex from Pliocene up to present, and followed a counter-clockwise rotation of the trend of compression from NE-SW to NW-SE (AHMAMOU and CHALOUAN, 1988), which is responsible for both paleostress ellipsoids observed at Ain Lorma (Fig. 8).

The deformations of the Rif Cordillera also propagate to the foreland, and were analyzed in the area of Rabat. There are brittle deformations since the Pliocene in this area, which had been considered stable from a tectonic point of view. This area is also affected by recent and active progressive deformations: the older rocks are more deformed than the younger ones, and syn-sedimentary faults are recognized (Fig. 5). Although the trends of the fracture sets are constant in an outcrop, they are variable on the scale of the entire area. This tectonic setting is typical of radial extension, with weak differences between the intermediate and minor stress axes, which may allow the local permutation of the stress axes. Also typical of this setting are the subvertical faults with vertical slips observed in Douar Doum. This setting is probably the consequence of the flexure of the Meseta; however a major difference with the Saïss Basin is that far from the mountain front, the trends of extension are not as well determined. In addition, there are intercalated events of transcurrence, with compression and horizontal extension. We have identified both NW-SE and NNE-SSW compression, which can be connected to the propagation towards the foreland of the compressive deformations, related to Eurasian and African plate convergence, although which in detail produces a heterogeneous stress field. The heterogeneity of the joints recognized in rocks of different ages of the Temara Quarry (Fig. 5) also testifies to this type of deformation. This setting evokes that of the Internal Zones, with permutation of the present-day stresses determined by seismicity (MEDINA, 1995). The Temara Quarry, located in the Meseta and south of the other outcrops (Fig. 5), shows the compression deformations better than the outcrops near Rabat, located at the edge of the Gharb basin.

Conclusions

The frontal region of the Rif Cordillera has undergone alpine inversion tectonics of the extensional and transcurrent structures that determine the position of the Mesozoic basins of the region. The location of the Prerif ridges in the mountain front, with ENE-WSW trends, is the consequence of the deep distribution of the Triassic rocks that constitute the main detachment level. This setting impedes the possible propagation of the deformation to the Saïss basin, which represents the foreland basin.

The Prerif ridges located at the northern edge of the Saïss basin are folds of southwards vergence associated with reverse faults that sometimes cut the topographic surface (Jbel Kefs and Jbel Kannoufa) or are less developed structures, with vergent folds probably associated with reverse blind thrusts (Jbel Trhat and Jbel Zalarh). The reverse faults have a S-SEwards tectonic sense of transport. These structures are probably active, as they deform up to Quaternary rocks. The striated pebbles signify that the stress field where these structures developed is characterized by N-S to NNW-SSE prolate compression.

The Saïss foreland basin underwent counter-clockwise rotation of the trend of compression since Pliocene times, from NE-SW to NW-SE with related orthogonal extension. The more recent deformations (microfaults and joints) suggest that the extensional setting is related with the compression at the Prerif ridges as a consequence of the crustal flexure of the Meseta. The N-S to NNE-SSW trend of extension is well defined near the ridges, with oblate stress ellipsoids, yet extension is more complex far from the topographic mountain front.

Deformation in the southernmost edge of the foreland basin and in the foreland itself—represented by the Gharb basin and Meseta in the area of Rabat—is weaker but also active. It develops sets of orthogonal tension joints and normal and transcurrent faults that deform up to Quaternary rocks. The stress field is generally characterized by radial extension, with permutations and progressive variations of the trend of extension in close outcrops. The reactivation of fractures as transcurrent faults indicates NW-SE and NNE-SSW compression, more often in the outcrops located south in the foreland.

The Rif Cordillera is characterized at present by the succession of zones in extension and compression: the Internal Zone in extension, compression in the External Zones and Prerif ridges, extension in the foreland basin and compression again in the foreland. The regional stresses of NW-SE compression related to Eurasian-African convergence are largely disturbed by the development of active structures in the Cordilleras, resulting in a locally heterogeneous stress field.

Acknowledgements

We thank J. Sanders for the English revision. This research was supported by project BTE-2000-1490-C02-01 and the Spanish-Moroccan Inter-Universities Cooperation Programs 29/PRO/99 of the A.E.C.I. and the Junta de Andalucía.

REFERENCES

AHMAMOU, M. and CHALOUAN, A. (1988), *Distension synsédimentaire plio-quaternaire et rotation anti-horaire des contraintes au Quaternaire ancien sur la bordure nord du bassin du Saïss (Maroc)*, Bull. Inst. Sci. Rabat *12*, 19–26.

Aït Brahim, L. and Chotin, P. (1989), *Genèse et déformation des bassins néogènes du Rif central (Maroc) au cours du rapprochement Europe-Afrique*, Geodinamica Acta *3*, 295–304.

Akil, M., *Les dépôts quaternaires littoraux entre Casablanca et Cap Beddouza (Meseta côtière marocaine): Etudes géomorphologiques et sédimentologiques*, [unpublished Ph. D. thesis] (Université Mohammed V, Rabat 1990).

Bourcart, J., Choubert, G., and Marçais, J. (1949), *Sur la stratigraphie du Quaternaire côtier de Rabat*, C.R.Acad. Sci. Paris *201*, 108–109.

Chalouan, A., Ouazani-Touhami, A., Mouhir, L., Saji, R., and Benmakhlouf, M. (1995), *Les failles normales à faible pendage du Rif interne (Maroc) et leur effet sur l'amincissement crustal du domaine d'Alboran*, Geogaceta *17*, 107–109.

Chalouan, A., Saji, R., Michard, A., and Bally, A.W. (1997), *Neogene tectonic evolution of the southwestern Alboran Basin as inferred from seismic data of Morocco*, AAPG Bull. *81*, 1161–1184.

Charrière, A., *Héritage hercynien et évolution géodynamique alpine d'une chaîne intracontinentale: le Moyen-Atlas au SE de Fès (Maroc)*, [unpublished Ph. D. thesis] (Université de Toulouse, Toulouse 1990).

Choubert, G. and Ambroggi, R. (1953), *Note préliminaire sur la présence de deux cycles dans le Pliocène marin du Maroc*, Notes Serv. Géol. Maroc *7*, 3–72.

Choubert, G. (1965), *L'étage Moghrébien dans le Maroc occidental*, Notes Serv. Géol. Maroc *25*, 47–55.

Cirac, P., *Le bassin sud rifain occidental au Néogène supérieur. Evolution de la dynamique sédimentaire et de la paléogéographie au cours d'une phase de comblement*, [unpublished Ph. D. thesis] (Université de Bordeaux I, Bordeaux 1985).

Colo, G. (1961), *Contribution à l'étude du Jurassique du moyen-Atlas septentrional*, Notes et Mém. Serv. Géol. Maroc *6*, 27–31.

deMets, C., Gordon, R.G., Argus, D.F., and Stein, S. (1990), *Current plate motions*, Geophys. J. Int. *101*, 425–478.

Durand-Delga, M., Rossi, P., Olivier, P., and Puglisi, D. (2000), *Structural Setting and Ophiolitic Nature of Jurassic Basic Rocks Associated with the Maghrebian Flyschs in the Rif (Morocco) and Sicily (Italy)*, C. R. Acad. Sci. Paris *331*, 29–38.

Faugères, J.C., *Les rides sud-rifaines: Evolution sédimentaire et structurale d'un bassin atlantico-mésogéen de la marge africaine*, [unpublished Ph. D. thesis] (Université de Bordeaux I, Bordeaux 1978).

Feinberg, H. and Lorenz, H. (1970), *Nouvelles données stratigraphiques sur le Miocène supérieur et le Pliocène du Maroc*, Notes Serv. Géol. Maroc *30*, 21–26.

Feinberg, H. (1986), *Les séries tertiaires des Zones Externes du Rif (Maroc). Biostratigraphie, paléoecologie et aperçu tectonique*, Notes et Mém. Serv. Géol. Maroc *315*, 192 pp.

Flinch, F.J., *Accretion and extensional collapse of the external Western Rif (Northern Morocco).* In *Peri-Tethys Memoir 2: Structure and Prospects of Alpine Basins and Forelands* (eds. Ziegler, P.A. and Horvath, F.) (Mém. Mus. Nation. Hist. Natur., Paris 1996) pp. 61–85.

Frizon de Lamotte, D., *La structure du Rif Oriental (Maroc)*, [unpublished Ph. D. thesis] (Université Pierre et Marie Curie, Paris 1985).

Frizon de Lamotte, D., Andrieux, J., and Guézou, J.C. (1991), *Cinématique des chevauchements néogènes dans l'Arc bético-rifain: Discussion sur les modèles géodynamiques*, Bull. Soc. Géol. de France *162*, 611–626.

Galindo-Zaldívar, J. and González-Lodeiro, F. (1988), *Faulting Phase Differentiation by Means of Computer Search on a Grid Pattern*, Annales Tectonicae *2*, 90–97.

Galindo-Zaldívar, J., González-Lodeiro, F., and Jabaloy, A. (1993), *Stress and Paleostress in the Betic-Rif Cordilleras (Miocene to Present)*, Tectonophysics *227*, 105–126.

González-Lodeiro, F., Aldaya, F., Galindo-Zaldívar, J., and Jabaloy, A. (1996), *Superimposition of Extensional Detachments During the Neogene in the Internal Zones of the Betic Cordilleras*, Geol. Rundsch. *85*, 350–362.

Haddaoui, Z., *Influence de la géometrie d'un bassin Jurassique sur la propagation des chevauchements néogènes: Geodynamique meso-Cenozoïque des rides sud-rifaines (Maroc). Modelisation geometrique et numerique*, [unpublished Ph. D. thesis] (Université Mohammed V, Rabat 2000).

Hassani, A., *La bordure nord de la chaîne hercynienne du Maroc, chaîne calédonienne des Shoul et plateforme nord mesetienne*, [unpublished Ph. D. thesis] (Strasbourg University, Strasbourg 1990).

LESPINASSE, P., *Géologie des zones externes et des flysch entre Chaouen et Zoumi(centre de la chaîne rifaine, Maroc)*, [unpublished Ph. D. thesis] (Université Pierre et Marie Curie, Paris 1975).

LITTO, W., JAAIDI, E., MEDINA, F., and DAKKI, M. (2001), *Seismic Study of the Structure of the Northern Margin of the Gharb Basin (Morocco) - Evidence for a Late Miocene Distension*, Eclogae geol. Helv. *94*, 63–74.

MEDINA, F. and CHERKAOUI, T.E. (1992), *Mécanismes au foyer des séismes du Maroc et des régions voisines (1959–1986). Consequences tectoniques*, Eclogae geol. Helv. *85*, 433–457.

MEDINA, F. (1995), *Present-day state of Stress in Northern Morocco from Local Mechanism Analysis*, J. Struct. Geol. *17*, 1035–1046.

MEGHRAOUI, M., MOREL, J.L., ANDRIEUX, J., and DAHMANI, M. (1996), *Pliocene and Quaternary Tectonics of the Tell-Rif Mountains and Alboran Sea, a Complex Zone of Continent Convergence*, Bull. Soc. Géol. de France *167*, 141–157.

MOREL, J.L. (1989), *Etats de contrainte et cinématique de la chaîne rifaine (Maroc) du Tortonien à l'actuel*, Geod namica acta *3*, 283–294.

MORLEY, C.K. (1987), *Origin of a Major Cross-element Zone: Moroccan Rif*, Geology *15*, 761–764.

OUAZANI-TOUHAMI, A. and CHALOUAN, A. (1995), *La distension de l'Oligocene supérieur à Burdigalien dans les nappes Ghomarides (Rif interne septentrional, Maroc)*, Geogaceta *17*, 113–116.

PLATT, J.P. and VISSERS, R.L.M. (1989), *Extensional Collapse of Thickened Continental Lithosphere: A Working Hypothesis for the Alboran Sea and Gibraltar Arc*, Geology *17*, 540–543.

TALTASSE, P. (1953), *Recherches géologiques et hydrogéologiques dans le bassin lacustre de Fès-Meknès*, Notes et Mém. Serv. Géol. Maroc, 300 pp.

VIDAL, J.C. and FAUGÈRES, J.C. (1975), *Une nouvelle interprétation de la structure des Rides sudrifaines (Rif, Maroc): rôle des décrochements*, C.R.Acad. Sci. Paris *281*, 1951–1954.

WERLNE, R. (1978), *La base du Moghrébien est d'âge pliocène moyen(zone à G. crassaformis)*, Arch. Sci. Genève *31*, 129–132.

ZIZI, M., *Triassic-Jurassic extension and Alpine inversion in Northern Morocco*. In *Peri-Tethys Memoir 2: Structure and Prospects of Alpine Basins and Forelands* (eds. Ziegler, P.A. and Horvath, F.) (Mém. Mus. nation. Hist. natur., Paris 1996a) pp. 87–101.

ZIZI, M., *Triassic-Jurassic Extensional Systems and their Neogene Reactivation in Northern Morocco (The Rides Prerifaines and Guercif Basin)*, [unpublished Ph. D. thesis] (Rice University, Houston 1996b).

ZOBACK, M.L. and BURKE, K., *World Stress Map* (EOS, Am. Geophy. Union, Washington D. C. 1993).

(Received February 11, 2002, revised February 25, 2002, accepted March 7, 2002)

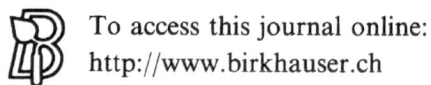

To access this journal online:
http://www.birkhauser.ch

Pure appl. geophys. 161 (2004) 541–563
0033–4553/04/030541–23
DOI 10.1007/s00024-003-2462-5

Pure and Applied Geophysics

Recent Tectonic Structures in a Transect of the Central Betic Cordillera

P. Ruano[1], J. Galindo-Zaldívar[1], and A. Jabaloy[1]

Abstract—The Betic Cordillera has undergone recent Alpine deformations related to the Eurasian-African plate interaction boundary. Most of the present-day relief has been built up since Tortonian times, and is related to the development of folds and faults that are overprinted on older deformations, and some of the faults may be considered as out-of-sequence. The combination of geophysical and geological data makes it possible to determine the main features of the recent tectonic structures. or those recently active, in its central transect. The main fault is a crustal detachment that separates a footwall constituted by the Iberian Massif and a hanging wall formed by the rocks of the Betic Cordillera. While the footwall is practically undeformed, the hanging wall has been folded and faulted. The folds are mainly E-W to NE-SW and have larger sizes and higher related relieves towards the South. The reverse faults are mainly concentrated in the northern mountain front. However, normal faults affect the southern part of the Cordillera and are associated with the development of large asymmetrical basins such as the Granada Depression. In this setting, the slip along the crustal detachment is variable and should increase southwards. The model of the recent tectonics in the central transect of the Cordillera is compatible with the presence of an active subduction in the Alboran Sea, and contrasts notably with the setting of the eastern Betic Cordillera, mainly deformed by transcurrent faults.

Key words: Betic Cordillera, recent tectonics, crustal detachment, uplift, normal faulting, large folds.

Introduction

The Betic-Rif Cordilleras are formed by the interaction of the Eurasian and African plates in the western Mediterranean. Available models of plate motion indicate a NW-SE convergence in recent times of about 4 mm/yr in the studied region (DeMets *et al.*, 1990). The seismicity and tectonic deformation change in character along the plate boundary. To both the East and West of the Betic-Rif Cordilleras, in Algeria and near the Azores Islands, the band of seismicity is very sharp (Udías and Buforn, 1991). However, in the area near the Betic-Rif Cordilleras, the seismicity and surface deformations are widespread along a band over 300-km wide related to the plate boundary (Udías and Buforn, 1991; Galindo-Zaldívar *et al.*, 1993). The extensive use of geophysical data (refraction

---

[1] Departamento de Geodinámica, Universidad de Granada, 18071 Granada. Spain.
E-mail: pruano@ugr.es, jgalindo@ugr.es or jabaloy@ugr.es

and multichannel seismic reflection profiles, gravity and magnetic data, heat flow, seismicity) in the last few years, has led to the proposal of a variety of crustal scale models that involve the Betic Cordillera.

The Betic-Rif Cordilleras are a complex region with a long history of deformation. Paleozoic rocks were deformed during the Variscan Orogeny. In Early Mesozoic times, the region underwent a crustal thinning, with the development of basins and the intrusion of igneous, mainly volcanic rocks (GARCÍA-HERNÁNDEZ et al., 1980). Since Late Cretaceous, several compressional and extensional stages of deformation have led to the present-day structure of the region. In this setting, the westward motion of the Alboran block (Alboran microplate of ANDRIEUX et al., 1971) has proved very relevant in the development of the major structures of the region.

The Betic Cordillera comprises the Internal and the External Zones (Fig. 1), respectively corresponding to the Alboran Domain and the South Iberian Domain (BALANYÁ and GARCÍA-DUEÑAS, 1987). While the units that make up the Internal Zones are partially metamorphosed during the Alpine Orogeny and may include Paleozoic rocks, the units that belong to the External Zones are formed essentially by Mesozoic and Cenozoic sedimentary rocks (FALLOT, 1948). Flysch units of Cretaceous to Miocene age are located between the External and Internal zones.

The Internal Zones are composed of several superposed nappe complexes (Fig. 1). The structures related to thrusting are unknown due to the transposition by later ductile penetrative deformations. The ages obtained for the high pressure metamorphism related to thrusting and crustal thickening vary between 91 21 Ma (NIETO et al., 1997) and 15 Ma (LÓPEZ SÁNCHEZ-VIZCAÍNO et al., 2001). Consequently this matter is at present under discussion. From bottom to top, it is possible to differentiate the Nevado-Filábride, Alpujárride and Maláguide complexes that form most of the outcrops of these zones. They are composed mainly of schists, quartzites, and marbles; the Maláguide Complex also contains slates, detritic rocks and a carbonate Mesozoic and Paleogene sequence. Moreover, the Dorsal, Predorsal and Alozaina complexes, comprising Mesozoic and Cenozoic carbonate rocks, are included in the Internal Zones of the Cordillera. The contacts between the main complexes are low-angle normal faults. The Nevado-Filábride/ Alpujárride contact shows a top-to-the-W-SW sense of motion during the Early

▶

Figure 1

Geological setting of the central part of the Betic Cordillera. General map of the Betic Cordillera (a), and detailed map of the central sector (b). 1. Upper Miocene-Quaternary sedimentary rocks. 2. Upper Subbetic Unit. 3. Intermediate Subbetic Unit (the whole Subbetic Zone in part a). 4. Lower Subbetic Unit. 5. Prebetic 6. Oligocene-Lower Miocene sedimentary rocks, including flysch and olistostroms. 7. Campo de Gibraltar Flyschs units. 8. Predorsal, Dorsal, Alozaina and Maláguide complexes. 9. Alpujárride Complex. 10. Nevado-Filabride Complex. 11. Iberian Massif Cover. 12. Iberian Massif. 13. Pedroches Batholith. 14. unconformity. 15. fault. 16. low-angle normal fault. 17. high-angle normal fault 18. reverse fault. 19. syncline 20. anticline. 2-5, External Zones. 8-10, Internal Zones. The location of seismic reflection profiles in Figures 3 and 4 are indicated.

Miocene (GALINDO-ZALDÍVAR *et al.*, 1989; JABALOY *et al.*, 1993). However, the Maláguides were displaced over the Alpujárrides with a top-to-the-ENE sense of motion during Late Oligocene-Early Miocene times (ALDAYA *et al.*, 1991). In addition, translations with a top-to-the-North sense of Early Miocene age have

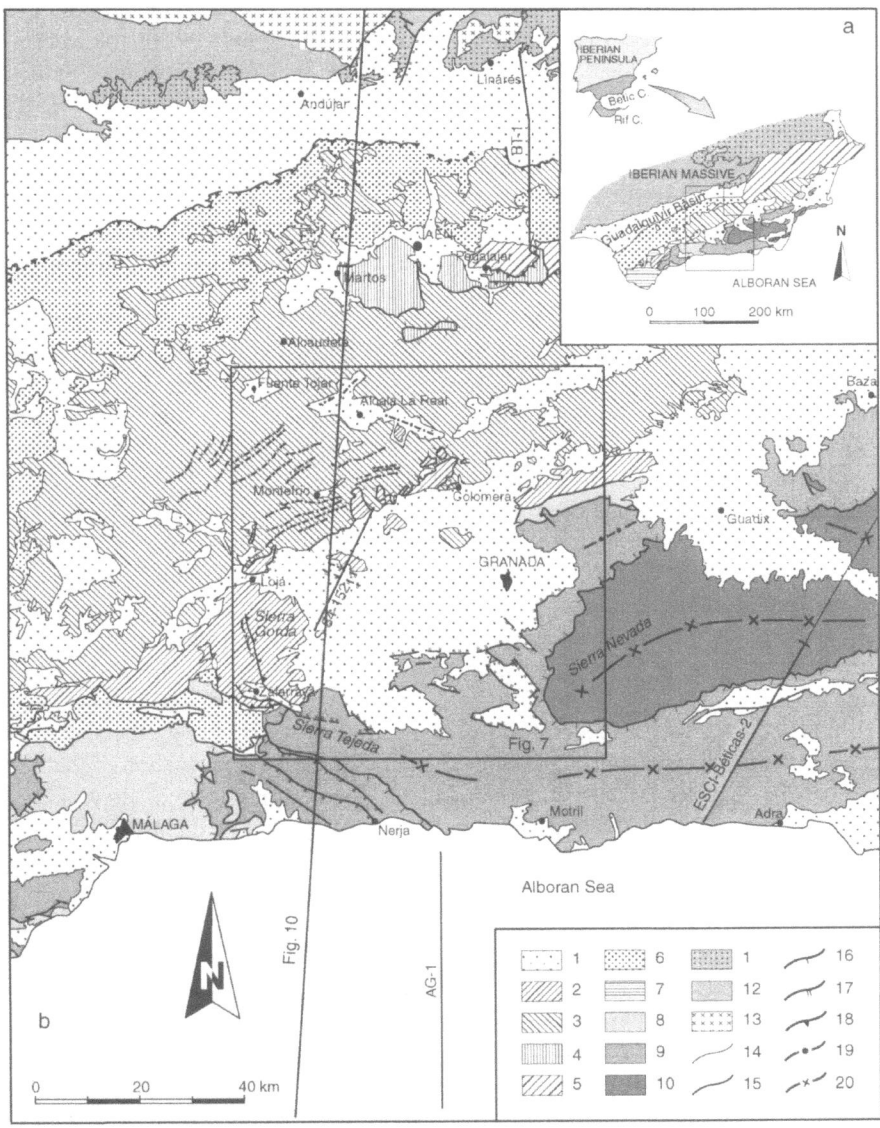

been documented in the Alpujarride complex (CUEVAS *et al.*, 1986) although their compressive or extensional character is under discussion (SIMANCAS and CAMPOS, 1993).

The External Zones are formed mostly by units with carbonate sequences of Mesozoic and Cenozoic age (GARCÍA-HERNÁNDEZ *et al.*, 1980). While the Prebetic Zone represents the deposits located closer to the South Iberian Margin, the Subbetic Zone comprises the deposits situated far from the margin. In the central sector of the Cordillera, three superposed tectonic sequences are differentiated in the Subbetic Zone (RUANO *et al.*, 2000) (Fig. 1). In this region, the compressive deformations probably started in Late Aquitanian. The evolution of the area includes the superposition of the External Zones over the Internal Zones during the Middle Burdigalian (LONERGAN, 1993). Later deformations involve translations of the tectonic units towards the WSW after the Burdigalian, and the activity of NW vergent thrusts after Late Tortonian (RUANO *et al.*, 2000; GALINDO-ZALDÍVAR *et al.*, 2000).

Most of the relief is built up from Tortonian times, when the Alboran Sea extended over the bulk of the Cordillera. During the uplift of the mountain chain, several Neogene depressions were differentiated. To the North, and separating the Cordillera from the Iberian Massif, is the Guadalquivir Depression, which constitutes a foreland basin. The Granada and the Guadix-Baza depressions are two of the largest intramountain depressions developed in the central Betic Cordillera. Calcarenites, marls, chalk, and detritic sediments with Late Miocene to present-day ages constitute the infill of these depressions. The Tortonian rocks of these depressions cover the contact between the External and Internal Zones. Other smaller depressions can be found in both the External and Internal Zones. Finally, the most important depression of the whole Cordillera is the Alboran Sea, developed in the Internal Zones and which continues below sea level to date.

The area has been surveyed by numerous researchers with the aim to highlight the structure and geological evolution of the region. In the External Zones of the central transect, it is necessary to underline the contributions of VERA (1969), GARCÍA-DUEÑAS (1969) and GARCÍA-HERNÁNDEZ *et al.* (1980) who, among others, describe the stratigraphic sequences of the main units and determine the location of the most important contacts. In the Internal Zones, the contributions of FERNÁNDEZ-FERNÁNDEZ *et al.* (1992) and ALONSO-CHAVES (1995) establish the features of large folds and main faults of Sierra Tejeda. The Neogene infill of the basins has been determined by RODRÍGUEZ-FERNÁNDEZ (1982). However, detailed studies concerning the recent deformations of the region are scarce. Deserving mention among them are the models based on the strike-slip faulting activity and related normal faults, such as those proposed in the Subbetic Zone by DE SMET (1984), or extrapolated to the whole Cordillera by SANZ DE GALDEANO (1983).

The study of paleostresses along the central transect of the Cordillera (GALINDO-ZALDÍVAR *et al.*, 1993), together with studies of the present-day stresses in the Granada Depression (GALINDO-ZALDÍVAR *et al.*, 1999; RUANO *et al.*, 2000) and the

Málaga region (MORALES et al., 1999) show them to be variable. Below the Malaga region, there are extensional stresses at a depth of around 100 km, and NW-SE compressional stresses in the lower crust and upper part of the mantle. In the upper crust of the Alboran Sea and Internal Zones of the Betic Cordillera, the stresses are predominantly extensional with a radial or a NE-SW preferred orientation. In the eastern part of the Cordillera front, the paleostresses show a mainly NW-SE trend of compression (GALINDO-ZALDÍVAR et al., 1993).

Most of the region has undergone fast uplift since the Miocene, which is a common feature of compressional regions. Despite the convergent character of the plate boundary and the presence of large folds, most of the faults of the studied region are normal and exhibit a generally extensional setting.

The aim of this contribution is to analyze the recent structures in the central transect of the Betic Cordillera in order to determine the distribution and character of the deformation. This transect provides insights on the relationship between structures deforming the External and Internal Zones and facilitates comparison of the lower and upper crustal level deformation. In addition, this transect manifests a different tectonic evolution to the one described in the Eastern Betic Cordillera, which has been studied by numerous international research groups (i.e. Groupe de Recherche Néotectonique de l'Arc de Gibraltar, 1977; Ott d'Estevou and Montenat, 1985; Montenat and Ott d'Estevou, 1996)

Geophysical Data

Geophysical data help establish the main features of the deep structure of the region. Refraction seismic profiles indicate the existence of a thin continental crust with a thickness usually varying between 16 and 20 km in the Alboran Sea (SURIÑACH and VEGAS, 1993). Below the Betic Cordillera, however, a relatively thick continental crust (up to 36–37 km) has been determined (BANDA and ANSORGE, 1980).

The Bouguer anomaly map (Fig. 2) shows a very high anomaly gradient parallel to the coastline, from negative values in the continent extending to positive values in the Alboran Sea. TORNÉ and BANDA (1992) and GALINDO-ZALDÍVAR et al. (1997), using Bouguer gravity anomalies, have described the transition between both crusts. The models of these Bouguer gravity anomalies indicate an abrupt thickening of the crust located close to the coastline.

The ESCI-Béticas-2 deep reflection seismic profile (GARCÍA-DUEÑAS et al., 1994) shows a flat Moho discontinuity below the southern part of the Cordillera. Also, a reflective lower crust and a transparent upper crust can be identified, separated by a subhorizontal Conrad discontinuity (Fig. 3). This crustal structure does not show any correlation with the surface structure recognized by field geological methods, which reveal several open folds of great size in the same transverse of the seismic profiles. Several authors (e.g. GALINDO-ZALDÍVAR et al., 1997; MARTÍNEZ-MARTÍNEZ

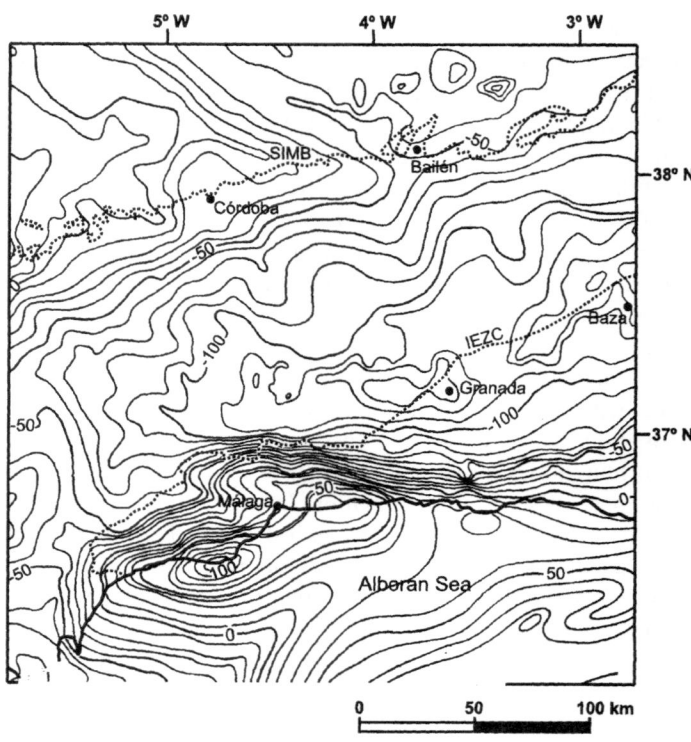

Figure 2
Bouguer anomaly map of the central sector of the Betic Cordillera and neighboring areas. Anomaly in mGals. Land data from Bouguer anomaly map, scale 1:500.000 (I.G.N., 1975). Alboran Sea data from Bonini *et al.* (1973). SIMB, South Iberian Massif Boundary. IEZC, Internal and External Zones Contact.

et al., 1997) have explained this lack of correlation as produced by a crustal detachment in the upper crust. The seismicity and seismic tomography of the region suggest that there is an active continental subduction of the Iberian Massif crust below the Internal Zones of the region in the western Alboran Sea (MORALES *et al.*, 1999; JABALOY *et al.*, in this volume).

Multichannel seismic profiles in the main basins of the transect—Guadalquivir basin, Granada Depression and Alboran Sea—allow us to observe the structure of the sedimentary infill in areas of poor exposure. The Alboran Sea profile near the Nerja region (Fig. 4), which cuts the prolongation to the sea of several main faults in the Alpujárride, shows that very continuous reflectors represent the Neogene infill. In addition, the southward tilting of the basement top and the thickening of the sedimentary infill can be associated with the progressive deformation of the region by folding, and not by faulting. The profile across the Granada Depression (Fig. 4) depicts the styles of deformation of the largest intramountain Neogene basin located

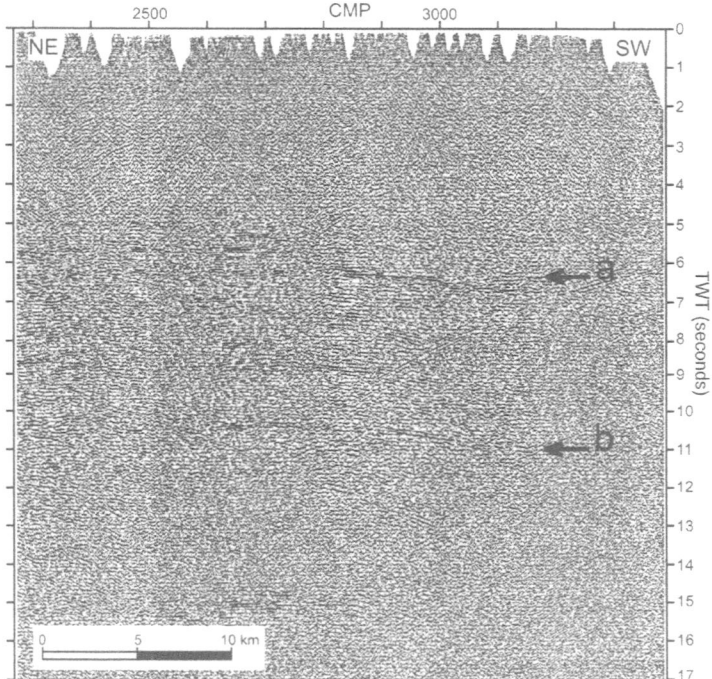

Figure 3

Segment of the ESCI-Béticas-2 deep seismic reflection profile. a. discontinuity between Lower and Upper crusts. b. Moho. Location in Figure 1.

in the studied transect. This profile illuminates the relationship between tilting and faulting in the Neogene evolution of the region. The Guadalquivir Basin is asymmetrical, with a maximum thickness of sediments towards the South (Fig. 4). A broad band of high-amplitude continuous reflectors, related to the Triassic rocks, is clearly identified along the seismic profile. This band of reflectors, and the overlying unit corresponding to the Miocene infill of the basin, is not deformed by faulting along the central and northern part of the Depression. However, in the southern part it is possible to identify a reverse fault superposing the Prebetic units with its Langhian and Serravallian cover over Tortonian rocks.

The aeromagnetic anomalies of southern Spain (Fig. 5) clearly demonstrate the continuity of bodies of basic rocks emplaced in the upper crust. The elongated anomalies observed in the southern part of the Iberian Massif, outside the Cordillera, continue with the same trend below the Guadalquivir basin, and can also be recognized below the greater part of the External Zones (GALINDO-ZALDÍVAR et al., 1997), although anomalous bodies do not outcrop in this area.

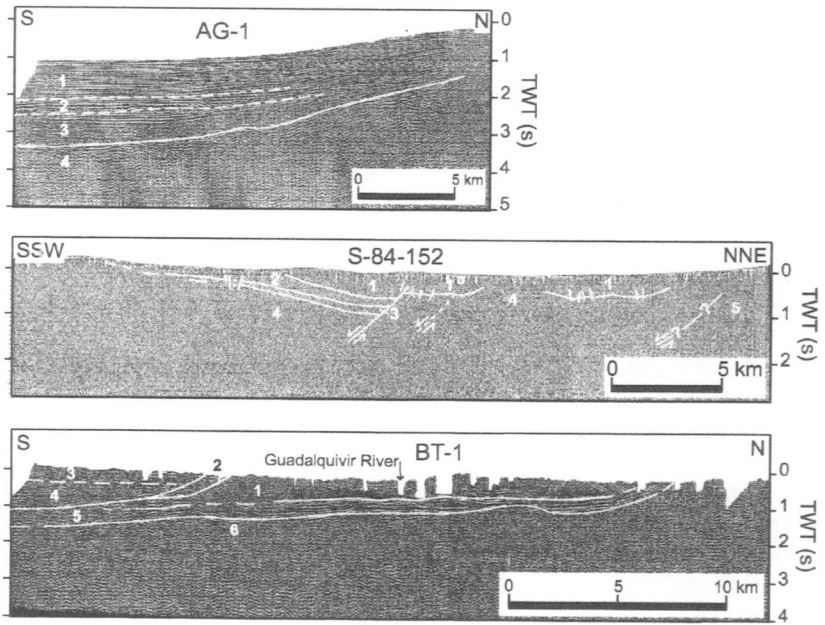

Figure 4
Multichannel seismic reflection profiles across the Alboran Sea (AG-1), Granada Depression (S-84-152) and eastern Guadalquivir basin (BT-1). Location in Figure 1. Profile AG-1: 1. Pliocene and Quaternary sediments. 2. Messinian rocks. 3. Tortonian- Lower Miocene rocks. 4. Betic Basement. Profile S-84-152: 1. Plio-Quaternary and Turolian sediments. 2. Messinian rocks. 3. Tortonian rocks. 4. Upper Subbetic Unit. 5. Intermediate Subbetic Unit. Profile BT-1: 1. Plio-Quaternary –Tortonian rocks. 2. Serravallian rocks. 3. Langhian rocks. 4. Internal Prebetic Unit? 5. Triassic rocks of the Iberian Massif Cover. 6. Iberian Massif Basement. 2-3, Olistostroms.

Field Observations

The field observation of structures across the transect indicate that folds and faults have developed since Late Miocene, and determine the present-day relief. However, they have different features and significance along the studied area. The central transect of the Betic Cordillera will be described from the South, at the border of the Alboran Sea, up towards the North, to the Iberian Massif.

Coast and Sierra Tejeda

Near the coastline, in the area around Nerja (Fig. 1) both Plio-Quaternary deposits and Alpujárride rock foliations are tilted southwards. The Pliocene beach deposits are formed mainly by calcarenites and show an internal unconformity that separates a lower part—dipping southwards generally more that 25°—from an upper

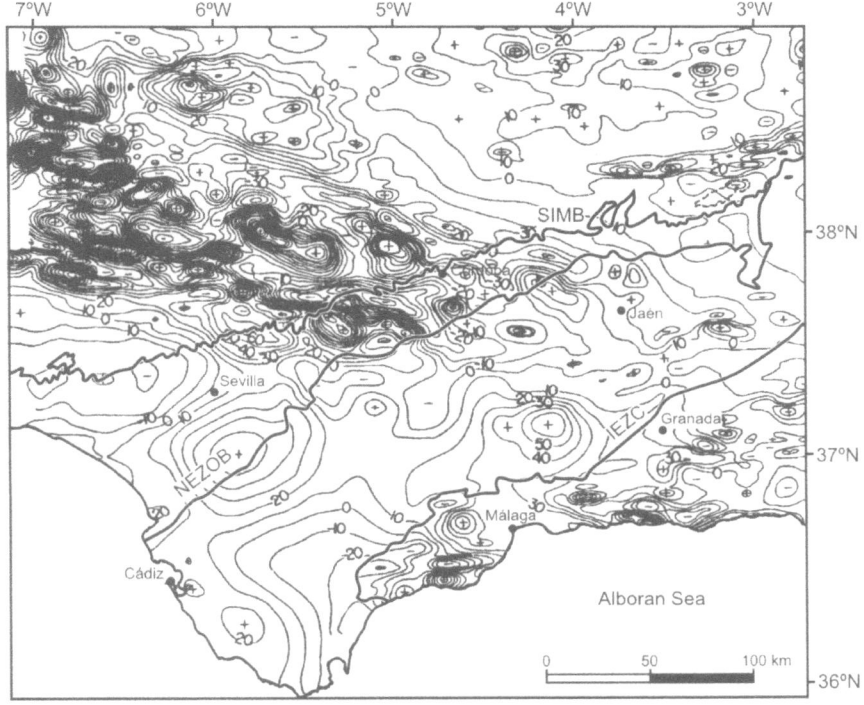

Figure 5

Total field aeromagnetic anomaly map of Central and Western Betic Cordillera and neighboring areas.
Anomaly in nT. Simplified from Ardizone et al. (1989). SIMB, South Iberian Massif boundary. NEZOB,
North External Zones and Olistostroms boundary. IEZC, Internal-External Zones contact.

part with low dips (Fig. 6a). This unconformity was recognized by FOURNIGUET
(1975); and more recently GUERRA-MERCHÁN and SERRANO (1993) relate its
development with the fracturation of the region. The general southwards dipping all
along the Alpujárride foliations indicates that it is related to large-scale folding
(Fig. 1) with a subvertical axial plane.

North of the coastline, the Sierra Tejeda constitutes a large antiform (Fig. 1)
with a deep incision of the fluvial network. This late antiform deforms the
Alpujárride Complex rocks and its internal structures that are composed of several
superposed units separated by normal faults with variable orientation of Early
Miocene age, and top-to-the-SW kinematics (FERNÁNDEZ- FERNÁNDEZ et al., 1992;
ALONSO-CHAVES, 1995). The late fold is approximately symmetrical with a
westward periclinal end and extends, with a WNW-ESE to E-W trend, along a
stretch of over 40 km. This fold is also responsible for the tilting of the Alpujárride
and Pliocene rocks near the coastline and for the uplift of Sierra Tejeda, which

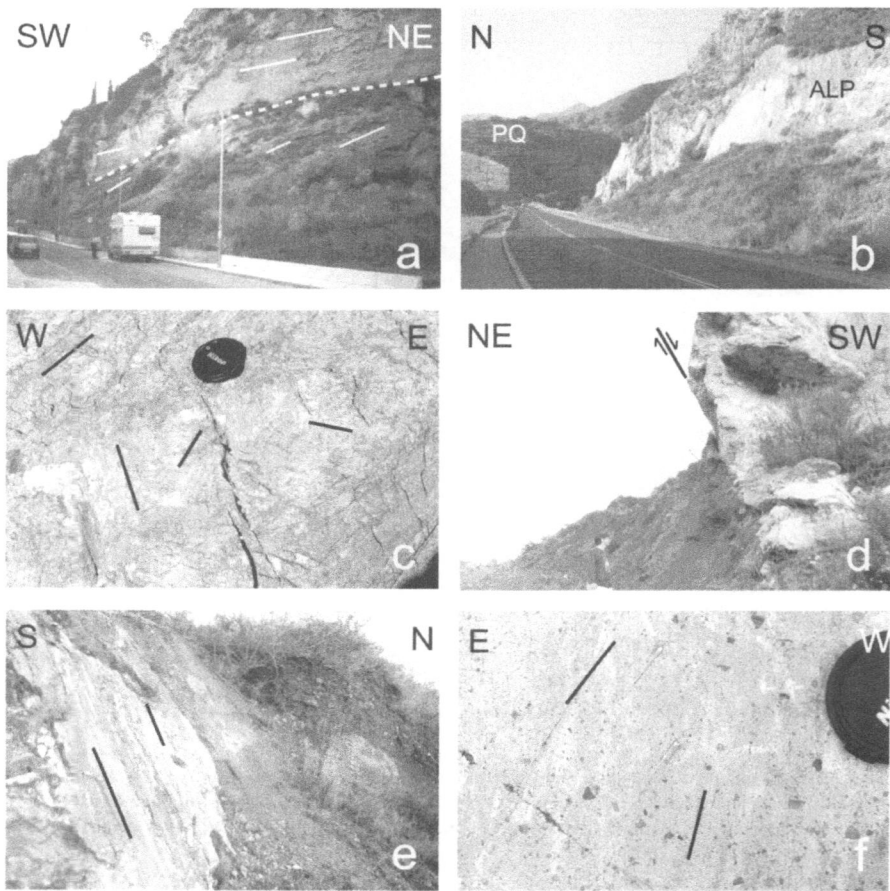

Figure 6
Field views of significant structures in the southern part of the transect. a. unconformity in Pliocene rocks, Burriana Beach (Nerja). b. NW-SE normal fault plane in Alpujárride rocks (ALP) covered by Plio-Quaternary sediments (PQ), east of Nerja. c. detailed view of fault in b, showing different striae orientat ons. d. NW-SE fault showing uplifted block with topography depression, east of Nerja. e. Zafarraya active normal fault cutting recent soils. f. detailed view of the Zafarraya fault plane with two oblique striations. Trails indicate normal regime.

runs parallel to the coast. However, Pliocene sediments are not tilted as much as the Alpujárride rocks.

In this area, there are several NW-SE trending faults, with variable dips towards the SW (Figs. 1 and 6b). These faults show several superposed striations (Fig. 6c) corresponding to different regimes that range from normal to transtensional, with dextral and sinistral characters for different events. They cut the Alpujárride complex rocks, and some joint detachment faults below, southwest of Sierra Tejeda (Fig. 1). The slip of these faults varies, but in most is of the order of hundreds of meters. They have a related scarp when they separate marbles and metapelites, yet in some cases the block tectonically uplifted constitutes a topographic depression, indicating the inactive character of the faults (Fig. 6d). In places they are covered by the Plio-Quaternary rocks. The fault system indicates a NE-SW trend of extension and was active later than 21 m.a. (the age of the end of metamorphism in the Alpujárride Complex, MONIÉ et al., 1991) and until the Pliocene. Although during Lower Miocene these normal faults may have been related to the low-angle normal faults, after Tortonian the NW-SE trending faults may have continued their activity. Faults with similar features continue to be active at present in the Cordillera, however, for instance in the Padul Fault, Western Sierra Nevada (Fig. 7) (ALFARO et al., 2001).

In the northern limb of the Sierra Tejeda Fold, the Alpujárride and Maláguide rocks extend under the Granada Depression. The contact between the Internal and External Zones is sealed by the Tortonian rocks. Flysch units located along this contact show moderate dips towards the N as a consequence of the activity of the Sierra Tejeda Fold. Furthermore, the area has undergone recent fault-related activity; the Zafarraya Fault being the most important one.

Zafarraya Fault

The Zafarraya fault (Figs. 6e, f and 7) is one of the most active faults in the area. It was involved in the earthquake of considerable magnitude (6.5–7, MUÑOZ and UDÍAS, 1981) which occured on Christmas Day, 1884. The fault has a total length exceeding 15 km (Fig. 7). Its trend varies along the strike, approximately E-W to the south of the Zafarraya Depression and curves to a NW-SE orientation at its western end, where deformed rocks of the External Zones can be seen. The fault plane in the Zafarraya region dips 60° northwards and its striations indicate that it is a normal fault (Fig. 6f). The Zafarraya Depression is an endoreic related basin located in the hanging wall of the fault (Fig. 7) and filled by sediments that range in age from Tortonian to the present. The fault plane cuts several recent soils intercalated with clastic wedges (Fig. 6e). The C^{14} dating of cut soils (REICHERTER et al., in press.) confirms repeated activity of the fault, with at least three events in the last 9000 years. However, in different outcrops it is possible to identify a variable fault slip between the clastic wedges, and the number of cut soils also varies, suggesting the segmented character of the fault.

To the East, the fault cuts the External/Internal Zones Contact and extends towards the southern border of the Granada Depression (Fig. 7). Research done in this sector indicates that the contact between the Alpujárride and the Plio-Quaternary alluvial fans in this sector of the Depression has been practically inactive in recent times. However, the presence of historical damage in this sector

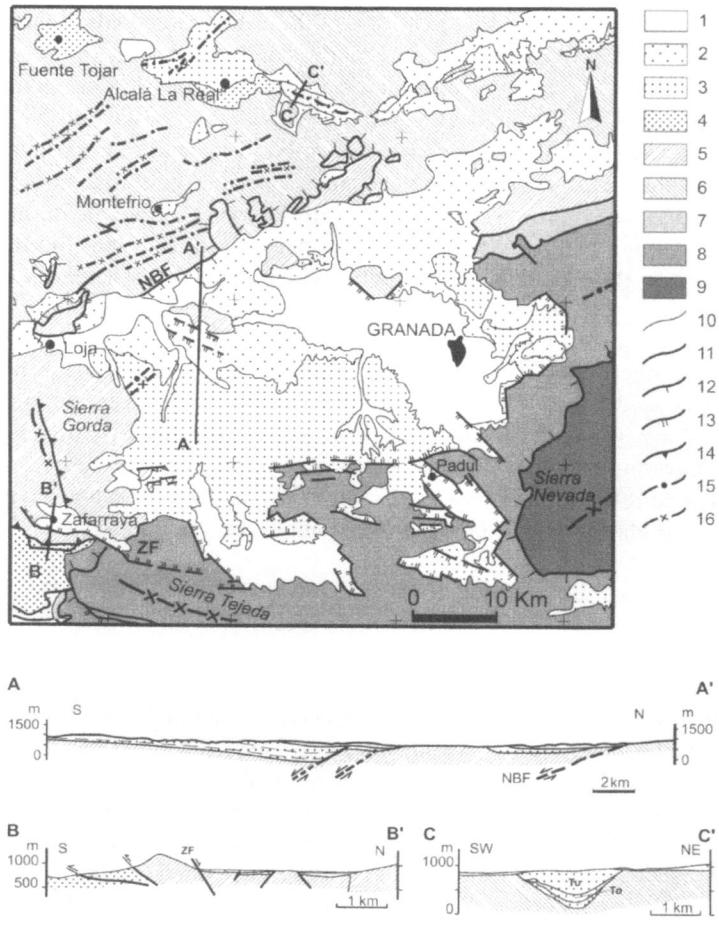

Figure 7

Detailed map of the Granada Depression and adjacent areas and cross sections. The geological A-A' cross section is based in the seismic profile on Figure 4. ZF, Zafarraya fault. NBF, North boundary fault of the Granada Depression. 1. Quaternary sediments. 2. Pliocene sediments. 3. Upper Miocene rocks. 4. Middle and Lower Miocene rocks. 5. Upper Subbetic Unit. 6. Intermediate Subbetic Unit. 7. Maláguide Complex. 8. Alpujárride Complex. 9. Nevado-Filábride Complex. 10. unconformity. 11. fault. 12. low-angle normal fault. 13. high-angle normal fault. 14. low-angle reverse fault. 15. syncline.16. anticline. 5-6, External Zones. 7-9, Internal Zones. B-B' cross section is based in LÓPEZ-CHICANO (1989).On C-C' cross section: Tu, Turolian sediments. To, Tortonian sediments.

during the 1884 Christmas earthquake (MUÑOZ and UDÍAS, 1981) may suggest that the active fault continues across the Plio-Quaternary alluvial fans. It is impossible to determine the precise location because of poor outcropping conditions.

Granada Depression

The Granada Depression is bounded eastwards by the Sierra Nevada, which constitutes a large antiform with outcropping rocks of the Internal Zones (Fig. 7). Similarly, the Sierra Gorda, which is a large dome of Subbetic rocks, constitutes the western boundary of the Depression. The Depression infill features a wedge structure opening toward the north (Figs. 4 and 7). This wedge is recognized in rocks ranging in age from Tortonian to Pliocene. The Depression is affected by several high angle normal faults, located mainly in its northern part and dipping southwards, producing a thoroughly complex structure. The Turolian rocks (equivalent in marine sequences to the Messinian) are affected by very open folds with a mean NW-SE trend (Fig. 8b). The Plio-Quaternary rocks lie unconformably over these folds. In the northern part of the Depression there are horsts where there are outcrops of Subbetic rocks (Fig. 7).

The northern boundary of the Granada Depression is determined by a large normal fault (RUANO et al., 2000) (Fig. 7). The normal character of this fault is indicated by the fault gouges kinematics. South dipping normal faults in the northwestern part of the Granada Depression are responsible for the wedge geometries of the depression infill which indicate that the faults have been active at least from the Tortonian to present-day, as demonstrated by the study of seismic profiles and gravity data used to build the cross sections (Figs. 4 and 7). The eastern boundary is affected by a set of NW-SE normal faults dipping mainly towards the SW (ALFARO et al., 2001) (Fig. 7). The Quaternary fluvial sediments are located along a curved band parallel to the northern and eastern boundaries of the Depression, where most of the active faults are concentrated.

Subbetic Ranges and Associated Basins

Northwards, in the Subbetic ranges, the large-sized folds developed since Early Tortonian determine the morphology of the relief. In this region, the faults with recent activity and affecting recent soils are mainly normal, very scarce, with short slips and have little importance at regional scale. Large fault gouge bands are only recognized at low-angle faults related to the Burdigalian-Tortonian nappe displacements (GALINDO-ZALDÍVAR et al., 2000). The synforms can be distinguished better than the antiforms because they concentrate Neogene deposits, for example in the depressions of Montefrío, Fuente Tojar and Alcalá la Real (Fig. 7). However, these depressions are practically not deformed by faulting, and the Quaternary deposits, constituted by clastic sediments and travertine, are undeformed and practically restricted to the present fluvial network.

In the Montefrío Depression, there is evidence of deformation during the Late Tortonian. ESTÉVEZ et al. (1982) describe the existence of an unconformity with the rocks of the lower part tilted towards the NW, indicating a NW-SE trend of compression (Fig. 8c).

Figure 8
Field views of significant structures in the northern part of the transect. a. incision of the fluvial network in the southern part of Granada Depression. b. NE-SW folds in Turolian limestones of the eastern Granada Depression. c. Intra-Tortonian unconformity in Montefrío. d. NW-SE synform in Turolian and Tortonian rocks of Alcalá la Real Depression. e. example of reverse fault in Tortonian rocks of the mountain front near Martos. f. P-C structure in the fault gouge of fault in e. indicating a top-to-the-N sense of motion of the hanging wall.

The largest Neogene rock outcrop in this transect on the Subbetic Ranges is the Alcalá la Real Depression, in which the WNW-ESE and NE-SW oriented folds determine the distribution of the sedimentary rocks since the Late Miocene (Fig. 7). The eastern end of the Alcalá la Real Depression is affected by an open symmetrical syncline with a WNW-ESE axis (Figs. 7 and 8d) that features a southeastward periclinal end. The fold affects both Tortonian and Turolian rocks, but is most intense in the Tortonian rocks. These data suggest that most of the folding was

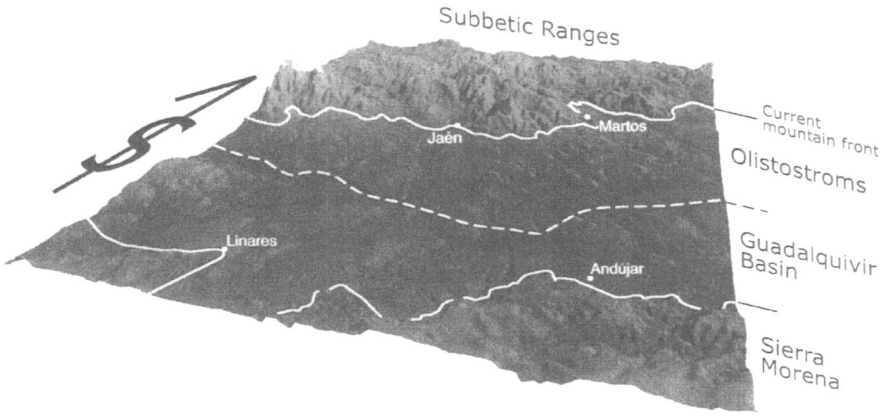

Figure 9
Topographic shaded view of the mountain front of the Betic Cordillera and the Guadalquivir basin. Southwards view.

produced during Late Miocene times. However, there is a large outcrop of Upper Tortonian calcarenites in the central part of the Depression, located along the westward prolongation of this fold, which is not deformed. Thus, this fold does not continue northwestwards and is probably less than 10 km in length.

The northwestern part of the Alcalá la Real Depression is deformed by NE-SW oriented folds that affect rocks up to Late Tortonian in age. The axes are sometimes longer than 5 km, and their shapes range from open and symmetrical to closed and asymmetrical, with vergence towards the SE and reverse limbs.

Mountain Front and Guadalquivir Basin

The present mountain front that separates the Betic Cordillera from the Guadalquivir foreland basin has sharp and irregular slopes and is well defined in the topography (Fig. 9). However, the presence of deformed olistostrom bodies of Early Miocene age north of the mountain front, along the southern border of the Guadalquivir basin (Fig. 1) indicates that the deformations related to the Betic Cordillera extended further northward in the past.

In the sector between Martos and Pegalajar, the mountain front is characterized by E-W to ENE-WSW reverse faults (Figs. 1 and 8e,f) that affect at least the Middle Tortonian rocks and probably younger rocks. However, the outcrops in this part of the Cordillera are scarce and poorly exposed, and their relationship with the more recent sediments has not been established. In addition, small sinistral E-W strike-slip faults, affecting up to Quaternary deposits, have been recognised in the Martos area at the present-day mountain front.

As observed in the seismic profiles (Fig. 4) and in the field outcrops, most of the infill of the Guadalquivir foreland basin, including the Quaternary sediments, is not deformed. Yet in the northern part of the Guadalquivir basin, between Andújar and Linares (Fig. 1) NE-SW oriented normal faults dipping towards the SE and NW deform this border. These faults are normal, with high-angle dips; and they produce two asymmetrical basins that connect with the Guadalquivir basin. In the southern margin, the olistostroms are superposed over the basin sediments.

Discussion

The acquisition of an extensive array of geophysical data over recent years has constituted a substantial contribution to the knowledge of the deep structure of the Betic Cordillera and the main features of the northern Alboran Sea. The existence of a detachment fault that affects the crust has been determined by studies of seismicity (SERRANO *et al.*, 1998) and seismic tomography (MORALES *et al.*, 1999) together with a deep seismic reflection profile (ESCI-Béticas–2, GALINDO-ZALDÍVAR *et al.*, 1997; MARTÍNEZ-MARTÍNEZ *et al.*, 1997). The ESCI-Béticas-2 seismic profile (Fig. 3) clearly shows subhorizontal reflectors in the Lower crust that are not deformed by the kilometre-sized folds of the Upper crust evidenced by field observations (Fig. 1). Hence, a detachment should be located between the two elements. Other geophysical methods, such as the gravity and aeromagnetic data, confirm this hypothesis. Indeed, the Iberian Massif aeromagnetic anomalies continue below the External Zones, pointing to a different nature of both blocks separated by the detachment fault. The footwall of this detachment corresponds to the southward extension of the Iberian Massif, while the rocks outcropping in the Betic Cordillera form the hanging wall (Fig. 10). This detachment is probably one of the most important active structures of the region. In the southern part of the Cordillera it is related to intense seismicity (SERRANO *et al.*, 1998). Moreover, it is located approximately at the compensation level (about 10 to 15 km) of the late large open folds that deform the Betic Cordillera, as Sierra Nevada fold, and its activity would have therefore been simultaneous with their development.

The field data of the region allow the discussion of the recent evolution of the upper crustal level of the Betic Cordillera. The presence of Tortonian rocks with similar shallow marine facies along the whole transect suggests that at the time of the deposition of these rocks, there were no large reliefs and the Alboran Sea probably extended north as far as the Iberian Massif. The Cordillera underwent previous deformations that advanced further to the North, as suggested by the presence of large bodies of olistostroms of Early Miocene age in the Guadalquivir basin.

The deformations since the Tortonian produced out-of-sequence thrusts. They affect a region that was already deformed, with frontal structures sealed by Late Tortonian to present-day deposits. These recent deformations consist of folds and

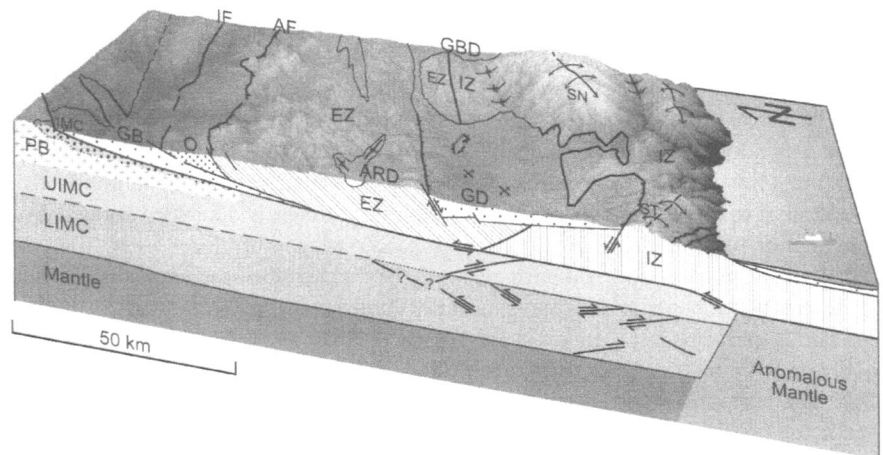

Figure 10

Tectonic sketch of the main structures of the central sector of the Betic Cordillera. IMC, Iberian Massif cover. PB, Pedroches Batholith. UIMC, Upper Iberian Massif Crust. LIMC, Lower Iberian Massif Crust. GB, Guadalquivir Basin. O, Olistostroms. EZ, External Zones of Betic Cordillera. IZ, Internal Zones of Betic Cordillera. ARD, Alcalá la Real Depression. GD, Granada Depression. GBD, Guadix-Baza Depression. SN, Sierra Nevada. ST, Sierra Tejeda. IF, inactive front of olistostroms. AF, active mountain front.

faults of essentially simultaneous development, responsible for the present-day relief. The folds are of variable size and orientation, but the largest ones are located in the southern part of the cross section (Fig. 10) and determine the location of the main basin and ranges.

Sierra Tejeda is clearly formed by the development of an antiform of E-W orientation, parallel to the coastline. This fold, in fact, determines the E-W trend of the coast, producing a general southward dipping of the foliations in Alpujárride rocks and the bedding in the Neogene sediments that are located in the southern limb. The development of the fold seems to have been discontinuous over time. The Pliocene unconformity in the Nerja region clearly indicates a sharp event during the fold development. The deep incision of the drainage system also confirms the recent age of the development of the relief. The normal faults recognized in this area have mainly NW-SE orientations and do not show present-day activity. Some of them are folded at the western end of Sierra Tejeda. Seismic profiles of the region (Fig. 4) corroborate that the upper part of the sedimentary infill of the Alboran Sea is not affected by faults. Large faults that run parallel to the coastline are absent in the southern limb of the Sierra Tejeda antiform. However, in the northern limb, the Zafarraya normal fault is active and indicates a N-S trend for present-day extension in the shallow levels of the crust. In this setting, the Sierra Tejeda fold may represent an accommodation fold developed in the hanging wall of the main detachment (Fig. 10). The location of the Zafarraya fault is compatible with the extensional stresses in the external arch of the main fold. Therefore, this normal fault could be a secondary-order feature with respect to the compression.

The recent uplift of Sierra Tejeda, evidenced as well by the presence of large alluvial fans in its northern border with the Granada Depression, could be related with the northward tilting of Granada Depression Neogene rocks. The uplift of the southern part of the Depression in Quaternary times is likewise evident from the incision of the drainage network (Fig. 8a), which produces erosion of the sedimentary infill of the southern part of the Depression and deposition of the Quaternary sediments in the central and northern part (Fig. 7).

The Granada Depression is deformed mainly by extensional structures, however compressional deformations, as ENE-WSW oriented folds affecting Turolian rocks, are also recognized. The main faults that determine the wedge geometry of the sediment infill are located in the north-central part and eastern borders. They are normal faults with a progressive activity since Tortonian times, simultaneous to the basin uplift, as shallow marine Tortonian deposits reach elevations at times over 800 m and reaching up to 1500 m in the eastern Granada Depression (SANZ DE GALDEANO, 1976). At present, the Depression mainly undergoes NE-SW and N-S extensional deformations. The simultaneous development of uplift and normal faulting may be the consequence of deformations in the shallow levels of the thick crust (Fig. 10). The scarce compressional deformations consist mainly of very open folds that deform the Turolian rocks and indicate a NW-SE trend of compression. This direction of compression coexists or may alternate with the extensional events, as shown by RUANO *et al.* (2000) that study striated pebbles in the northern sector of the Granada Depression. Both normal and reverse earthquake focal mechanism in the detachment (GALINDO-ZALDÍVAR *et al.*, 1999), that constitutes a first-order structure with a probably complex present-day kinematics, are related to the shortening and extension deformations observed at surface.

North of the Granada Depression, most of the Subbetic rocks are deformed by folds, and faulting is concentrated in the mountain front. In the Alcalá la Real region there is evidence of the progressive development of folds during the Tortonian and Turolian, less deformed in the most recent sediments. The uplift of the Neogene rocks, (the latter being preserved only in small synforms), together with the development of folds, would be compatible with the compressive character of the deformation along the crustal detachment. Taking into account the variability of the fold trends, the slip along the detachment should also be heterogeneous in small sectors, probably with minor splays with different slip, although N-S to NW-SE component of translations would predominate (Fig. 10). The virtual absence of Quaternary deposits in the basins along the Subbetic ranges may be a consequence of the relief rising.

The mountain front (Fig. 9) is deformed by strike-slip and reverse faults that affect the sedimentary infill of the Guadalquivir basin (Figs. 4 and 10), and may represent the northward outcropping of the main crustal detachment. In this area the kinematics of the faults clearly indicate a N-S to NW-SE compression. However, its irregular shape suggests a low recent tectonic activity.

The offsets along the reverse faults of the mountain front are very small and do not justify great translations along the crustal detachment front. However, the development of large folds—recognized by field observations in the hanging wall of the detachment but absent in the footwall, as indicated by deep seismic reflection profiles (Figs. 3 and 10)—could be responsible for the increase of the slip southwards along the crustal detachment. In fact, the large folds and the higher reliefs are located near the coastline.

The hanging wall shows variable internal deformations, and the detachment is likely to evidence different kinematics in each sector. In the Málaga region, westwards from the studied area, the detachment may connect with a continental subduction zone, described by MORALES et al. (1999), consisting of a slab that dips southeastwards, and shows NW-SE compressive focal mechanisms associated with it. However, in the Granada Depression, the main deformation of the hanging wall consists of NE-SW and NW-SE extensional faulting and recent folds with predominantly E-W to NE-SW trends. These observations suggest that the detachment has undergone complex kinematics, ranging mainly from top-to-the-NW movement during compressional events, to top-to-the-SW motion during extensional events. However, to date there is no detailed data to determine the exact kinematics of this structure.

Stresses in the transect of the Betic Cordillera are very heterogeneous. In the front and at depth, NW-SE to N-S compression predominates, as it does between the Eurasian and African plates. However, the upper part of the Internal Zones undergoes variable deformations, mainly NE-SW recent extension simultaneous to and as a consequence of the uplift. There are sectors with overprinted extensional and compressional events over time.

The available geophysical and geological data on the central transect of the Cordillera clearly indicate that the simultaneous or alternating activity of compressional and extensional deformations in orthogonal orientations practically do not develop in association with large strike-slip faulting. These features contrast sharply with the setting in the eastern Betic Cordillera, which is widely deformed by transcurrent faults (e.g., SANZ de GALDEANO, 1983; OTT D'ESTEVOU and MONTENAT, 1985; MONTENAT and OTT D'ESTEVOU, 1996). The available data also suggest that the southern boundary of the Eurasian plate (Iberian Massif) is located at the main detachment fault that is below the crust of the Betic Cordillera (Fig. 10), and is not a vertical fracture as was proposed in the earliest models of the region (ANDRIEUX et al., 1971). The Betic Cordillera belongs to the wide deformation area related to this plate boundary.

Conclusions

The aforementioned data and observations show that the main structures in the Cordillera related to the shortening from Late Miocene are kilometer-size folds whose trends are predominantly E-W to NE-SW. These folds might develop in the hanging wall of a major detachment that separates the Betic Cordillera rocks and the Iberian Massif basement. The development of theses structures would be responsible for the crustal thickening and uplift of the Cordillera since the Tortonian. These structures may be considered out-of-sequence because they deform a pre-Tortonian orogen and are compatible with the NW-SE Eurasian-African plate convergence.

The positive relief of the Internal Zones favors the development of normal faults in the upper crust, which do not extend down to the lower crust. The reverse faults are preferently located at the front of the Cordillera. The short slip on these reverse faults suggests that most of the shortening in the upper crust would be accommodated by the large-sized folds, and that slip increases southwards along the main crustal detachment.

The paleostress determinations also confirm the predominance of E-W to NE-SW extension in most of the Internal Zones, locally simultaneous with the development of N-S to NW-SE compression in the same sectors, and with a compression ranging from NW-SE to N-S in the northern part of the External Zones. These paleostresses could develop in the same general stress field. Nonetheless, since different incompatible deformations develop simultaneously in close areas, the stress field must be complex in detail. All these features suggest that the deformation since the Late Miocene was continuous throughout the time, but had distinctive regional features. This makes it impossible to establish isolated tectonic phases in the evolution of the Cordillera.

The proposed model for the structures of recent activity in the central transect of the Cordillera contrasts with the evolution of the eastern Betic Cordillera, clearly determined by the activity of transcurrent faults.

Acknowledgements

We thank M. J. Román-Alpiste for the assistance in the topographic models and J. Sanders for the English revision. The comments of two anonymous referees have improved the quality of the paper. This research was supported by project BTE-2000-1490-C02-01 of the CICYT.

References

ALDAYA, F., ALVAREZ, F., GALINDO-ZALDÍVAR, J., GONZÁLEZ-LODEIRO, F., JABALOY, A., and NAVARRO-VILÁ, F. (1991), *The Malaguide-Alpujarride Contact (Betic Cordilleras, Spain): A Brittle Extensional Detachment*, C. R. Acad. Sci. Paris *313, Série II*, 1447–1453.

ALFARO, P., GALINDO-ZALDÍVAR, J., JABALOY, A., LÓPEZ-GARRIDO, A.C., and SANZ DE GALDEANO, C. (2001), *Evidence for the Activity and Paleoseismicity of the Padul Fault (Betic Cordillera, Southern Spain)*, Acta Geológica Hispánica 36, 283–295.

ALONSO-CHAVES, F. M., Evolución tectónica de Sierra Tejeda y su relación con procesos de engrosamiento y adelgazamiento corticales en las Cordilleras Béticas, [unpublished Ph.D. thesis] (Univ. Granada, Granada 1995).

ANDRIEUX, J., FONTBOTÉ, J. M., and MATTAUER, M. (1971), *Sur un modèle explicatif de l'Arc de Gibraltar*, Earth Planet. Sci. Lett. 12, 191–198.

ARDIZONE J., MEZCUA J., and SOCIAS I. *Mapa aeromagnético de España Peninsular, escala 1:1.000.000* (Instituto Geográfico Nacional, Madrid 1989).

BALANYÁ, J. C., and GARCÍA-DUEÑAS, V. (1987), *Les directions structurales dans le Domaine d'Alborán de part et d'autre du Détroit de Gibraltar*, C. R. Acad. Sci. Paris 304, Série II, 929–932.

BANDA, E., and ANSORGE, J. (1980), *Crustal Structure under the Central and Eastern Part of the Betic Cordillera*, Geophys. J. R. Astr. Soc. 63, 515–532.

BONINI, W. E., LOOMIS, T. P., and ROBERTSON, J. D. (1973), *Gravity Anomalies, Ultramafic Intrusions, and the Tectonics of the Region around the Strait of Gibraltar*, J. Geophys. Res. 78, 1372–1382.

CUEVAS, J., ALDAYA, F., NAVARRO-VILA, F., and TUBÍA, J. M. (1986), *Caractérisation de deux étapes de charriage principales dans les nappes Alpujarrides centrales (Cordillères Bétiques, Espagne)*, C. R. Acad. Sci. Paris 302, Série II, 1177–1180.

DEMETS, C., GORDON, R. G., ARGUS, D. F., and STEIN, S. (1990), *Current Plate Motions*, Geophys. J. Int. 101, 425–478.

DE SMET, M. E. M. (1984), *Wrenching in the External Zone of the Betic Cordilleras, Southern Spain*, Tectonophysics 107, 57–79.

ESTÉVEZ, A., RODRÍGUEZ-FERNÁNDEZ, J., SANZ DE GALDEANO, C., and VERA, J. A. (1982), *Evidencia de una fase compresiva de edad Tortoniense en el sector central de las Cordilleras Béticas*, Estudios Geol. 38, 55–60.

FALLOT, P. (1948), *Les Cordilleres Bétiques*, Estudios Geol. 4, 83–172.

FERNÁNDEZ-FERNÁNDEZ, E., CAMPOS, J., and GONZÁLEZ-LODEIRO, F. (1992), *Estructuras extensionales en los materiales alpujárrides al E de Málaga (Sierra Tejeda, Cordilleras Béticas)*, Geogaceta 12, 13–16.

FOURNIGUET, J., *Néotectonique et Quaternaire marin sur le littoral de la Sierra Nevada. Andalousie (Espagne)*, [unpublished Master Thesis] (Univ. Orleans, 1975).

GALINDO-ZALDÍVAR, J., GONZÁLEZ-LODEIRO, F., and JABALOY, A. (1989), *Progressive Extensional Shear Structures in a Detachment Contact in the Western Sierra Nevada (Betic Cordilleras, Spain)*, Geodinamica Acta 3, 73–85.

GALINDO-ZALDÍVAR, J., GONZÁLEZ-LODEIRO, F., and JABALOY, A. (1993), *Stress and Paleostress in the Betic-Rif Cordilleras (Miocene to Present-day)*, Tectonophysics 227, 105–126.

GALINDO-ZALDÍVAR, J., JABALOY, A., GONZÁLEZ-LODEIRO, F, and ALDAYA, F. (1997), *Crustal Structure of the Central Sector of the Betic Cordillera (SE Spain)*, Tectonics 16, 18–37.

GALINDO-ZALDÍVAR, J., JABALOY, A., SERRANO, I., MORALES, J., GONZÁLEZ-LODEIRO, F., and TORCAL, F. (1999), *Recent and Present-day Stresses in the Granada Basin (Betic Cordilleras): Example of a Late Miocene-present-day Extensional Basin in a Convergent Plate Boundary*, Tectonics 18, 686–702.

GALINDO-ZALDÍVAR, J., RUANO, P., JABALOY, A., and LÓPEZ CHICANO, M. (2000), *Kinematics of Faults between Subbetic Units during the Miocene (Central Sector of the Betic Cordillera)*, C. R. Acad. Sci. Paris 331, Série II, 811–816.

GARCÍA-HERNÁNDEZ, M., LÓPEZ GARRIDO, A. C., RIVAS, P., SANZ DE GALDEANO, C., and VERA, J. A. (1980), *Mesozoic Paleogeographic Evolution the External Zones of the Betic Cordillera*, Geol. en Mijnbouw 59, 155–168.

GARCÍA-DUEÑAS, V. (1969), *Les unités allochtones de la Zone Subbétique, dans la transversale de Grenade (Cordillères Bétiques, Espagne)*, Rev. Gèograp. Phy. Gèol. Dinam. XI, 211–222.

GARCÍA-DUEÑAS, V., BANDA, E., TORNÉ, M., CÓRDOBA, D., and ESCI-Béticas Working Group (1994), *A Deep Seismic Reflection Survey across the Betic Chain (Southern Spain): First Results*, Tectonophysics 232, 77–89.

GROUPE DE RECHERCHE NÉOTECTONIQUE DE L'ARC DE GIBRALTAR (1977), L'Histoire tectonique récent (Tortonien à Quaternaire) de l'Arc de Gibraltar et des bordures de la mer d'Alboran, Bull. Soc. géol. France 19, 575–614.

GUERRA-MERCHÁN, A. and SERRANO, F., Análisis estratigráfico de los materiales neógenos-cuaternarios de la región de Nerja, In Geología de la Cueva de Nerja (ed. Carrasco Cantos, F.) (Patronato de la Cueva de Nerja Nerja 1993) pp. 55–90.

I.G.N., Mapa de España de Anomalía de Bouguer, escala 1:500.000 (Instituto Geográfico Nacional, Madrid 1975).

JABALOY, A., GALINDO ZALDÍVAR, J., and GONZÁLEZ LODEIRO, F. (1993), The Alpujárride Nevado-Filábride Extensional Shear Zone, Betic-Cordilleras, SE Spain, J. Struct. Geol. 15, 555–569.

LONERGAN, L. (1993), Timing and Kinematics of Deformation in the Malaguide Complex, Internal Zone of the Betic Cordillera, Southeast Spain, Tectonics 12, 460–476.

LÓPEZ-CHICANO, M., Geometría y estructura de un acuífero kárstico perimediterráneo: Sierra Gorda (Granada y Málaga) [unpublished Master Thesis] (Univ. Granada, 1989).

LÓPEZ SÁNCHEZ-VIZCAÍNO, V., RUBATTO, D., GÓMEZ-PUGNAIRE, M. T., TROMMSDORFF, V. and MÜNTENER, O. (1997), Middle Miocene High-pressure Metamorphism and Fast Exhumation of the Nevado-Filábride Complex, SE Spain, Terra Nova 13, 327–332.

MARTÍNEZ-MARTÍNEZ, J. M., SOTO, J. I., and BALANYÁ, J. C. (1997), Crustal Decoupling and Intracrustal Flow beneath Domal Exhumed Core Complexes, Betic (SE Spain), Terra Nova 9, 223–227.

MONIÉ, P., GALINDO-ZALDÍVAR, J., GONZÁLEZ-LODEIRO, F., GOFFÉ, B., and JABALOY, A. (1991), First Report on 40Ar/39Ar Geochromology of Alpine Tectonism in the Betic Cordilleras (Southern, Spain), J. Geol. Soc. London 148, 289–297.

MONTENAT, CH. and OTT D'ESTEVOU, P. Late neogene basins evolving in the eastern betic transcurrent fault zone; an illustrated review. In Tertiary Basins of Spain: The stratigraphic Record of Crustal Kinematics (eds. P.F. Friend and C.J. Dabrio) (Cambridge University Press 1996) pp.372–386.

MORALES, J., SERRANO, I., JABALOY, A., GALINDO-ZALDÍVAR, J., ZHAO, D., TORCAL, F., VIDAL, F., and GONZÁLEZ-LODEIRO, F. (1999), Active Continental Subduccion beneath the Betic Cordillera and the Alborán Sea, Geology 27, 735–738.

MUÑOZ, D. and UDÍAS, A., El Terremoto de Andalucía del 25 de Diciembre de 1884 (Instituto Geográfico Nacional, Madrid 1981).

NIETO, J. M., PUGA, E., MONIÉ, P., DÍAZ DE FEDERICO, A. and JAGOUTZ, E. (1997), High pressure metamorphism in metagranites and orthogneisses from the Mulhacén Complex (Betic Cordillera, Spain), Terra Nova 1, 22–23.

OTT D'ESTEVOU, P. and MONTENAT, C., (1985) Evolution structurale de la zone bétique orientale (Espagne) du Tortonien à l'Holocène. C. R. Acad. Sc. Paris 300, 363–368.

REICHERTER, K. R., JABALOY, A., GALINDO-ZALDÍVAR, J., RUANO, P., BECKER-HEIDMANN, P., MORALES, J., REISS, S. and GONZÁLEZ-LODEIRO, F. (in press), Repeated palaeoseismic activity of the Ventas de Zafarraya Fault (S-Spain) and its Relation with the 1884 Andalusian Earthquake, Int. J. Earth Sci.

RODRÍGUEZ-FERNÁNDEZ, J., El Mioceno en el sector central de las Cordilleras Béticas. [unpublished Doctoral Thesis] (Univ. Granada, 1982).

RUANO, P., GALINDO-ZALDÍVAR, J., and JABALOY, A. (2000), Evolución geológica desde el Mioceno del sector noroccidental de la depresión de Granada (Béticas Cordilleras), Rev. Soc. Geol. España 13, 143–155.

SANZ DE GALDEANO, C., Datos sobre las deformaciones neogenas y cuaternarias del sector del Padul (Granada), In Reunión sobre la geodinámica de las Cordilleras Béticas y el Mar de Alborán (Univ. Granada, Granada 1976) pp. 197–218.

SANZ DE GALDEANO, C. (1983), Los accidentes y fracturas principales de las Cordilleras Béticas, Estudios Geol. 39, 157–165.

SERRANO, I., MORALES, J., ZHAO, D., TORCAL, F., and VIDAL, F. (1998), P-wave Tomographic Images in the Central Betics-Alboran Sea (South Spain) Using Local Earthquakes: Contribution for Continental Collision, Geophys. Res. Lett. 25, 4031–4034.

SIMANCAS, J. F. and CAMPOS, J. (1993), Compresión NNW-SSE tardi a postmetamórfica y extensión subordinada en el Complejo Alpujárride (Dominio de Alborán, Orógeno Bético), Rev. Soc. Geol. España 6, 23–35.

SURIÑACH, E. and VEGAS, R. (1993), *Estructura general de la corteza en una transversal del Mar de Alborán a partir de los datos de sísmica de refracción-reflexión de gran ángulo: Interpretación geodinámica*, Geogaceta *14*, 126–128.

TORNÉ, M. and BANDA, E. (1992), *Crustal Thinning from the Betic Cordillera to the Alboran Sea*, Geo-mar. Lett. *12*, 76–81.

UDÍAS, A. and BUFORN, E. (1991), *Regional Stresses along the Euroasia-Africa plate Boundary Derived from Focal Mechanisms of Large Earthquakes*, Pure Appl. Gephys. *136*, 433–448.

VERA, J. A., *Estudio geológico de la zona Subbética en la transversal de Loja y sectores adyacentes* (Mem. Inst. Geol. Min., Madrid 1969).

(Received February 3, 2002, revised September 10, 2002, accepted February 28, 2003)

To access this journal online:
http://www.birkhauser.ch

Pure appl. geophys. 161 (2004) 565–587
0033–4553/04/030565–23
DOI 10.1007/s00024-003-2463-4

❙ Pure and Applied Geophysics

Neogene Through Quaternary Tectonic Reactivation of SW Iberian Passive Margin

N. Zitellini[1], M. Rovere[2], P. Terrinha[3], F. Chierici[4],
L. Matias[5], and Bigsets Team[6]

Abstract — Southwest Portugal, the Gulf of Cadiz and Morocco are under the potential threat of natural hazards linked to seismicity and tsunami generation. We report the results of two multi-channel seismic (MCS) surveys carried out in 1992 and 1998 along the continental margin and oceanic crust of SW Iberia. This MCS data set shows the evidence of the compressional deformation which involves both the continental and the oceanic crust of the study area. The area of deformation extends from the southern border of the Tagus Abyssal Plain to the Seine Abyssal Plain, encompassing the continental margin of SW Portugal. Most of the structures observed are probably related to a Mid-Miocene phase of Africa-Europe plate convergence. In this paper we discuss the recent advances on the identification of the tectonic structures that are still active and that may generate great earthquakes and tsunamis. The tectonic structures identified are located respectively at the Guadalquivir Bank, along the eastern border of the Horseshoe Abyssal Plain and along the southern continental slope of SW Portugal.

Key words: Southwest Iberia, seismic reflection data, active faulting, compressional tectonic, earthquake, tsunami.

Introduction

Since historical times a number of destructive earthquakes/tsunamis has been reported to have occurred in SW Iberia like the tsunami of 60–63 B.C., which devastated the city of Cadiz, and the 1531 and 1722 events that struck the coasts of SW Portugal. This area was also the source of the famous 1755 Lisbon Earthquake

---

[1] Istituto di Scienze del Mare, Sezione di Bologna CNR, Via P. Gobetti, 101, 40129 Bologna, Italy. E-mail: nevio.zitellini@bo.ismar.it

[2] Dipartimento di Scienza della Terra e Geologico-Ambientali, Università di Bologna, Via Zamboni, 67, 40120 Bologna, Italy.

[3] Instituto Geológico e Mineiro, Departamento de Geologia Marinha, 2721–866 Amadora, Portugal; also at Faculdade de Ciências da Universidade de Lisboa, Departamento de Geologia, LATTEX 1749–016 Lisboa, Portugal.

[4] Istituto Radio Astronomia, CNR, Matera, Italy.

[5] Centro de Geofisica da Universidade de Lisboa, Portugal.

[6] BIGSETS Team: L. Mendes Victor, C. Corela, A. Ribeiro, D. Cordoba, J. J. Danobeitia, E. Grácia, R. Bartolomé, R. Nicolich, G. Pellis, B. DellaVedova, R. Sartori, L. Torelli, A. Correggiari, L. Vigliotti.

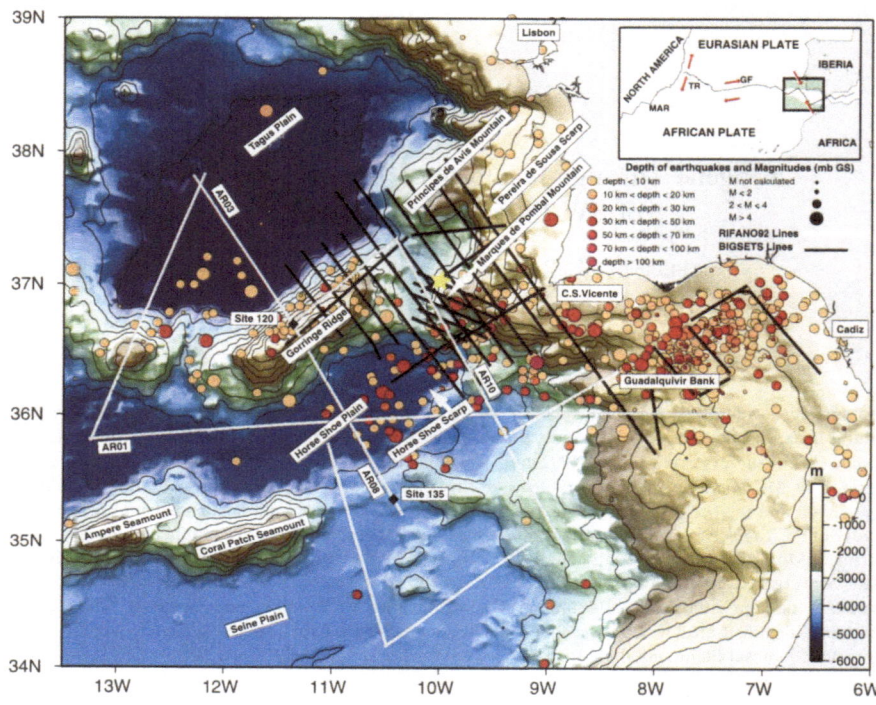

Figure 1
Bathymetry map of the study area (data from GEBCO97 Digital Atlas Web Site: www.nbi.ac.uk) with location of earthquakes ipocenters recorded during 1973 to Present (data retreived from USGS National Earthquake Information Center, World Data Center for Seismology, Denver, PDE catalog at http://www.neic.cr.usgs.gov). Also shown the location of MCS lines acquired during RIFANO92 and BIGSETS98 surveys. The yellow star indicates the position of the 1755 A.D. event (from UDIAS *et al.*, 1976). Inset sketches the main elements of plate boundaries (data from NOAA Global Relief Data), the stippled patch indicates the absence of a well-defined plate boundary: MAR: Mid Atlantic Ridge; TR: Terceira Ridge; GF: Gloria Fault. Solid arrows display relative plate kinematics, after BUFORN *et al.* (1988).

(yellow star in Fig. 1), the most terrifying cataclysm to have occurred since historical times in Western Europe with an estimated earthquake magnitude of 8.5–8.7 (MARTINS and MENDES VICTOR, 1990). That day, in Lisbon alone, between 10,000 to 60,000 people died, 10% of them due exclusively to the tsunami. The earthquake was so strong as to be felt throughout Europe as far as Belgium, Finland and Northern Africa The generated tsunami wave reached the Caribbean Sea and the coasts of Great Britain (see for a review BAPTISTA *et al.*, 2003). The tsunami wave heavily damaged the coasts of SW Portugal, Western and Northern Morocco and the Gulf of Cadiz, reaching a height of 30 m in C. S. Vicente (SW Portugal).

During the Nineties two multi-channel seismic (MCS) surveys, RIFANO92 (SARTORI *et al.*, 1994) and BIGSET98 (ZITELLINI *et al.*, 2001), have been carried out offshore SW Iberia (Fig. 1) with the aim to localize and to map the tectonic structures related to the source area of the 1755 Lisbon Earthquake. This research led to the identification of the most probable source area (ZITELLINI *et al.*, 1999, 2001), the Marquês de Pombal (Fig. 1), an abyssal mountain of tectonic origin located on the continental margin off SW Iberia, about 100 km WSW of Cape San Vincente (Fig. 1). This localization was strengthened by the backward ray-tracing tsunami modeling and shallow-water simulations performed by BAPTISTA *et al.* (1998). The two geophysical investigations have shown that besides the Marquês de Pombal Mountain, other active tectonic structures are present in the area. The major aim of this work is to provide their documentation as well as a more extensive record of the structural setting of the Marquês de Pombal structure. The progress in the knowledge of the tectonic evolution during Neogene through recent times of the seismogenic area and the detailed study of the largest tectonic structures is a necessary step for a better assessment of the seismic/tsunami hazard of the area.

Geological Setting

The region off the SW Iberian Peninsula and the Gulf of Cadiz shows an important natural seismic activity and puzzling tectonics (Fig. 1). The southern margin formed during the Jurassic continental breakup between North America and Africa, the western margin is the result of Cretaceous separation of Iberia with respect to North America. In the Gulf of Cadiz, the South Iberia and Moroccan passive margins are partially overlain by a thick accretionary wedge related to the neogenic westward migration of the Alboran Microplate and by huge olistostromes, discharged gravitationally from the tectonic mélange itself (TORELLI *et al.*, 1997). According to plate tectonic reconstruction the Eurasia/Africa plate boundary owns its present configuration only since late Oligocene. The Europe-Africa plate boundary (inset of Fig. 1) trends roughly E-W, connecting the Azores-Triple Junction to the Gibraltar Strait. Along this boundary the present plate motion is divergent east of the Azores with a dextral strike-slip component, transform in the middle segment, and convergent to the east of the Tore-Madeira Ridge. In the area between the Gorringe Ridge and the Gibraltar Strait, compressive stress trends mainly NNW-SSE with plate convergence rate of 4 mm yr[1] in the last three million years. In this area the plate boundary consists of diffuse, active compressive deformation, distributed over a 200–300 km wide area (SARTORI *et al.*, 1994; HAYWARD *et al.*, 1999; TORTELLA *et al.*, 1997). The seismicity pattern reflects this kind of plate interaction with the presence of scattered hypocenters spanning from shallow in the oceanic area to intermediate depth in the continental domain (Fig. 1). Focal mechanisms display a mixture of thrust and strike-slip motion (BUFORN *et al.*, 1988). The plate convergence is responsible for the reactivation of the older rift faults, and a number of large, active, tectonic structures have been detected along the continental margin and within the oceanic domain (ZITELLINI *et al.*, 2001). Apparently, the tectonic reactivation began in the Miocene. This is supported by the age of the inversion at the Lusitanian Basin located along the western Portuguese coast, approximately 150 km North of Cape S. Vicente (RIBEIRO *et al.*, 1990; KULLEERG *et al.*, 2000) and by the existence of a Middle Miocene major unconformity observed within the sedimentary cover in the Tagus Abyssal Plain (MAUFFRET *et al.*, 1989, Fig. 10). In the southern Portuguese Margin, the Miocene compression is milder and it is superimposed on an older important tectonic inversion that took place during latest Cretaceous through Paleogene times (TERRINHA, 1998). Two major questions arise to assess the seismic/tsunami hazard of the area. One is related to the dimension of the rupture area of the 1755 Lisbon Earthquake. ZITELLINI *et al.* (2001) suggested that the most probable location of the source area of this great earthquake is the Marquês de Pombal Thrust Fault (MPTF in Fig. 3) which, during the earthquake, may rupture simultaneously with a subsidiary thrust fault acting as a back thrust fault. The resulting earthquake magnitude for a mean displacement of 15 meters of the hypothesized rupture system is of the same order of magnitude of the 1755 Lisbon Earthquake although it does not reach the 8.5–8.7 value inferred by MARTINS and MENDES VICTOR (1990). This implies that the hypothesized rupture system could be part of a larger generator involving an area twice as large as the one calculated in ZITELLINI *et al.* (2001). This fact is stressed by the recent paper of TERRINHA *et al.* (2003), which suggests a northward continuation of the fault system associated with the Marquês de Pombal structure. The second question is related to the possible presence of other potential seismogenic/tsunamigenic structures offshore SW Iberia. This is testified by the occurrence of other historical large earthquakes and tsunamis, such as the 1531, which was probably as strong as the 1775 Lisbon event, and the 1722 event, that were mostly likely located offshore SW Portugal (UDÍAS *et al.*, 1976). An additional question arises concerning the reciprocal interaction of the various active tectonic structures during an earthquake activity. This last point is far beyond the scope of the present work because we lack the key data relative to the architecture of the décollement surfaces at depth and the knowledge of the rheology and the thermal state of the crust.

MCS Acquisition and Processing

The first survey (RIFANO92 lines in Fig. 1) was sponsored by the Italian Consiglio Nazionale delle Ricerche. The survey was performed during 1992 on board the R/V OGS-Explora and collected 2,000 km of seismic lines across the Africa-Europe plate boundary: from Seine to Tagus Plains and from Gibraltar Arc to Horseshoe Plain. The energy source was generated by a 32 airgun array with a total volume of 4880 cubic inches and 50 m shooting interval. The data were acquired with a 3000-m-long, 120-channel streamer, with 25 m group interval. The second survey was performed in 1998 and was funded by the European Community, project BIGSETS (BIG Sources of Earthquake and Tsunami in SW Iberia, ENV4-CT97–9547). BIGSETS survey was focused in the area between the Gorringe Ridge and the continental margin of SW Portugal to investigate the tectonic deformation related to the "1755 Lisbon Earthquake" (BIGSETS98 lines in Fig. 1). This second survey collected 1,715 km of seismic lines and it was carried out on board the R/V Urania. The data were acquired using a 48-channel, 1,200-m-long streamer made available by the Institut de Ciències de la Terra Jaume Almera of Barcelona (Spain). The energy source was generated by two GI guns with total volume varying from 150 to 465 cubic inches and shooting interval of 25, 37 and 50 meters. The processing of the MCS lines shown in this paper was performed at the Institute of Marine Geology of Bologna using the commercial DISCO package of PARADIGM Inc. The processing sequence was: decimation from 2 ms to 4 ms, common depth point gathering, spiking deconvolution in RIFANO92 lines, velocity analysis every 2.5 km, normal move out correction, CDP staking, spherical divergence correction, finite-difference wave-equation migration using stacking velocities with reduction of 10%.

Data Description

The first investigation, the RIFANO92 survey (Fig. 1), was designed to acquire geophysical information, on a regional scale, about the plate interaction between Africa and Europe across the plate boundary in the region of the Gorringe Ridge. The MCS lines were shot in E-W and NNW-SSE direction, roughly parallel and perpendicular to the present day NNW-SSE slip vectors. One 500-km-long, east-west oriented, line AR01 (Fig. 1) was acquired from Gibraltar Arc to the Horseshoe Plain. One NNW-SSE line was shot from the Tagus to the Seine Plain throughout the deformed area of the Gorringe Ridge (line AR03-AR08 in Fig. 1). This line was located entirely in oceanic domain, from the undeformed Cretaceous oceanic crust of the Tagus Plain to the undeformed Jurassic oceanic crust of the Seine Plain. Another NNW-SSE MCS line AR10 (Fig. 1) was shot closer to SW Iberia, entirely in continental domain. The main results from this study have been reported by TORELLI et al. (1997) for line AR01, SARTORI et al. (1994) for line AR03-AR08 and ZITELLINI et al. (1999) for line AR10. The whole MCS data set showed that the area is dominated by the presence of compressive tectonic structures, most of them still active. In particular the discovery of a large tectonic structure 100 km WSW of Cape S. Vicente, thought as the possible generator of the 1755 Earthquake (ZITELLINI et al., 1999), led to the acquisition campaign BIGSETS98. This campaign was designed to investigate in detail the structure found offshore Cape S. Vicente and to search for other possible, undetected tsunamigenic sources along the margin. The first results of BIGSETS98, regarding the structure found offshore Cape S. Vicente, were cescribed by ZITELLINI *et al.* (2001), who named the structure "Marquês de Pombal." Here we report a more comprehensive documentation on the Marquês de Pombal and on the other active structures found along the SW Iberia continental margin integrating the RIFANO92 and BIGSETS98 data sets. The data show that all the major morphological features, i.e., the Marquês de Pombal Mountain, the Horseshoe Scarp, the Guadalquivir Bank, the Gorringe Ridge, the Princes de Avis Mountains and the Pereira de Sousa Scarp (Fig. 1) are of tectonic origin. For clarity we will describe them individually.

Marquês de Pombal Mountain

The continental slope of the SW Portuguese Margin displays two promontories before reaching the abyssal depths, the Marquês de Pombal Mountain (Fig. 1 and MPM in Fig. 2) and the Principes de Avis Mountains (Fig. 1 and PAM in Fig. 2), both described in this paper. The Marquês de Pombal Mountain stands out morphologically as an elevated rectangular smooth surface bounded by the Pereira

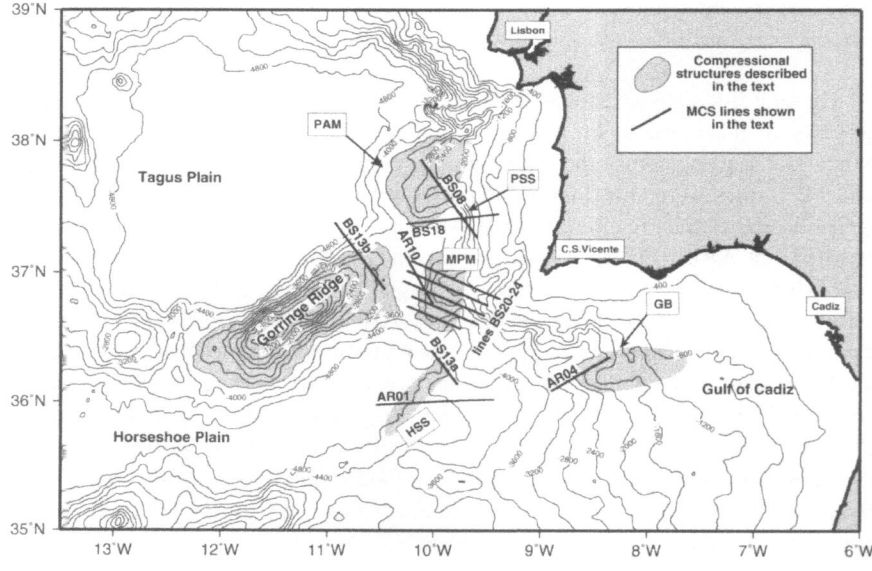

Figure 2
Bathymetry map of the study area. Contours are every 400 m (data from GEBCO97). Heavy lines represent the portions of the MCS lines shown in the following figures. Grey shaded areas indicate compressional structures. MPM: Marquês de Pombal Mountain; HSS: Horseshoe Scarp; GB: Guadalquivir Bank; PSS: Pereira de Sousa Scarp; PAM: Principes de Avis Mountains.

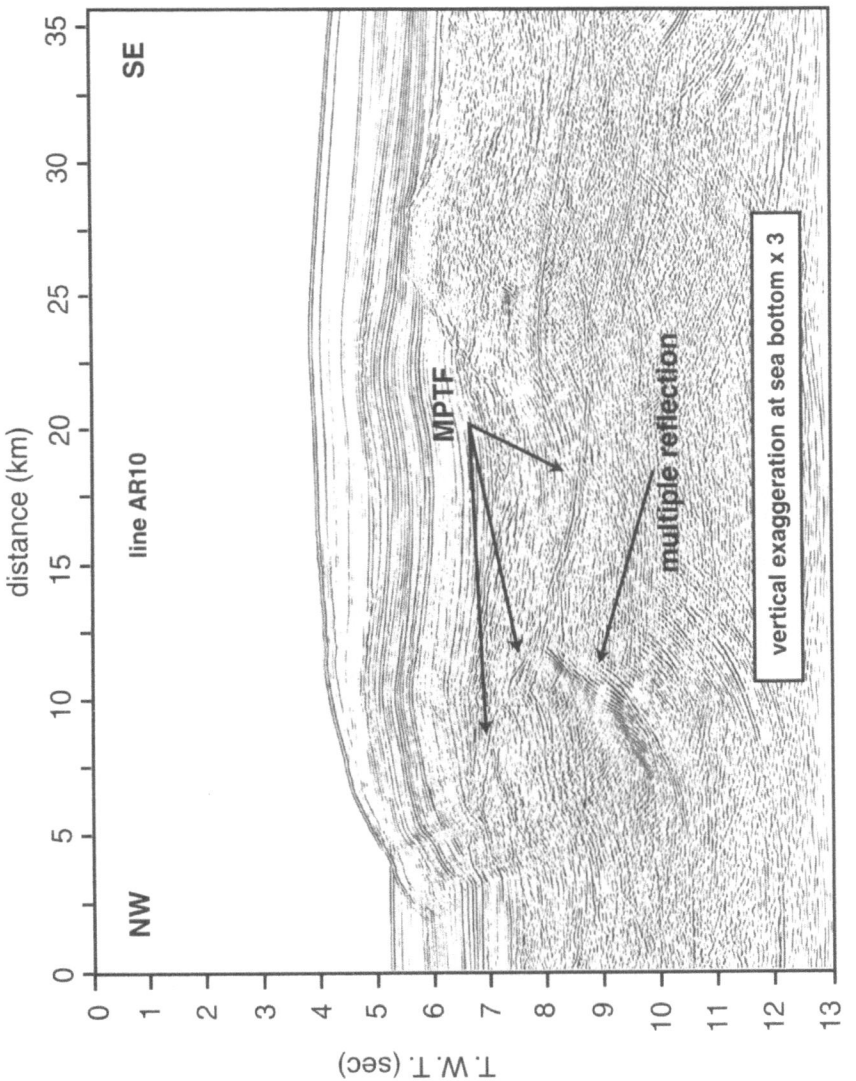

Figure 3
Post-stack time migration of line AR10, for location see Figures 2 and 5.

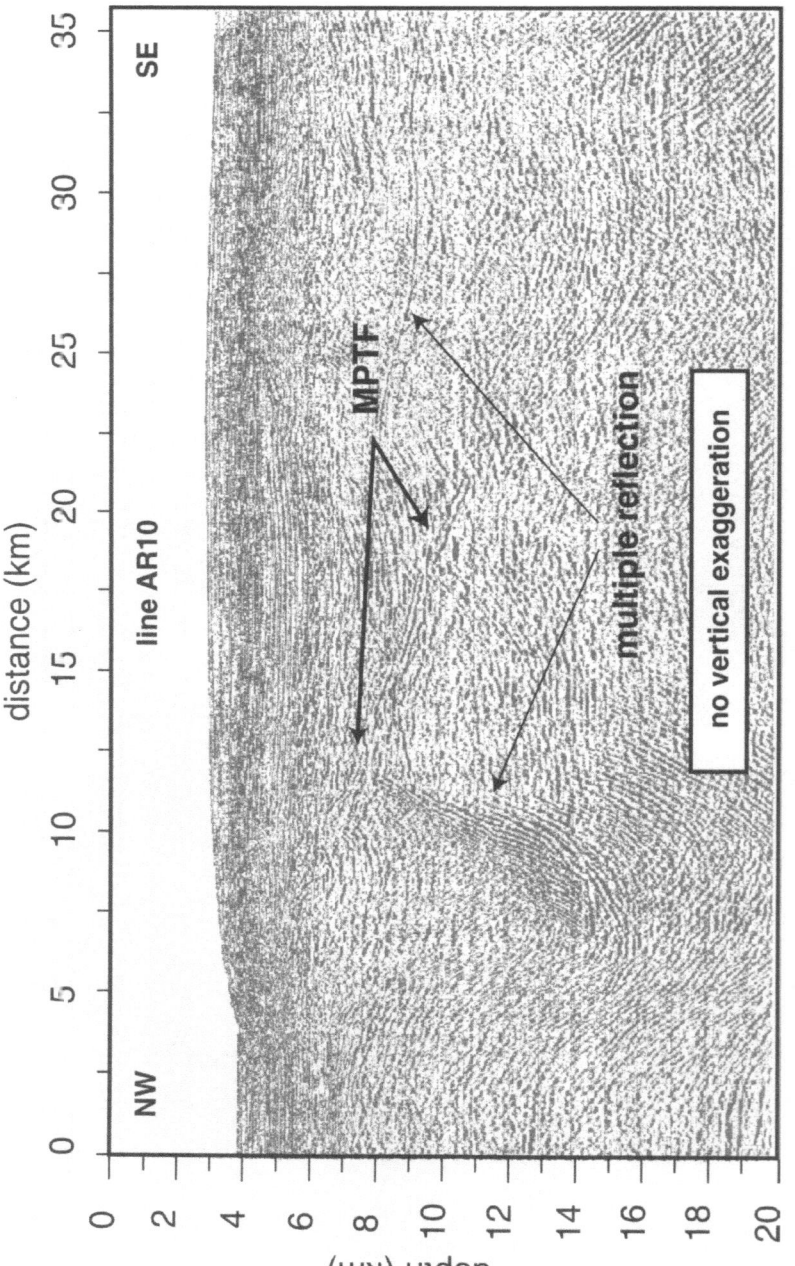

Figure 4

Line AR10 depth converted after post-stack time migration using interval velocities derived by smoothing the stack velocities.

de Sousa Scarp (Fig. 1 and PSS in Fig. 2) and the Cape S. Vicente Canyon (Fig. 5). A network of 10 MCS of lines (Fig. 1), 6 of them presented in this paper, could be used to trace the deformation associated with the Marquês de Pombal thrust fault (MPTF in Fig. 3) and constrain its geometry and kinematics. The whole set of existent MCS lines show that the Marquês de Pombal Thrust Fault consists of an approximately 55-km long NNE trending blind thrust, breaching through the youngest sediments of the continental slope only at line BS22 (Fig. 6c). Altogether the MPTF deforms a 100-km long narrow area that is being transported westwards on top of a landward dipping thrust whose depth conversion (Fig. 4) revealed an apparent average dip angle of 24 ° in the upper 11 km, according to ZITELLINI et al. (2001).

To the North, the MPTF does not breach the surface (line BS20, Fig. 6a) and we observe a simple flexure of the sediment cover on the hanging-wall (line BS20, Fig. 6a; line BS21, Fig. 6b) of MPTF. Inspection of Figs 6b and 6c shows that the maximum uplift and steepest scarp along the MPTF occur on lines BS21 and BS22. To the South, lines BS23 and BS24 (Figs 6d, 6e) show the presence of two anticlines in the hanging-wall of MPTF. These observations indicate that the style of deformation in the northern and southern parts of the MPTF is different. This, together with the different strike of the fault between these two, led to propose that the MPTF constitute two independent fault segments connected by a relay fault zone (Fig. 5), as suggested by TERRINHA et al. (2003).

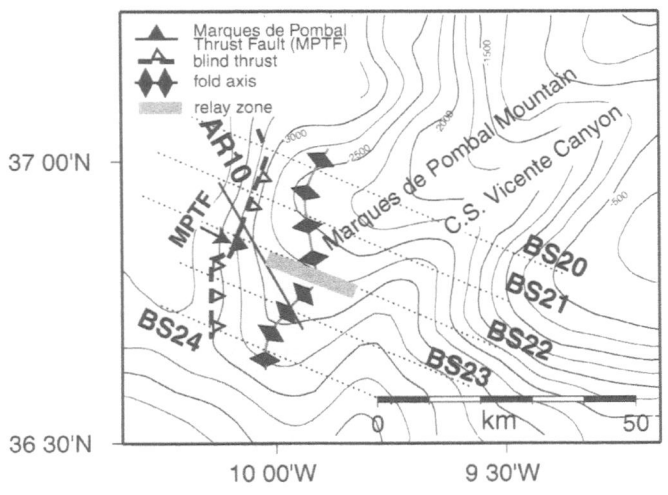

Figure 5
Tectonic sketch of the Marquês de Pombal Mountain.

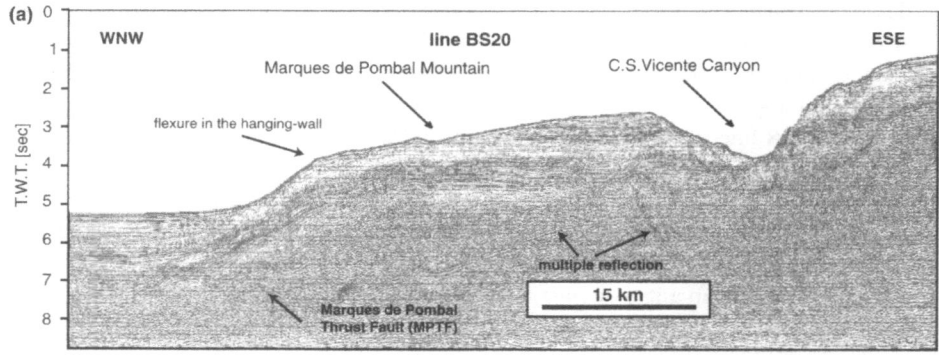

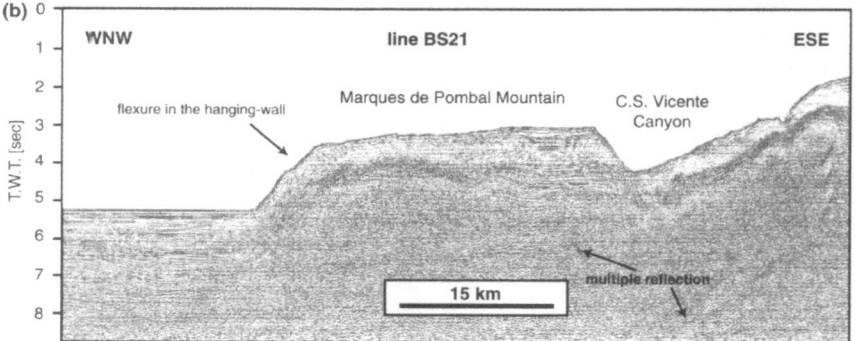

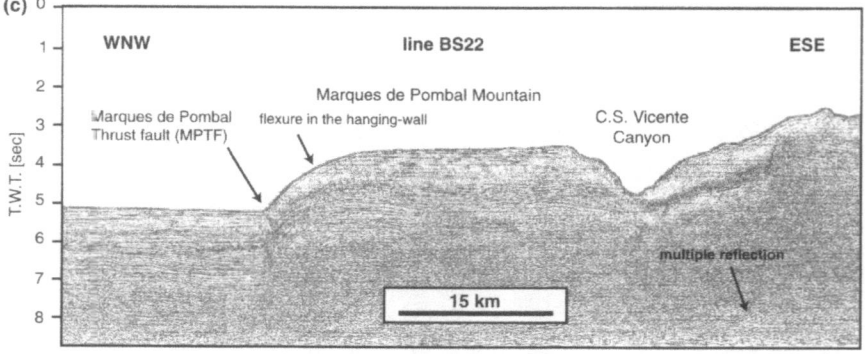

Figure 6

a) Time migration of line BS20, b) time migration of line BS21, c) time migration of line BS22 showing the Marquês de Pombal Thrust Fault (MPTF) breaching the sea bottom, d) time migration of line BS23, e) time migration of line BS24; all the lines have been acquired across the Marquês de Pombal structure, until the Cabo de São Vicente Canyon.

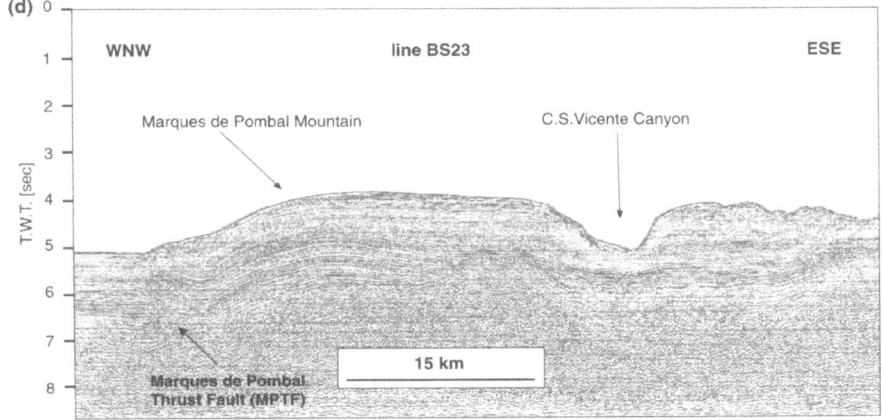

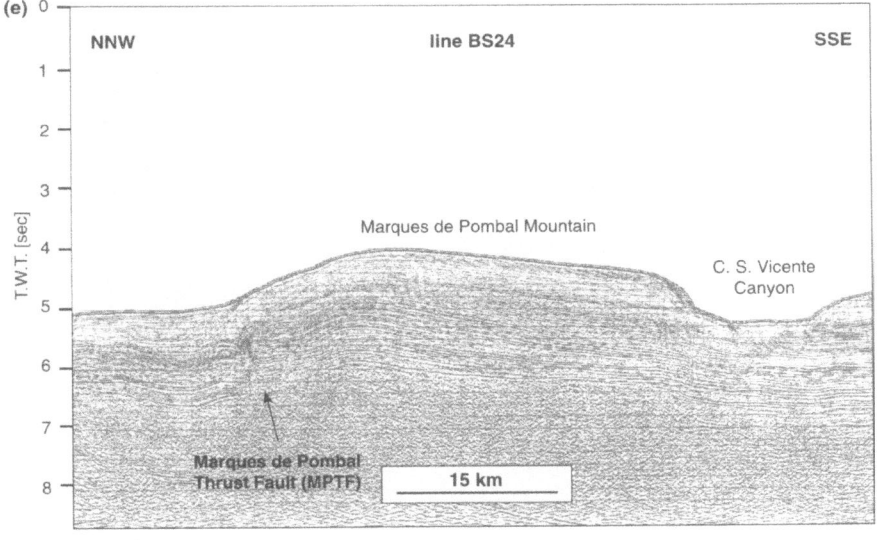

Figure 6
(Contd.)

The NNE-SSW striking faults of the western Portuguese Margin are generally inherited from the late orogenic Variscan fracturing event of Permian age (ARTHAUD and MATTE, 1977). It is somewhat possible that the MPTF was originally formed in the late stages of the orogenic event and that it played an active role during the Mesozoic tectonic extension and was reactivated again in the Cenozoic.

Horseshoe Scarp

The Horseshoe Scarp is a prominent feature that bounds eastward the Horseshoe Plain (Fig. 1). The Horseshoe Plain lies at a depth of 5000 m and is filled by about 2–5 km of sediments, where the oldest sediments are probably of Late Jurassic age (SARTORI *et al.*, 1994; HAYWARD *et al.*, 1999). Underneath the Horseshoe Scarp develops the Horseshoe Fault (HSF in Fig. 7) which consists of a NE-SW trending blind thrust. The chaotic body present within the sedimentary section formed by down-slope flow of olistostromes and debris discharged from the tectonic mélange associated with the subduction system of the Betic and Rifain chain (TORELLI *et al.*, 1997). The emplacement of this huge chaotic body formed by

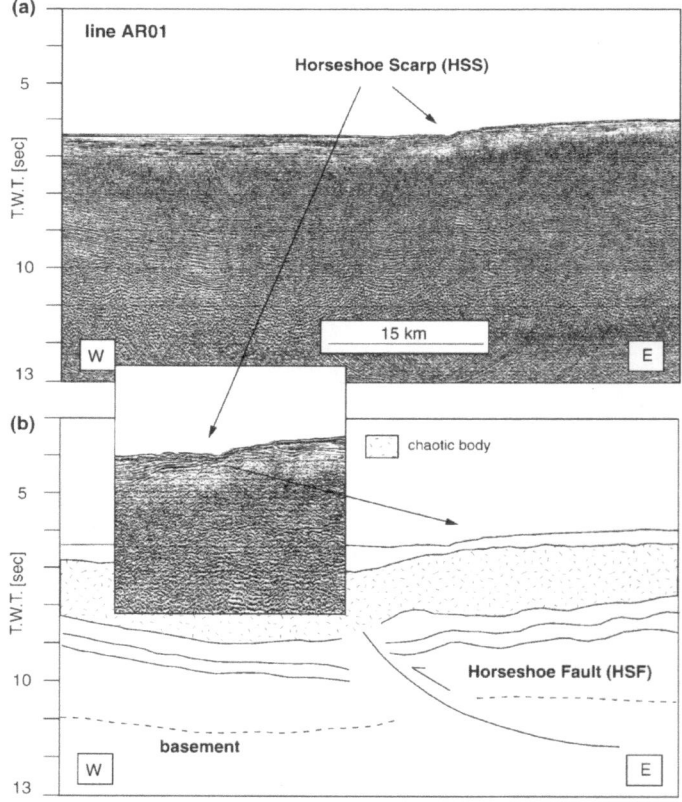

Figure 7
a) Time migration of line AR01 across the Horseshoe Scarp (HSS), for location of the MCS line see Figure 2. b) Interpretative sketch of the seismic line.

mass transport processes is certainly contemporaneous with the main uplift phase of the Betic-Rifain Cordillera of Middle to Late Miocene times, whilst the tectonic mélange associated with the Africa-Iberia subduction is older, possibly of Paleogene age, also the age of tectonic inversion of the Algarve Basin, South Portugal (TERRINHA, 1998). The chaotic body is capped by plane-parallel, laterally continuous reflections related to turbiditic sedimentation as suggested by the coring of the uppermost sediment layers (THOMSON and WEAVER, 1994). The chaotic body in the Horseshoe Abyssal Plain is an important stratigraphic marker, because it acts as a reference level for the post Middle Miocene tectonic deformation. The Horseshoe Fault is very continuous laterally, some 180-km long, it is almost rectilinear and it extends from the Ampère-Coral Patch Seamounts to the southern termination of the Cape S. Vicente canyon (Fig. 1). This fault yields a clear morphological expression (Fig. 1) and it is shown by MCS lines AR01 (Fig. 7), BS13 (Fig. 8) and also by the reflection profile PQ published in HAYWARD *et al.* (1999). The vertical displacement, using the chaotic body as reference level, varies along the fault increasing from south to north, about 0.7 s TWT in line AR01 (Fig. 7) and 1.6 s TWT in line BS13 (Fig. 8). This is also evident from the bathymetry map (Fig. 1). AR01 (Fig. 7) and BS13 (Fig. 8) MCS lines show that movements along the Horseshoe Fault post-date the deposition of the chaotic body. Moreover the fault is active at present since it deforms the uppermost sediments which, by lateral correlation with other BIGSETS MCS lines, can be dated as Quaternary. From our data set we cannot infer the southern and northern termination of the fault.

Guadalquivir Bank Structure

The Guadalquivir Bank shown by line AR04 (Fig. 9) is an elongated ENE-WSW seamount 50 km long and 30 km wide, (Fig. 1). It formed initially as a basement horst during the Mesozoic stretching of the southern Iberia continent and it represents the southern boundary of the Algarve Basin, whose rift phases lasted from Triassic to Albian-Cenomanian times (TERRINHA, 1998). Dredges taken from this seamount show that the basement crops out at the sea floor and consists of the low grade metamorphic sediments of Carboniferous age of the South Portuguese Zone (TERRINHA, 1998). MCS lines acquired across this basement high show that it is partially covered by sediments of Miocene through Quaternary age. The Guadalquivir Bank behaved as a rigid block during the main phases of tectonic inversion of the Algarve Basin of Late Cretaceous-Paleogene age during which it suffered uplift and compression. During Neogene times it subsided more than 1 km during the formation of the flexural Miocene Algarve Basin (TERRINHA, 1998).

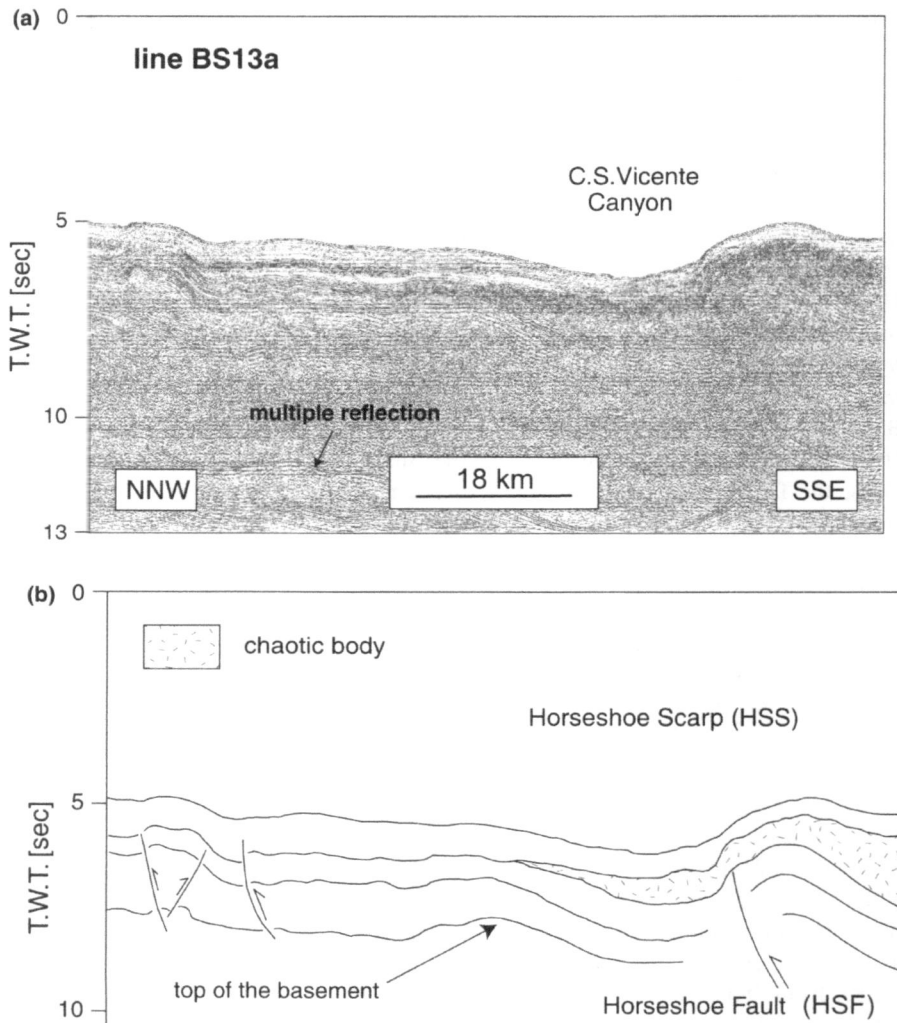

Figure 8

a) Time migration of line BS13a showing the northern termination of the Horseshoe Scarp and associated Fault (HSS and HSF), for location of the line see Figure 2. b) Interpretative sketch of the seismic line.

Line AR04 (Fig. 9) shows that the Guadalquivir Bank is presently bounded by reverse faults (T1 and T2 in Fig. 9), which are responsible for the subsequent vertical uplift of the basement. The displacement of the most recent reflectors shows the

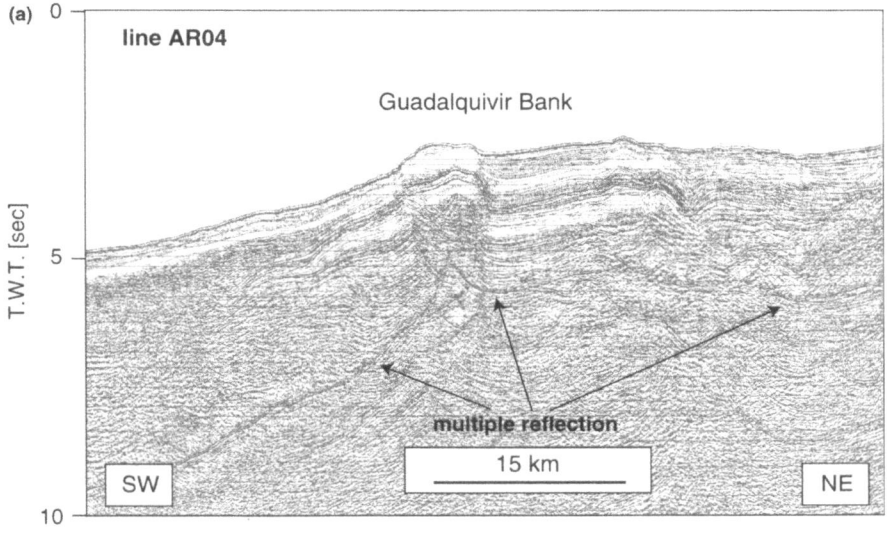

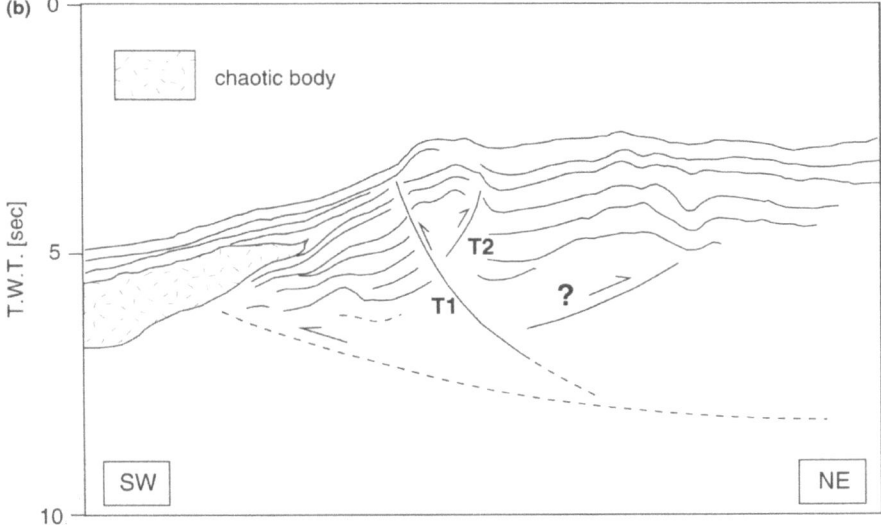

Figure 9

a) Time migration of line AR04, shot on the Guadalquivir Bank, for location of the seismic line see Figure 2. b) Interpretative sketch of the seismic line AR04, showing the possible deep décollement surface, accomodating the shallower deformation, expressed as thrust faults (T1 and T2).

present-day activity of the reverse faults, and the displacement of more than 1.0 s TWT of the chaotic body indicates the importance of the post Middle Miocene tectonic shortening in this sector. This is also testified by the presence of a cluster of shallow seismic activity just underneath and northward of the structure (Fig. 1). The paucity of MCS lines in this sector does not allow the determination of the fault trace around the structure.

Gorringe Ridge

The Gorringe Ridge (Fig. 1) is the most important topographic feature of the area. It is an uplifted block of crustal and upper mantle rocks (AUZENDE *et al.*, 1978; AUZENDE *et al.*, 1982; RYAN *et al.*, 1973; FÉRAUD *et al.*, 1986; GIRARDEAU *et al.*, 1998) forming one of the most prominent gravimetric anomalies of the world. The Gorringe Ridge separates the 5000 m deep Tagus Abyssal Plain from the Horseshoe Abyssal Plain reaching a water depth of 25 m at its top. The Gorringe Ridge trends NE-SW, it is 200-km long and 60-km wide and it is asymmetric with the northern flank steeper than the southern. The Gorringe Ridge was considered the most likely tectonic source for the 1755 Lisbon Earthquake (FUKAO, 1973) until the work of BAPTISTA *et al.* (1998), who demonstrated that the source zone should trend N-S and have proximity to the Portuguese coast. SARTORI *et al.* (1994) showed that the Gorringe Ridge overthrusted the Tagus Abyssal Plain for 4–5 km. RYAN et al. (1973) suggested that the uplift of the Gorringe Ridge occurred during post-Langhian, pre-early Pliocene time. Additional details of the structure can be inferred from the BIGSETS98 lines (Fig. 1). These MCS lines were collected on the eastern side of the Gorringe, presumably on thinned continental crust. Here the presence of a thicker sedimentary cover allows a better understanding of the deformation patterns and the tectonic timing. In Figure 10 MCS line BS13b summarizes the characteristics of this sector. Here the vertical uplift is still considerable, of the order of 3.0 secs TWT, and the flanks of the Gorringe Ridge are still asymmetric with the northern side steeper than the southern. This asymmetry is reflected by the different degree of tectonic deformation which is very intense in the northern side and on the apex of the seamount while it is minor in the southern flank. A relevant unconformity is detectable in the Tagus Abyssal Plain (Fig. 10). This unconformity marks the end of the northwestwards directed over thrusting of the Gorringe Ridge. This unconformity has a regional significance being widespread all over the Tagus Abyssal Plain and it corresponds to the unconformity found by MAUFFRET *et al.* (1989) in the area. Unfortunately there is no precise time calibration of this unconformity. MAUFFRET *et al.* (1989) assigned a Mid-Miocene age to it, based on seismic correlation with DSDP site 398 (SIBUET *et al.*, 1979).

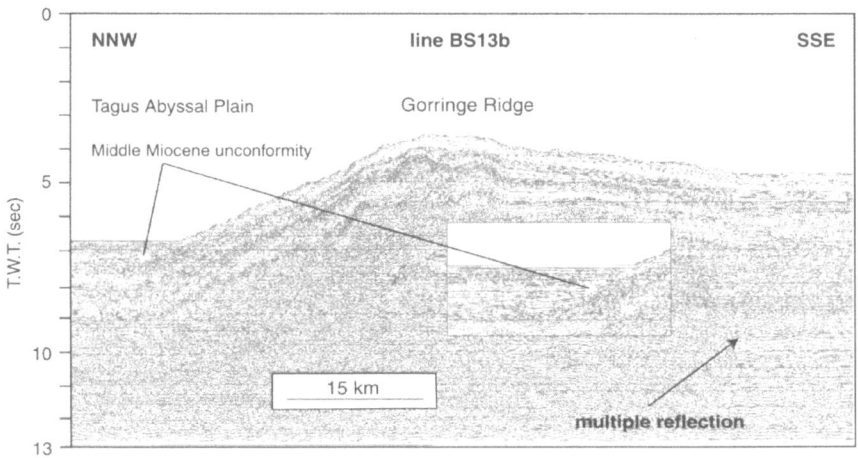

Figure 10

Time migration of line BS13b showing fold-related structures on the northeastern termination of the Gorringe Ridge and the Middle-Miocene unconformity in the Tagus Abyssal Plain, for location see Figure 2. The inset shows a magnification of the Tagus Abyssal Plain sedimentary sequence at the foot of the Gorringe Ridge.

Principes de Avis Mountains

The Principes de Avis Mountain is located on the continental slope (Fig. 1) and it lies on the continuation of the Gorringe Ridge and has the same NE-SW trend. MCS line BS08 (Fig. 11) crosses the structure normal to the major axis and displays the presence of two folds trending parallel to the mountain. The two folds are asymmetric and their geometry indicates northwestward tectonic transport direction. The southern fold is a nice example of ductile deformation of the sedimentary cover on top of a blind thrust dipping to the southeast. Shortening and uplift caused the progressive southeastward migration of the basin's depocenter, which is presently closer to the Pereira de Sousa Scarp (Fig. 11). Within the sedimentary cover of the modern Rincão do Lebre Basin a major unconformity separates a lower sequence of plane-parallel reflectors from an upper one of progressive on-lapping reflectors ("Y" in Fig. 11). This unconformity separates the pre-deformation from the syn-deformation or syntectonic sedimentary units; unfortunately there is no time calibration available for the unconformity. A second unconformity located in the upper part of the younger sequence marks the end of the contraction phase ("X" in Fig. 11). This unconformity is less clear and separates a lower "wedged" sequence from an upper one made up of more parallel, laterally discontinuous reflectors, which are related to re-sedimentary processes such as mass flow and debris flow originating from the flank of the topographic high. These processes are probably the cause of the

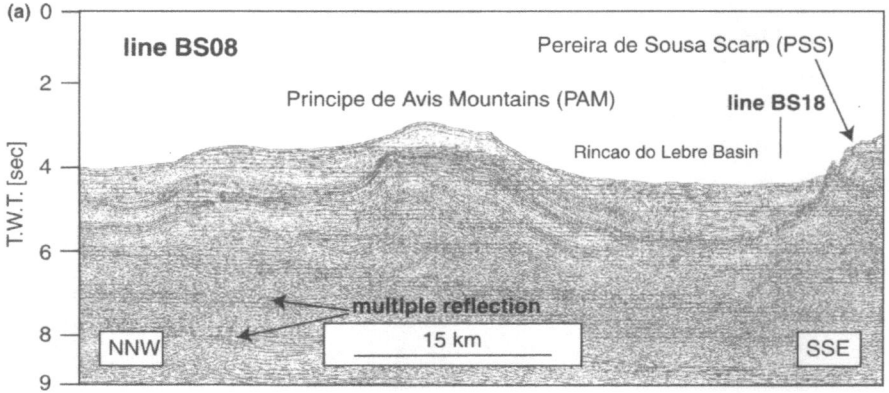

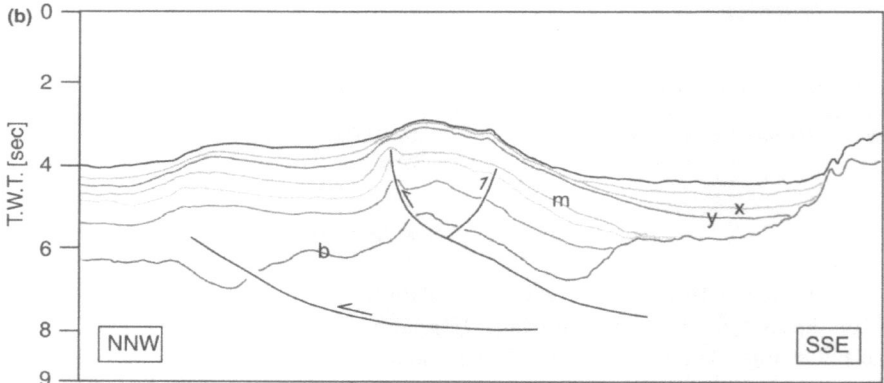

Figure 11

a) Time migration of line BS08 showing the basin Rincão do Lebre between the Principes de Avis Mountains and the Pereira de Sousa Scarp (see Figs. 1 and 2 for location). Also shown is the intersection with line BS18, shown in Figure 12. b) Interpretative sketch of the seismic line: letters indicate the major unconformities recognized within the basin and b = basement top. m = Middle Miocene unconformity; y, x, more recent angular unconformities, see text for a more exhaustive discussion.

local thickening of sediments on both sides of the basin. This unconformity is also poorly constrained in time.

Pereira de Sousa Scarp

The Pereira de Sousa Scarp (Fig. 1) is the morphological expression of a fault formed initially as a rift fault during the Mesozoic continental separation between

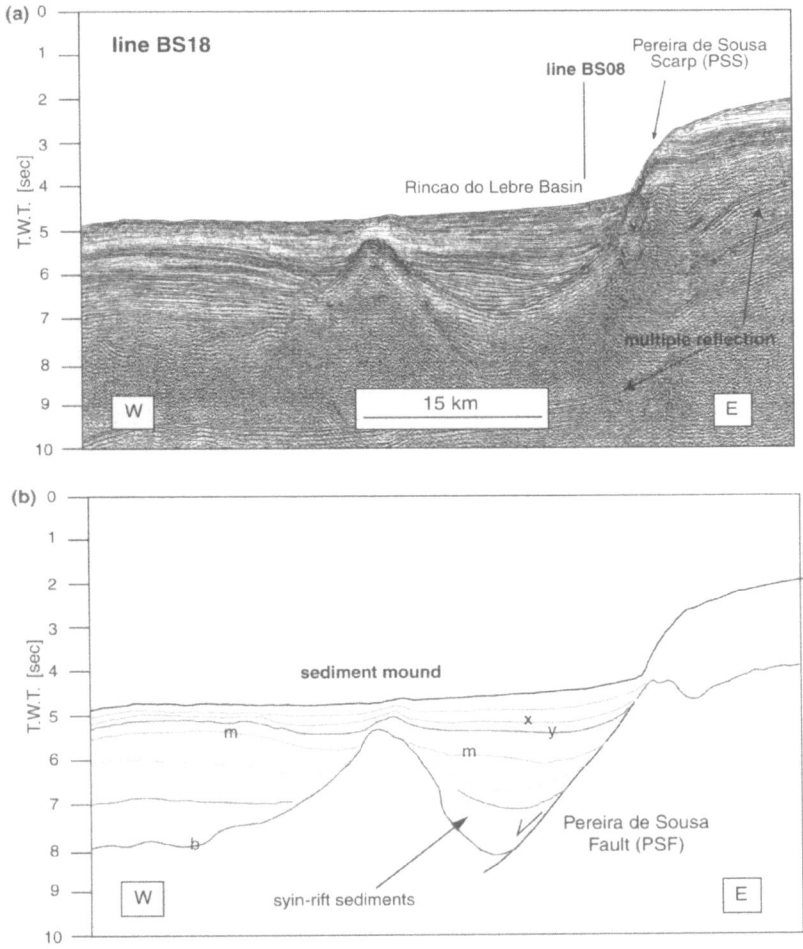

Figure 12
a) Time migration of line BS18 shot normal to the Pereira de Sousa Scarp (PSS), showing the scarp and the intersection of line BS08, for location see Figure 2. b) Interpretative sketch of the seismic line, showing the same angular unconformities presented in Figure 11.

Iberia and North America (PSF in Fig. 12). This fault originated one of the most prominent morphological features of the continental margin off SW Portugal, even recognizable on the bathymetry map (Fig. 1). The fault trends approximately N-S, offsets the basement by more than 4.5 s TWT and exhibits a steep well preserved fault scarp 600-m high (Fig. 12) and 45-km long (Fig. 1). The Pereira de Sousa Fault is described in this paper because of its following singular characteristics: 1) it remains an extensional fault from a geometrical point of view, despite the fact of being located in an area where compressive deformation is widespread, 2) it is laterally bounded by two compressional structures, the Marquês de Pombal Mountain and the Principes de Avis Mountain also described in this paper, 3) it represents the best developed morphological scarp of the area and 4) it is associated with the development of an active sedimentary system that produces sedimentary ridges that can reach 20 km of run out distance (TERRINHA *et al.*, 2003). The geometry of the PSF is best shown on MCS line BS18, which is normal to it (Fig. 12).

The Pereira de Sousa fault limits landward the lower continental slope where a thick sedimentary basin, 3.0 s TWT, is present (Fig. 12). The fault is listric as testified by the marked rotation of a large basement block of the hanging-wall and by the wedging of the Mesozoic syn-rift sediments (Fig. 12). Line BS18 is parallel to the direction of the Cretaceous extension while it is almost perpendicular to the direction of transport of the most recent compressive event. This fact makes the identification of the tectonic structures, and associated unconformities, more difficult regarding the last event. We can overcome the difficulties analyzing line BS08 (Fig. 11), which is nearly parallel to the direction of crustal shortening. On this MCS line we can see that the major shortening event affecting the post-rift sediments is confined below unconformity "X" (Fig. 11). Once the unconformity "X" is calibrated on line BS18 (Fig. 12) we infer that above "X" the sedimentation is dominated by sedimentary processes rather than tectonic movements. At the foot of the Pereira de Sousa Scarp the post "X" sequence appears related to the interplay of turbiditic with mass and debris flow deposition filling and smoothing a pre-existent topographic low. Seaward, in the other side of the basin, the sediment's mound is possibly related to velocity drop of turbiditic currents against a preexistent topographic high, with consequent local deposition. In conclusion, differently from the Marquês de Pombal Mountain, which has a similar trend and a similar Mesozoic history, the Pereira de Sousa Fault appears only slightly reactivated in compression during pre "X" time. In any case, at the present stage of knowledge, we cannot rule out that a minor contraction is also taking place today.

TERRINHA *et al.* (2003) in fact argue that the PSF is presently being uplifted, back-rotated, i.e., it is getting steeper and dissected by wrench faults from which can be inferred a WNW-ESE direction of compression. The completely different behavior of the Marquês de Pombal Mountain and Pereira de Sousa Scarp is still puzzling and may be related to the different orientation of the MPTF (roughly N10°E) and PSF (N10°W) with respect to the main stress field.

Discussion and Conclusions

In the investigated area, in addition to the Marquês de Pombal, there are at least two other remarkable, active, tectonic structures related to crustal shortening and showing considerable tectonic uplift: the Horseshoe Fault and the Guadalquivir Bank. The Gorringe Ridge and the Principes de Avis Mountains are also contractional structures although they appear deformed mostly during the Middle Miocene phase and present activity is almost absent or not detectable at the resolution of our data set. The most active structures are located where the present seismicity is higher, i.e., between latitude 35.5°N and 37.0°N and mostly within the continental crust. The lack of good stratigraphic control of the sedimentary cover prevents a more precise definition for the start and for the end of the Middle Miocene phase as well as for the beginning of the inversion of the Marquês de Pombal Thrust Fault, the Horseshoe Fault and the Guadalquivir Bank. The deformation style associated with the Neogene compressive episodes is very uniform and is expressed mainly by blind thrusts, i.e., thrust faults that do not reach the surface of the sea bottom. This deformation style produces folds in the sedimentary cover which are the most common tectonic features of the area and are easily recognizable in the high-resolution bathymetric map (GRÁCIA et al., 2003). Only the largest thrust faults reach the surface as in the Marquês de Pombal structure. The structural trends related to the Neogene compression found in the oceanic lithosphere are mainly NE-SW as in the Gorringe Ridge and in the Horseshoe Fault. The structural trends vary considerably along the continental margin, this is due to the local response to the shortening of pre-existing E-W Mesozoic and N-S Cretaceous extensional faults of the South and West Iberian passive margins respectively, as it is testified by the Marquês de Pombal and the Guadalquivir Bank strikes. Locally, along the continental margin, the preexisting faults may not be, or are only slightly, reactivated as for the Pereira de Sousa Fault. SARTORI et al. (1994) and HAYWARD et al. (1999) have shown that the deformation associated with the NE-SW compressive stress is spread over an area 200-km wide. From our investigations we can attest that the shortening is accommodated, on the upper crust, mainly by few major but large tectonic structures, of the order of tens of kilometers. The existence of few major active structures accommodating most of the deformations and their precise localisation should help greatly for future monitoring of the structure for risk assessment and forecast. The possible lateral prolongation of the Marquês de Pombal fault system, to obtain a larger rupture area for the 1755 Lisbon Earthquake, has a limited number of solutions being restricted to the structures found at the Horseshoe Fault and the Guadalquivir Bank. This solution does not role out the hypothesis of TERRINHA et al. (2003) which needs further proof to be demonstrated. The way different tectonic structures interact during a very large earthquake is at the moment speculative because we do not know the position of the décollement surfaces that may connect at depth the faults observed in the first few kilometres of crust. The existence at depth of a common décollement surface allows the contemporaneous, or almost contemporaneous, rupture of more than one fault. To enlighten this last aspect it will be necessary to perform additional investigations on the deep structure architecture of the SW Iberian margin.

Acknowledgements

The research was funded by the EC Environment and Climate Program 1994–1998, Technologies to Forecast, Prevent and Reduce Natural Risks (contract n. ENV4-CT97-0547, project BIGSETS). Additional funding was supported by the Portuguese Project MATESPRO (PDCTM/P/MAR/15264/1999). IGM contribution No. 1321.

REFERENCES

ARTHALD, F. and MATTE, P. (1977), *Late Paleozoic Strike-slip Faulting in Southern Europe and Northern Africa: Result of a Right-lateral Shear Zone between the Appalachians and the Urals*, Geol. Soc. Am. Bull. *88*, 1305–1320.

AUZENDE, J.-M., OLIVET, J. L., CHARVET, J., LEHANN, A., LE PICHON, X., MONTEIRO, J. C., NICOLAS, A., and RIBEIRO, A. (1978), *Sampling and Observation of Oceanic Mantle and Crust on Gorringe Ridge*, Nature *273*, 45–49.

AUZENDE, J.-M. and CYAGOR GROUP II (1982), *The Gorringe Ridge: First Results of the Submersible Expedition CYAGOR II*, Terra Cognita *2*, 2, 123–130.

BAPTISTA, M. A., MIRANDA, J. M., CHIERICI, F., and ZITELLINI, N. (2003), *New Study of the 1755 Earthquake Source Based on Multi-channel Seismic Survey and Tsunami Modeling*, Natural Hazards and Earth System Sciences *3*, 333–340.

BAPTISTA, M. A., MIRANDA, P. M. A., and MENDES VICTOR, L. (1998), *Constraints on the Source of the 1755 Lisbon Tsunami Inferred from Numerical Modelling of Historical data*, J. Geod. *25 (2)*, 159–174.

BUFORN, E., UDÍAS, A., and COLOMBÁS, M. A. (1988), *Seismicity, Source Mechanisms and Tectonics of the Azores-Gibraltar Plate Boundary*, Tectonophys. *152*, 89–118.

FÉRAUD, G., YORK, D., MEVEL, C., CORNEN, G., HALL, C. M., and AUZENDE, J.-M. (1986), *Additional 40Ar/39Ar Dating of the Basement and the Alkaline Volcanism of Gorringe Bank (Atlantic Ocean)*, Earth Plan. Sci. Lett. *79*, 255–269.

FUKAO, Y. (1973), *Thrust Faulting at a Lithospheric Plate Boundary, The Portugal Earthquake of 1969*, Earth plan. Sci. Lett. 18, 205–216.

GIRARDEAU, J., CORNEN, G., BESLIER, M.-O., LEGALL, B., MONNIER, C., AGRINIER, P., DUBUISSON, G., PINHEIRO, L., RIBEIRO, A., and WHITECHURCH, H. (1998), *Extensional Tectonics in the Gorringe Ridge Rock:, Eastern Atlantic Ocean: Evidence of an Oceanic Ultra Slow Mantellic Accreting Center*, Terra Nova *10*, 330–336.

GRÁCIA, E., DAÑOBEITIA, J. J., VERGÉS, J., CORDOBA, D., and PARSIFAL CRUISE PARTY (2003), *Mapping Active Faults at the SW Iberia Margin (38°–36°) from High-resolution Swath-Bathymetry Data. Implications for Earthquake Hazard Assessment*, Geology *31*(1), 83–86.

HAYWARD, N., WATTS, A. B., WESTBROOK, G. K., and COLLIER, J. S. (1999), *A Seismic Reflection and GLORIA Study of Compressional Deformation in the Gorringe Bank Region, Eastern North Atlantic*, Geophys. J. Int. *138*, 831–850.

KULLBERG, M. C., KULLBERG, J. C., and TERRINHA, P. (2000), *Tectónica da cadeia da Arrábida*, Mem. Geociências, Museu Nac. Hist. Nat. Univ. Lisboa *2*, 35–84.

MAUFFRET, A., MOUGENOT, D., MILES, P. R., and MALOD, J. A. (1989), *Cenozoic Deformation and Mesozoic Abandoned Spreading Center in the Tagus Abyssal Plain (West of Portugal): Results of the Multichannel Seismic Survey*, Can. J. Earth Sci. *26*, 1101–1123.

MARTINS, L. and MENDES VICTOR, L. A. (1990), *Contribuicao para o estudo da sismicidade de Portugal continental*. Univ. De Lisboa, Instituto Geofisico do Infante D. Luis, publicacao *18*, 1–70.

RYAN, W. B. F. *et al.* (1973), *Initial Reports of the Deep Sea Drilling Project*, *13*, pp. 19–41, US Gov. Printing Office, Washington, DC.

RIBEIRO, A., KULLBERG, M. C., KULLBERG, J. C., MANUPPELLA, G., and PHIPPS, S. (1990), *A review of Alpine Tectonics in Portugal: Foreland Detachment in Basement and Cover Rocks*, Tectonophysics *184*, 357–366.

SARTORI, R., TORELLI, L., ZITELLINI, N., PEIS, D., and LODOLO, E. (1994), *Eastern Segment of the Azores-Gibraltar Line (Central-eastern Atlantic): An Oceanic Plate Boundary with Diffuse Compressional Deformation*, Geology *22*, 555–558.

SIBUET, J.-C., RYAN, W. B. F. *et al.* (1979), *Initial Reports of the Deep Sea Drilling Project*, 47, US Gov. Printing Office, Washington, DC.

TERRINHA, P., PINHEIRO, L. M., HENRIET, J.-P., MATIAS, L., IVANOV, M. K., MONTEIRO, J. H., AKHMETZHANOV, A., VOLKONSKAYA, A., CUNHA, T., SHASKIN, P., ROVERE, M., and the TTR10 SHIPBOARD SCIENTIFIC PARTY (2003), *Tsunamigenic-seismogenic Structures, Neotectonics, Sedimentary Processes and Slope Instability on the Southwest Portuguese Margin*, Marine Geology *195*, 55–73.

TERRINHA, P. (1998), *Structural Geology and Tectonic Evolution of the Algarve Basin, South Portugal*, Ph.D. Thesis, Imperial College, London.

THOMSON, J. and WEAVER, P. P. E. (1994), *An AMS Radiocarbon Method to Determine the Emplacement Time of Recent Deep-sea Turbidites*, Sed. Geol. *89*, 1–7.

TORELLI, L., SARTORI, R., and ZITELLINI, N. (1997), *The Giant Chaotic Body in the Atlantic Ocean off Gibraltar: New Results from a Deep Seismic Reflection Survey*, Mar. Pet. Geol. *14*, 125–138.

TORTELLA, D., TORNÉ, M., and PEREZ-ESTAUN, A. (1997), *Geodynamic Evolution of the Eastern Segment of the Azores-Gibraltar Zone: The Gorringe Bank and the Gulf of Cadiz Region*, Mar. Geophys. Res. *19*, 211–230.

UDÍAS, A., LOPEZ-ARROYO, A., and MEZCUA, J., (1976), *Seismotectonic of the Azores-Alboran Region*, Tectonophys. *31*, 259–289.

ZITELLINI, N., CHIERICI, F., SARTORI, R., and TORELLI, L. (1999), *The Tectonic Source of the 1755 Lisbon Earthquake and Tsunami*, Annali di Geofisica *42*, 1, 49–55.

ZITELLINI, N., MENDES, L. A., CORDOBA, D., DANOBEITIA, J., NICOLICH, R., PELLIS, G., RIBEIRO, A., SARTORI, R., TORELLI, L., BARTOLOMÉ, R., BORTOLUZZI, G., CALAFATO, A., CARRILHO, F., CASONI, L., CHIERICI, F., CORELA, C., CORREGGIARI, A., DELLA VEDOVA, B., GRACIA, E., JORNET, P., LANDUZZI, M., LIGI, M., MAGAGNOLI, A., MAROZZI, G., MATIAS, L., PENITENTI, D., RODRIGUEZ, P., ROVERE, M., TERRINHA, P., VIGLIOTTI, L., and ZAHINOS-RUIZ, A. (2001), *Source of 1755 Lisbon Earthquake and Tsunami Investigated*, EOS, Transactions, Am. Geophys. Union *82*, 26, 285–282.

(Received June 5, 2002, revised February 2, 2003, accepted February 26, 2003)

 To access this journal online:
http://www.birkhauser.ch

Pure appl. geophys. 161 (2004) 589–606
0033–4553/04/030589–18
DOI 10.1007/s00024-003-02464-3

Pure and Applied Geophysics

GEOALGAR Project: First Results on Seismicity and Fault-plane Solutions

F. Carrilho[1], P. Teves-Costa[2,3], I. Morais[3,4],
J. Pagarete[2], and R. Dias[5]

Abstract — The project GEOALGAR, initiated in May 2000, is devoted to the geodynamic monitoring and seismic characterization of the Algarve region. A brief description of the project goals, as well as the first results concerning the analysis of the recent seismic digital data, from 1999 and 2000, are presented. After simultaneous inversion of the seismic data and the velocity model parameters, the relocation of the hypocenters was performed for two selected areas. Twenty-five earthquakes were used for Area 1 and 125 earthquakes for Area 2, selected from the period 1999–2000 and with magnitude $M_L \geq 2.0$. The results show that there are two main regions where there are more events: the Monchique region (inland) and another one in the area of Guadalquivir Bank (ranging from 36.4°N, 08°W to 36.8°N, 7.2°W); for Area 1, the hypocentral corrections are relatively small, with the focus slightly deeper than that in the old solution; for the more regional events, the hypocenters corrections are bigger, with the focus becoming more shallow.

Fault-plane solutions for the recent events were also estimated, showing that best solutions are dominated by strike-slip movement consistent with a stress controlled by a horizontal compression in the NW-SE to NNW-SSE direction, with two exceptions showing reverse mechanisms, with maximum horizontal stress orientation slightly rotated to a N-S direction. These results are also in agreement with those presented in previous studies performed by different authors.

The new epicentral locations show a more organized spatial distribution that could indicate a possible correlation with some known tectonic features. However the fault-plane solutions are considerably more difficult to correlate with the neotectonic features.

Key words: Algarve-Portugal, hypocentres relocation, fault-plane solutions.

1. Introduction

The Algarve region (southern Portugal) has been permanently affected by an important seismic activity due to its particular location close to the boundary between the Eurasian and the African plates. The main activity is located offshore, at the northern edge of a wide seismic belt that extends approximately from the

---

[1] Instituto de Meteorologia, Rua C do Aeroporto, 1700 Lisboa, Portugal.
E-mail: Fernando.Carrilho@meteo.pt
[2] Faculdade de Ciências da Universidade de Lisboa, Campo Grande – Edifício C8, Lisboa, Portugal.
[3] Centro de Geofísica da Universidade de Lisboa, Campo Grande – Edifício C8, Lisboa, Portugal.
[4] Instituto de Ciências da Terra e do Espaço, Rua da Escola Politécnica 58, 1269-102 Lisboa, Portugal.
[5] Instituto Geológico e Mineiro, Estrada da Portela – Zambujal, 2720 Alfragide, Portugal.

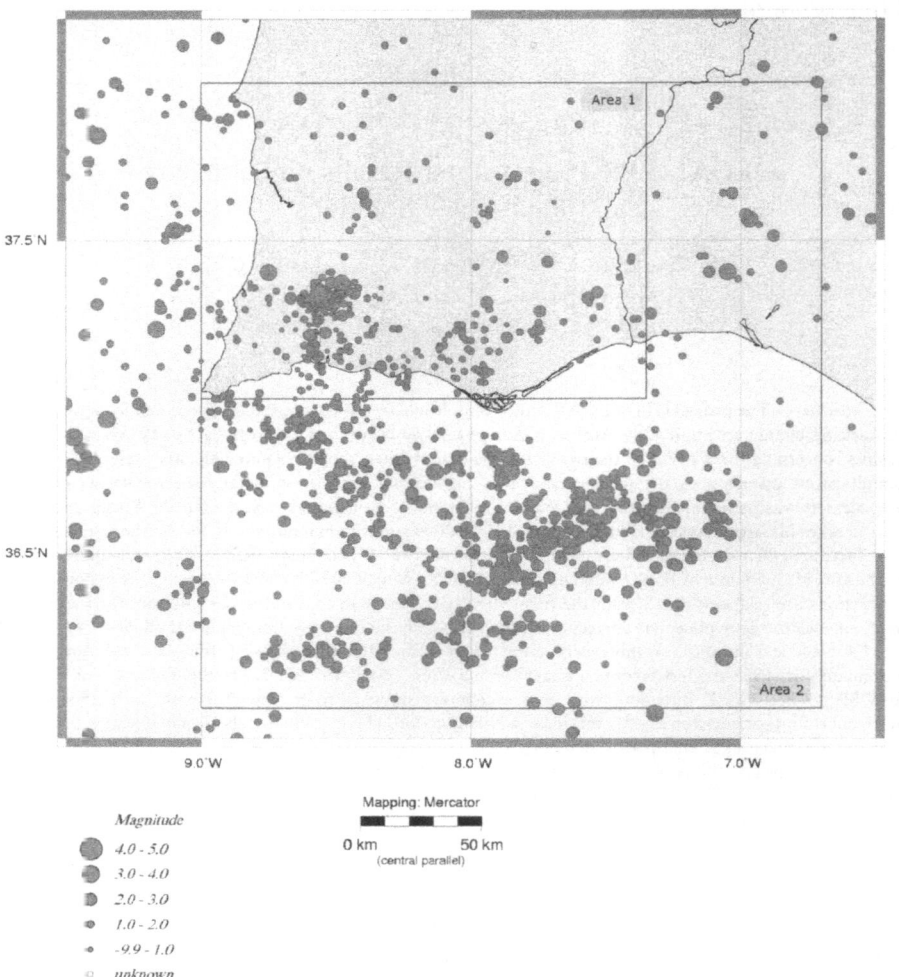

Figure 1
Seismic activity for southern Portugal and adjacent margins, from 1996 to 2000 (data from Instituto de
Meteorologia, Portugal).

Gorringe bank (SW of mainland Portugal) to the Straits of Gibraltar (Fig. 1). The
seismicity is presently monitored with a set of seismic stations belonging to the
national seismic network and the Algarve regional network.

Inland and offshore historical seismicity has been reported since the year 63
BC (Fig. 2). The big 1755 Lisbon earthquake (M = 8.75) (RICHTER, 1958; ABÈ,

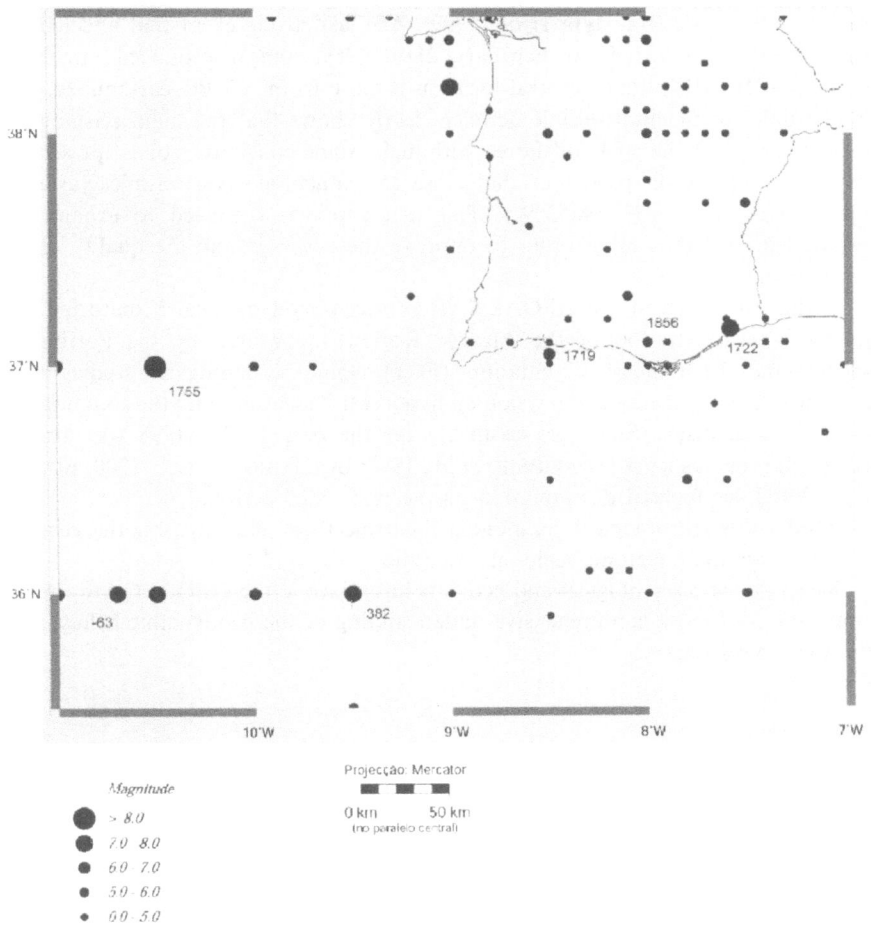

Figure 2
Regional historical and instrumental seismicity from 63BC to 1950 (data from MARTINS and MENDES-VICTOR, 1990).

1979), occurred SW of mainland Portugal and generated a big tsunami. Other reported tsunamigenic earthquakes occurred in 63 BC and in 382. Inland, or near the coast, the epicentral areas of the biggest events were located close to Portimão (1719), Tavira (1722) and Loulé (1856). These earthquakes produced intensities up to X (Modified Mercalli scale — MM). In spite of its epicentral location onshore, in the MARTINS and MENDES-VICTOR (1990) catalogue, it is now believed that the source of the 1722 earthquake is located offshore, due to significant evidence of a tsunami occurrence (BAPTISTA *et al.*, 1999). The last strong event that affected the Algarve region occurred on February 28th, 1969, and it produced intensities reaching VIII (MM); its epicentral location is close to the 63 BC earthquake. The present-day instrumental seismic activity clearly shows that the main seismogenic sources are still located offshore, although some onshore zones present a significant activity, in particular that close to Monchique (see seismic cluster in Fig. 1, close to 37.3°N; 08.53°W). This fact supports the need to expand the regional seismic network in order to improve the coverage and the quality of the acquired data.

Within the scope of the GEOALGAR project (Geodynamical Monitoring and Seismic Characterization of the Algarve Region), sponsored by the Portuguese Science and Technology Foundation (FCT), some seismological studies were developed, namely the careful revision on hypocentral locations and the computation of focal mechanisms. Since the seismicity for the period 1996–1998 was already studied in previous works (CARRILHO *et al.*, 1997; BEZZEGHOUD *et al.*, 2000; BORGES *et al.*, 2001), we focused our attention on the 1999–2000 period.

Besides this seismological component the project also encompasses the geodetic control of the main tectonic regional structures.

Based on the seismological and geodetic information, the GEOALGAR aims to contribute to a more comprehensive understanding of the geodynamic behavior of the Algarve region.

2. Geodynamic Environment

The Algarve region is located in the Eurasian lithospheric plate, near the crossing of the N-S West-Iberia continental margin with the E-W Eurasian-African plate boundary (Fig. 3). According to CABRAL and RIBEIRO (1989) and RIBEIRO *et al.* (1996), this N-S West-Iberia continental margin is probably in a transitional state to a convergent plate boundary. This tectonic environment is responsible for important regional neotectonic and seismic activities, which are evidenced by Pliocene to Pleistocene deformation (Fig. 4) and by a significant historical and instrumental seismicity (Figs. 1 and 2).

The neotectonic activity in the Algarve region is evidenced by vertical crustal movements and by the presence of several macroscale and mesoscale active faults, a large number of joints and a few mesoscale folds and soft sediments deformation structures. These two latter ones affect the younger outcropping sedimentary units, mostly the Faro-Quarteira Pliocene to Quaternary sands, as well as paleoseismites showing, too, the occurrence of high seismic intensities (DIAS and CABRAL, 1995, 1997; DIAS, 2001). Recent neotectonic studies (DIAS and

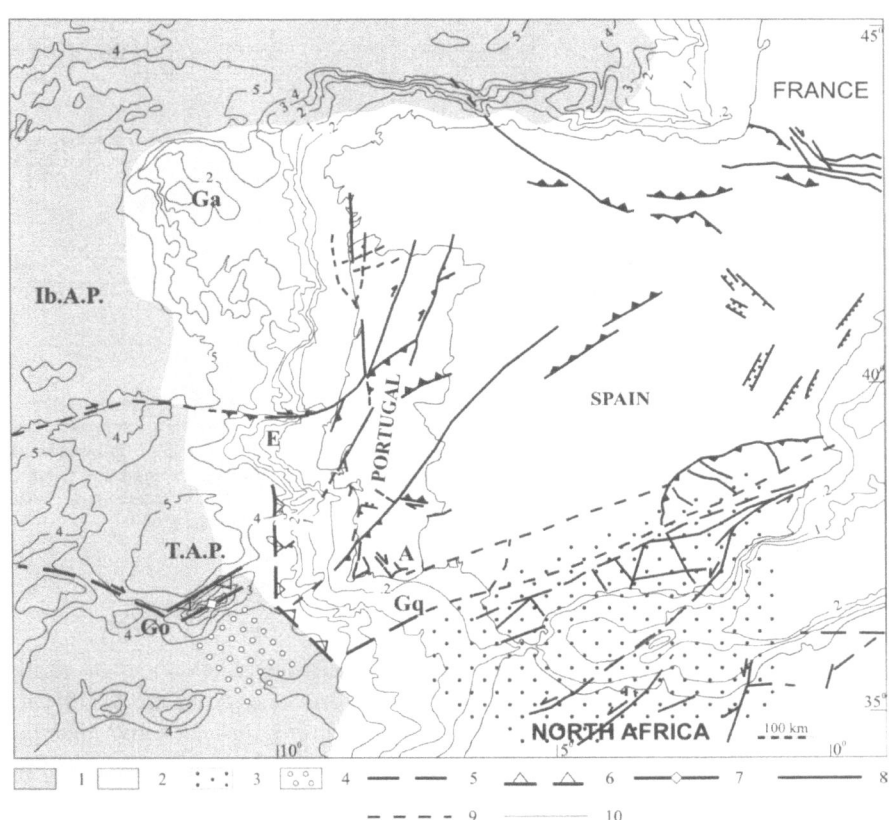

Figure 3

Regional geodynamical framework for the region under study (modified from RIBEIRO *et al.*, 1996). 1—Oceanic crust; 2—thinned continental crust; 3—diffuse plate boundary (continental collision); 4—zone of distributed plate deformation by buckling and thrusting; 5—plate boundary (approximate location); 6—incipient subduction along the Southwestern Iberian continental margin; 7—active antiformal fold; 8—active fault; 9—probable active fault; 10—bathymetric contour; Ga—Galiza bank; Ib.A.P.—Iberia Abyssal Plain; E—Extremadura high; T.A.P.—Tagus Abyssal Plain; Go—Gorringe bank; Gq—Guadalquivir bank. Bathymetric lines are in kilometers.

CABRAL, 2000; DIAS, 2001) led to a map of the main active faults for the Algarve region (Fig. 4).

However, there are some main structures affecting the Faro-Quarteira sands that are poorly understood. It is therefore important to monitor possible displacements between them, using geodetic and seismic techniques.

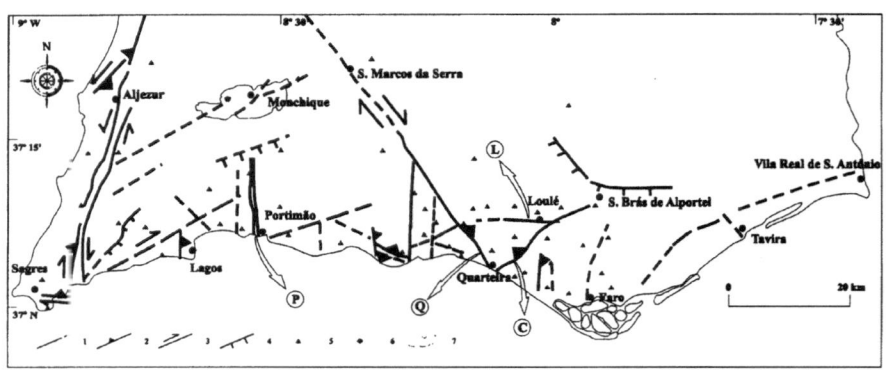

Figure 4

Synthet c map of the main active faults recently recognized in the Algarve region. The predominance of variably trending faults showing reverse movement component suggests the action of a compressive stress field that produces a regional constrictive finite strain in the last 3 My. 1—Active fault; 2—reverse fault (teeth cn hanging wall); 3—strike-slip fault; 4—fault with vertical movement component (teeth on downthrown block); 5—reference GPS; 6—EUREF GPS station (IPCC); 7—Monchique massif; Q—Quarteira fault; L—Loulé fault; C—Carcavai fault; P—Portimão fault.

3. Seismic and Geodetic Networks

In the beginning of 1996, the Instituto de Meteorologia (IM) and the British Geological Survey installed a short-period seismic network covering the Algarve area (Fig. 5). The network comprises 7 stations, transmitting through radio links to a central point that is accessed by a dialup system from the IM headquarters, and was installed in the framework of the *European Seismic Rapid Data Exchange Project* (WALKER *et al.*, 1997). Also three other stations from the national digital seismic network exist in the area.

In order to assist the characterization of the complex geological structure of this region, a GPS control network has been installed, with the goal of determining the geodynamic characterization of the main active faults already shown in Figure 4. Parts of this geodetic network were observed in 1998, 2001, and 2002, but the GPS data had not been treated until now.

4. Seismic Analysis

4.1 Hypocenter Relocations

In order to improve the hypocenter locations, we started to revise the local crust velocity models. We used a technique of simultaneous inversion of model parameters and hypocenter locations (KISSLING *et al.*, 1995).

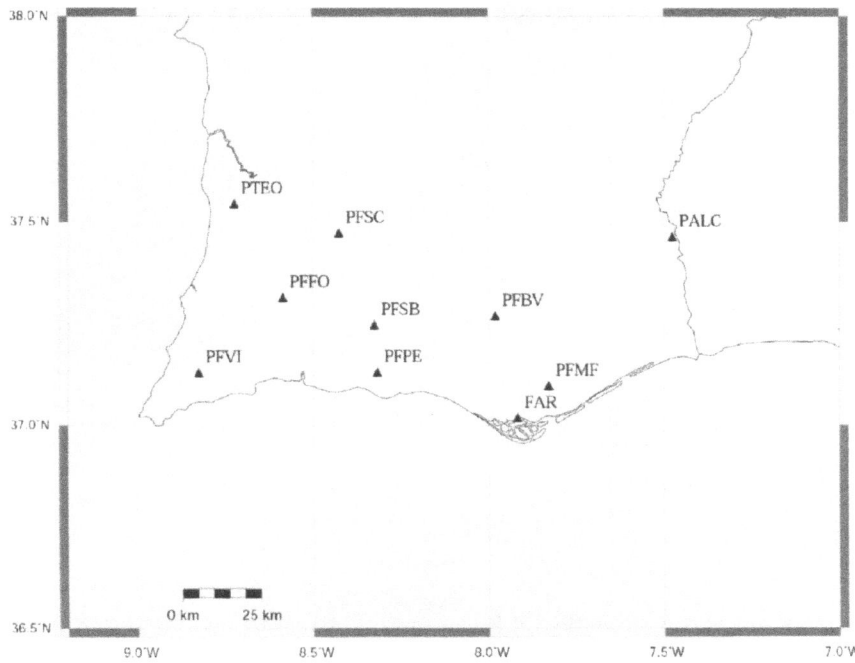

Figure 5
Distribution of the seismic stations in the Algarve area (regional and national network, Instituto de Meteorologia).

The following methodology was used:
- We selected two areas from the study zone in order to test a local inland model and a regional model for the whole zone (see figure 1):
  - Area 1 – inland area, from 37° to 38° N and 7.35° to 9° W;
  - Area 2 – regional area, between 36° to 38° N and 6.7° to 9° W
- Two different local crust velocity models were initially introduced, with the same number of layers for each area:
  - For Area 1, the model I1 was adapted from the model proposed by CARRILHO *et al.* (1997), and the model I2 was a model with the same number of layers but with higher P-wave velocity (Fig. 6);
  - For Area 2, initial models were taken from the structure of the regional model for Portugal, currently in use to locate the earthquakes generated in this area, IMG model (Fig. 7): one with the same velocities and another with higher velocities.
- The layers of these two models were split and several combinations of V_P/V_S ratios were used to build successive initial models for the inversion;

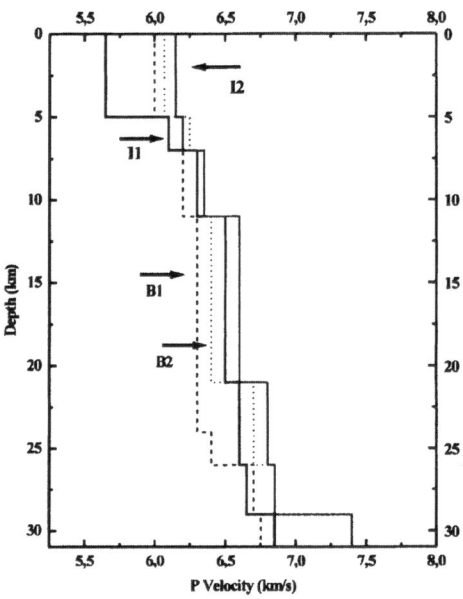

Figure 6
Final local crust velocity models (B1 and B2) for area 1, which resulted from two identical inversion procedures of the same data, with two models of different initial local crust velocity models (I1 and I2).

• The solution with the smallest root mean square (rms) was taken. Fig. 6 shows the two best models for Area 1, B1 and B2, obtained, respectively, from I1 and I2 initial models.

Two preliminary models were obtained for the two different areas (see Fig. 7): a local inland model ("earth model" – B1) and a new regional model.

In the computation of the models, 25 earthquakes were used for Area 1 and 125 earthquakes for Area 2. The selected earthquakes occurred during the period 1999–2000 and have magnitude $M_L \geq 2.0$. For Area 1, the global rms obtained with the IMG model is 0.234, whereas for the new model it is 0.228. For Area 2, the rms improvement is also not significant (0.232 versus 0.231)

As can be seen in Figure 8, the hypocentral locations derived with the new models, for Area 1 and Area 2, present different changes for each area. Regarding epicentral changes, the mean epicentral correction is 0.31 km (with a relatively large standard deviation of 0.28) for Area 1 (inland), whilst for Area 2 the mean epicentral correction is 3.39 km (also with a large standard deviation of 3.10). Concerning depth corrections, it could be seen that for Area 1 the corrections are smaller, with a mean of 0.82 km and a standard deviation of 0.81, than for Area 2, where the mean correction is −1.24 km with a standard deviation of 3.30. This mean depth correction for Area 2 indicates that the events tend to become slightly shallower with the new

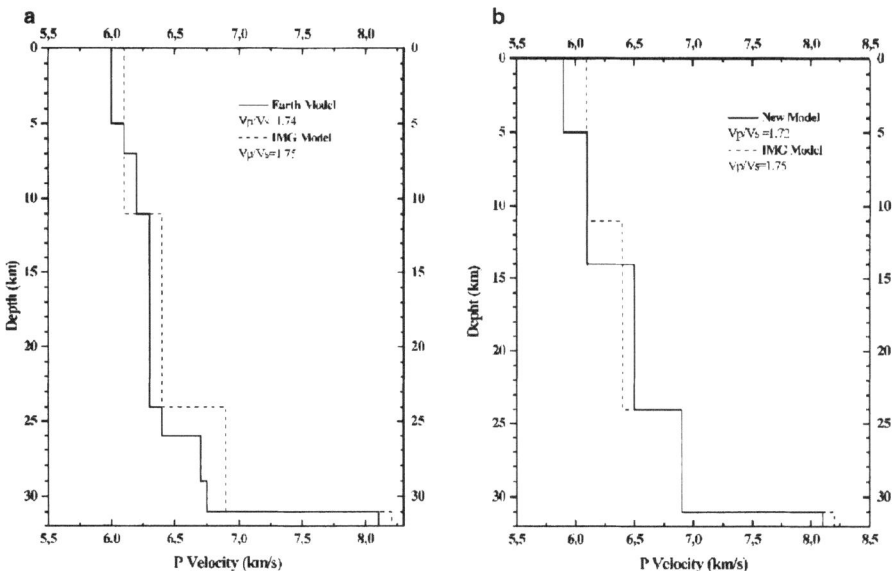

Figure 7

Obtained crust *P*-wave velocity model for Area 1 (a) and for Area 2 (b) and existent IMG model. New models obtained with data from 1999 to 2000.

model. Those are expected results because, if we consider the configurations of the epicentral locations—station locations, we can see that for Area 1 the epicentral distances and the azimuthal gaps are much smaller, resulting in better constrained solutions. The detailed depth changes for Area 1 are illustrated in Figure 9, where a vertical West to East section is presented.

Regarding Area 2, it is possible to see that there are two main regions where there are more events: the already mentioned Monchique region and another one in the area of Guadalquivir Bank (ranging from 36.4°N, 08°W to 36.8°N, 7.2°W). This corresponds to two main epicentral distance classes. Analyzing Figure 8b it can be verified that there are two main peaks: the first one, with corrections less than 1 km, corresponding to the first region (smaller epicentral distances and better constrained locations), and a second one, with a peak with corrections around 4–5 km, corresponding to the events where larger epicentral distances dominate.

Figure 10 presents the new locations for events in Area 1, together with the regional neotectonic map. It is possible to observe epicentral distribution patterns probably related with tectonic features: in the Portimão-Lagos region, an approximate N-S lineament is visible, close to the Portimão Fault (Fig. 4); in the eastern part of the Algarve, two other lineaments with SW-NE direction are suggested; also, a concentration of epicenters around Monchique mountain, aligned WSW-ENE, is present. However, the seismicity seems to have a deeper origin, with

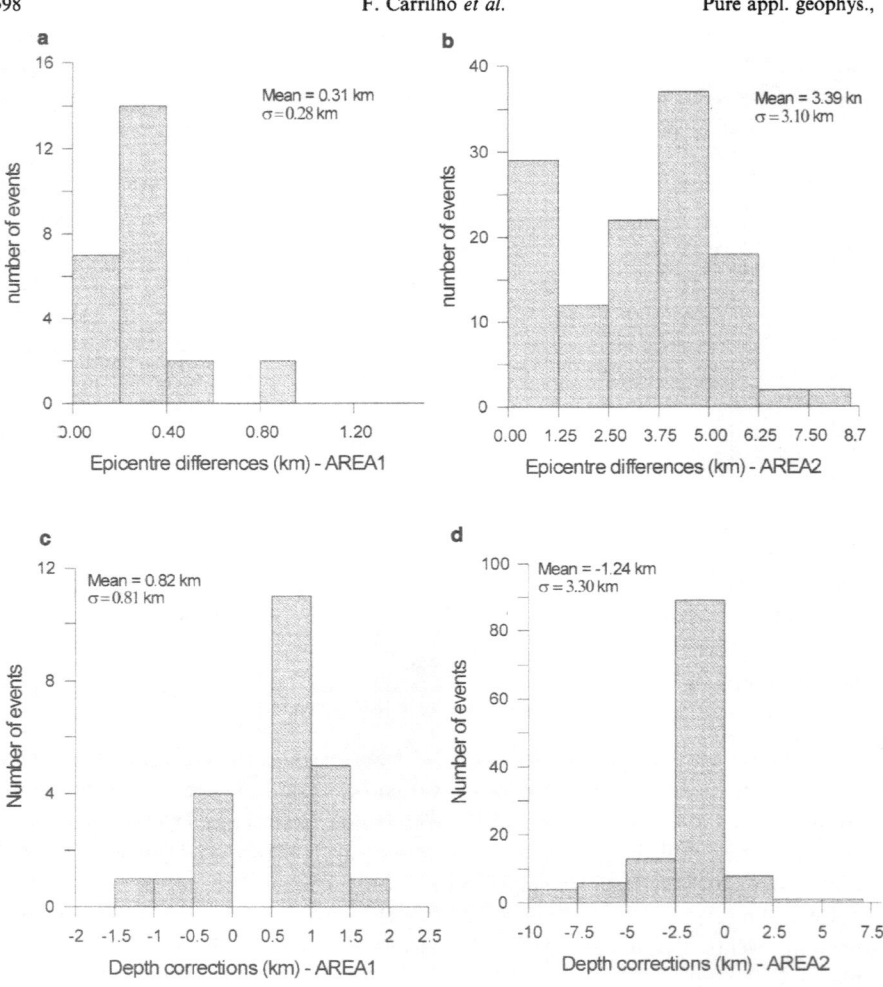

Figure 8

Hypocentral changes between the new models and the regional IMG model (a) Epicentral changes to the new inland model (Area 1); (b) Epicentral changes to the new regional model (Area 2); (c) Depth changes to the new inland model (Area 1); (d) Depth changes to the new regional model (Area 2).

most hypocenters located at depths greater than 5 km (Fig. 9), indicating that seismic events are probably generated in deeper pre-existing tectonic features, in the Variscan basement (DIAS, 2001).

4.2. Fault-Plane Solutions

To compute the focal mechanisms we used the FOCMEC program, included in SEISAN package programs (HAVSKOV and OTTEMOLLER, 2000), where all

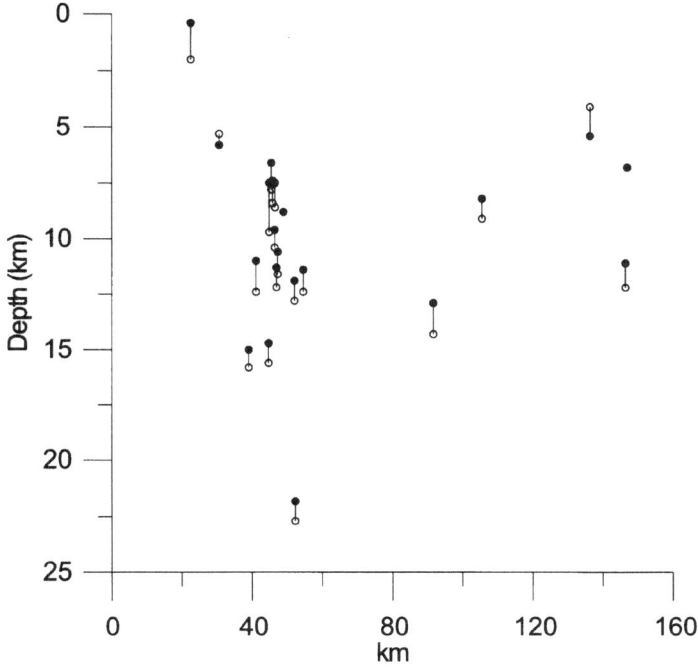

Figure 9

Vertical section, running West to East, comparing hypocentral depths computed with models IMG (closed circle) and AREA1 (open circle); the approximate location of this profile is presented in Figure 10.

solutions compatible with a given polarity distribution are obtained. Subsequently, the best solutions were used as start solutions for the MECSTA program (BRILLINGUER *et al.*, 1980) for the determination of the maximum likelihood focal mechanism (nodal planes orientation and also the orientation of the principal stress axes (P and T)).

For the calculation of these focal mechanisms we used all available polarities, reported in the seismic bulletins from the Portuguese, Spanish and Morocco seismic stations, together with the ones measured on the local network. We selected the events according to the following criteria (at least one of the items must be satisfied): a) earthquakes with $M_L \geq 1.9$; b) events with at least 7 polarities reported; c) phases selected preferentially from digital records; d) good covering of the focal sphere.

For these Portuguese, Spanish and Morocco data, we decided to use the IMG regional crust velocity model for the calculation of the focal mechanisms, because the two new models for Areas 1 and 2 are local ones and were derived using only Portuguese data. Table 1 shows the location, magnitude and fault plane solutions for the studied events. Individual fault plane solutions with polarity distribution are shown in Figure 11.

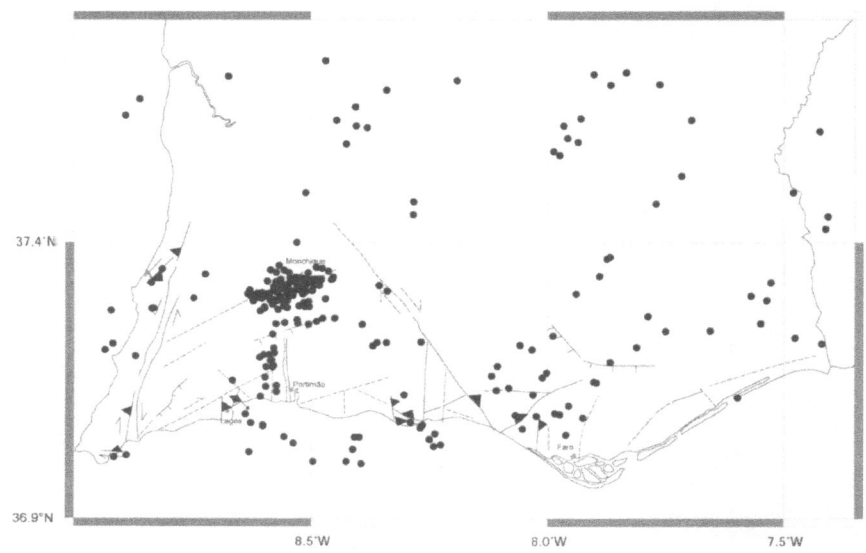

Figure 10
Epicenters computed with the new model for Area 1 plotted over the Algarve neotectonic model

Globally we can say that solutions are not well constrained, mainly due to the lack of polarities. Events 20 and 23 are of the reverse mechanism type with small differences in fault-plane azimuths. Despite some rotation between them, we can see that events 17, 19 and 22 are very similar with dominating strike-slip component. Events 15, 16, 18 and 21 have similar solutions, with a vertical nodal plane and another oblique, differing only in the azimuths of the nodal planes (Fig. 11).

Results for 1999–2000 are displayed in Figure 12, together with other mechanisms computed in previous works (BEZZEGHOUD *et al.*, 2000; BORGES *et al.*, 2001; MOREIRA, 1991; BUFFORN *et al.*, 1988), and in Table 2.

As we can observe in Figure 12, in general, the group of focal mechanisms on the left is in agreement with the focal mechanisms calculated previously. In spite of the poor constraint of most of the new mechanisms, it can be seen that the best solutions are dominated by strike-slip movement consistent with a stress controlled by a horizontal compression in the NW-SE to NNW-SSE direction (Fig. 13). The exceptions to this are the events 20 and 23, which mechanisms are reverse, with maximum horizontal stress orientation slightly rotated to a N-S direction. Both results are also in agreement with the discussion presented by BORGES *et al.* (2001).

Table 1

Hypocentral data and fault plane solutions for the studied events (Φ, Θ-azimuth and dip for the P and T axes; N—Number of polarities; S-score)

Event	Date/Origin Time Yyyy-mm-dd hh:mm	Lat. (°N)	Lon. (°E)	Depth (km)	Mag (M_L)	Nodal plane Strike	Dip	Rake	P Axes Φ	Θ	T Axis Φ	Θ	N	S
15	1999-02-13 11:24	37.261	-8.475	22	2.2	46	88	-52	348	36	105	32	10	0.90
16	1999-04-12 11:39	37.334	-8.475	12	2.0	186	81	-44	135	37	343	22	8	1.00
17	1999-06-11 23:53	37.322	-8.551	8	1.9	347	79	-9	303	14	213	2	7	1.00
18	1999-07-29 04:00	37.191	-7.869	5	2.3	178	61	-1	131	21	39	19	10	0.90
19	1999-10-21 14:45	37.332	-8.562	15	2.0	184	87	-23	137	18	232	14	7	1.00
20	2000-03-27 21:50	36.670	-7.305	19	3.7	250	64	72	353	17	127	66	16	1.00
21	2000-04-18 19:42	37.201	-8.597	11	2.1	210	81	-36	162	31	263	17	8	0.75
22	2000-04-26 12:05	36.782	-8.056	27	2.7	356	80	23	127	9	221	23	9	1.00
23	2000-07-30 17:11	36.646	-7.334	21	2.7	226	70	84	321	25	126	65	17	1.00

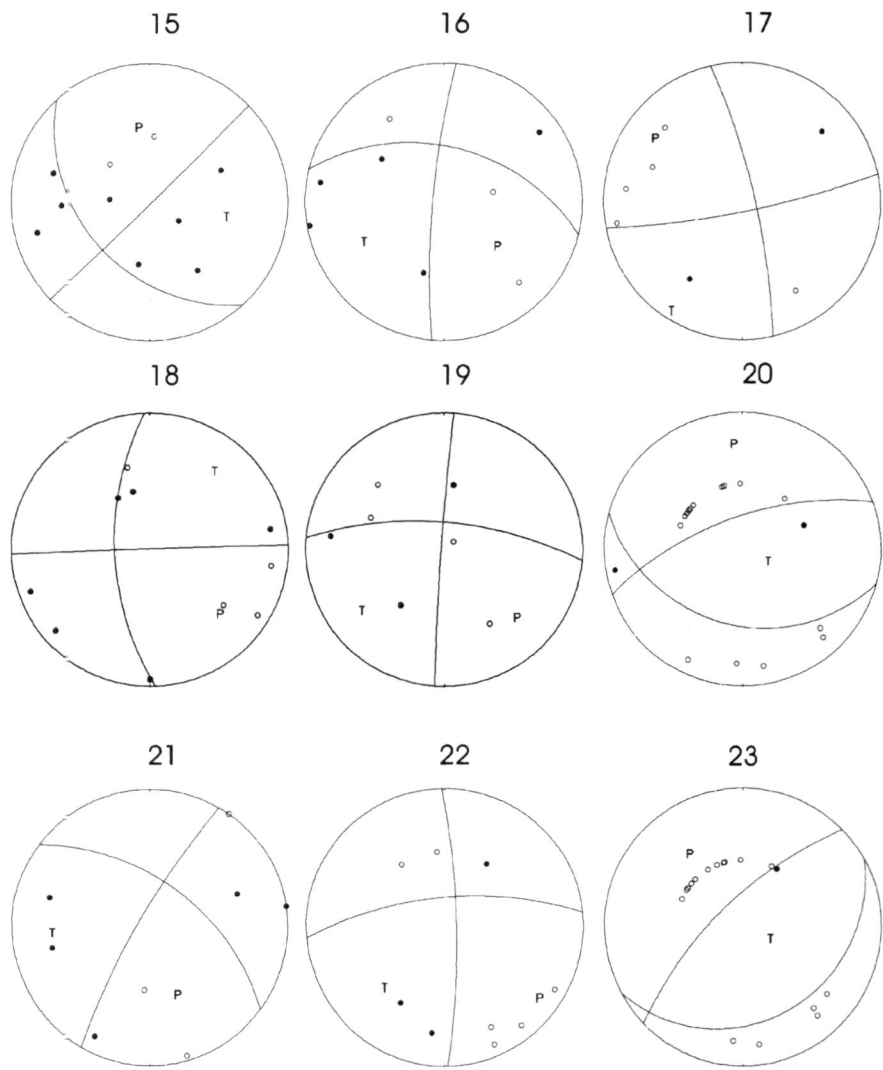

Figure 11
Computed individual fault plane solutions, with polarity distribution.

5. *Final Considerations and Future Work*

The overall goal of the GEOALGAR project is to contribute to the understanding of the complex geological structure and evaluate the seismic hazard for the

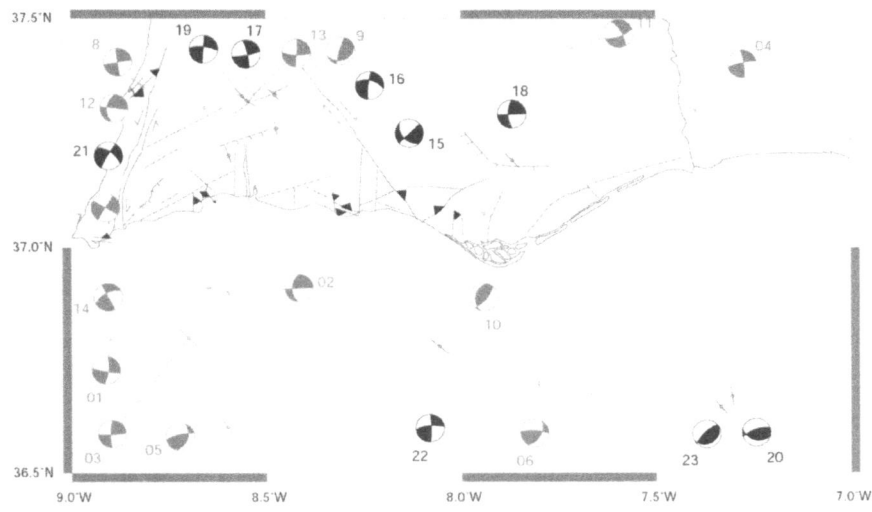

Figure 12

Focal mechanisms: gray—computed in previous works (BEZZEGHOUD *et al.*, 2000; MOREIRA, 1991; BORGES *et al.*, 2001; BUFFORN *et al.*, 1988); black — computed in this study (Algarve neotectonic model in background); ϕ-SH$_{max}$ represents the horizontal projection of the pressure axis.

Algarve region. In order to accomplish this goal, the integration of geodetic, geological and seismological studies is essential.

We started with the analysis of the seismic data, which enabled a better definition of the regional velocity model, also improving the quality on the hypocentral locations. The new epicentral locations show a distribution pattern that could indicate a possible correlation with some tectonic features. However, the fault plane solutions appear to be difficult to correlate with the neotectonic features, nonetheless the horizontal direction of the main stress axis exhibited a predominant NW-SE to NNW-SSE trend that is in agreement with the previous results published for this region.

Next steps will be (i) the joint analysis of the whole set of digital seismic data (from 1998 to 2001), in order to improve the regional crust velocity model, and (ii) to include the *S*-wave information in the estimation of the fault plane solutions.

Regarding the geodetic component, the project is still under development and a set of work packages is planned for the near future. In particular, planning is underway to treat now the GPS data collected between 1998 and 2002 and in the near future to carry out a new GPS campaign based on all the stations already observed. The entire set of geodetic data will be further analyzed, in order to detect possible evidence of displacements of tectonic origin.

Table 2

Hypocentral data and fault plane solutions for the other events previously reported. MOR: MOREIRA (1991); BUF: BUFFORN et al. (1988); BOR: BORGES et al. (2001); IGN: Instituto Geografico Nacional (Spain); BEZ: BEZZEGHOUD et al.) 2000). (Symbols as in Table 1)

Event	Date/Origin Time Yyyy-mm-dd hh:mm	Lat. (°N)	Lon. (°E)	Depth (km)	Mag (M$_L$)	Nodal plane			P Axes		T Axis		Ref
						Strike	Dip	Rake	Φ	Θ	Φ	Θ	
1	1986-09-25 06:31	36.8	-8.9	–	4.3	7	70	-10	325	21	232	7	MOR
2	1986-10-20 14:48	36.9	-8.6	37	4.8	180	37	3	147	33	29	36	BUF
3	1989-11-02 06:00	36.8	-8.7	40	4.5	180	75	8	135	5	43	16	BOR
4	1989-12-20 04:15	37.3	-7.4	23	5.0	351	77	10	305	2	215	16	IGN
5	1993-02-16 03:11	36.6	-8.6	26	4.3	17	33	34	326	22	202	54	BOR
6	1994-09-24 10:50	36.7	-7.8	52	4.3	274	70	126	338	17	226	51	BOR
7	1996-04-23 10:56	37.107	-8.544	12	2.6	301	90	-179	166	1	76	1	BEZ
8	1996-07-08 16:03	37.354	-8.818	5	3.4	81	90	179	126	1	36	1	BEZ
9	1996-07-17 12:31	37.339	-8.479	9	2.2	196	78	74	299	31	86	54	BEZ
10	1996-08-30 04:44	37.074	-7.942	22	2.4	39	65	102	120	19	332	68	BEZ
11	1997-06-11 11:16	37.437	-7.717	27	2.7	71	65	-153	290	36	20	0	BEZ
12	1997-07-05 20:50	37.305	-8.656	18	2.6	276	90	-125	156	35	36	35	BEZ
13	1998-03-04 13:08	37.334	-8.531	12	3.2	93	72	179	317	12	50	13	BEZ
14	1998-04-24 09:09	36.911	-8.648	23	2.3	332	90	36	101	25	203	25	BEZ

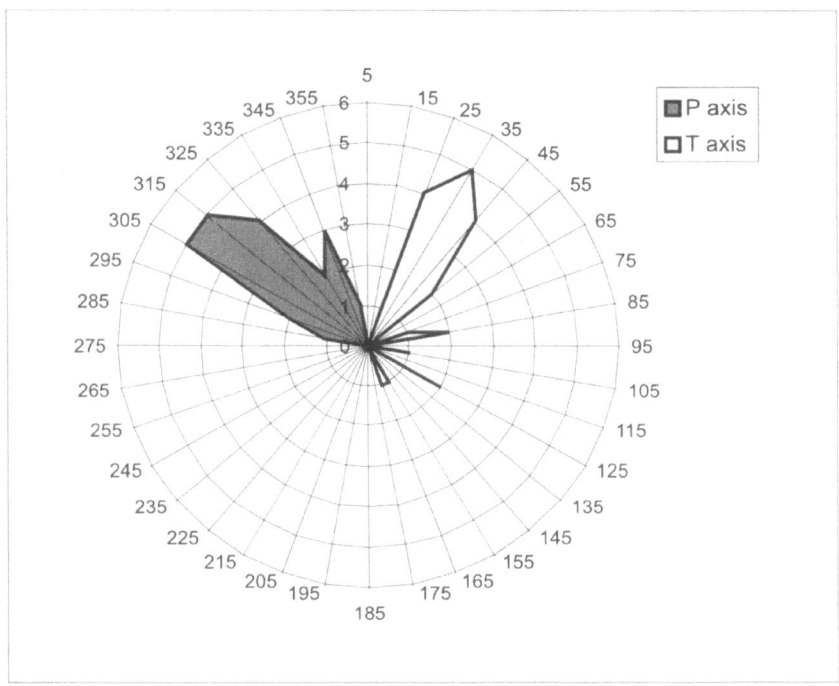

Figure 13
Orientation of the horizontal projection of *P* and *T* axis for the focal mechanisms shown in tables 1 and 2.
Labelled azimuths correspond to the central value for each class.

Acknowledgements

This paper was performed under the project **PRAXIS/CTE/14257/98** financed by the Portuguese Science and Technology Foundation (FCT), which supported a grant for one of the authors, and was also partially supported by the project "Caracterização do Potencial Sismogenético de Falhas na Região do Algarve Ocidental Meridional," sponsored by Instituto Geológico e Mineiro and the FCT.

REFERENCES

ABE, K. (1979), *Size of Great Earthquakes of* 1837–1974 *Inferred from Tsunami Data*, J. Geophys. Res. *84* (B4), 1561–1568.

BAPTISTA, M.A., LEMOS, C., and MIRANDA, J. M. (1999), T*he Tsunami of* 1722.12.27, *Algarve, Portugal*, IUGG99 Abstracts, 133.

BEZZEGHOUD, M., FITAS, A., BORGES, J. F., CARRILHO, F., and SENOS, M. L. (2000), *Seismicity, Focal Mechanisms and Source Parameters in South Portugal*, 2nd Spanish-Portuguese Assembly on Geodesy and Geophysics, Lagos, Portugal, 129–130.

BORGES, J. F., FITAS, A. J. S., BEZZEGHOUD, M., and TEVES-COSTA, P. (2001), *Seismotectonics of Portugal and its Adjacent Atlantic area*, Tectonophysics *337*, 373–387.

BRILLINGER, D. R., UDIAS, A., and BOLT, B. A. (1980), *A probability Model for Regional Focal Mechanism Solutions*, Bull. Seismol. Soc. Am. *70*, 149–170.

BUFFORN, E., MÉZCUA, J., and UDIAS, A. (1988), *Mecanismo focal del terramoto del Cabo San Vicente de 20 de Outubre de 1986*, Rev. Geofis., 109–112.

CABRAL, J. and RIBEIRO, A. (1989), *Incipient Subduction along the West-Iberia Continental Margin*, 28th International Geological Congress, Washington, D.C., USA, 1/3, 223.

CARRILHO, F., SENOS, M. L., FITAS, A., and BORGES, F. (1997), *Estudo da Sismicidade do Algarve e da Região Atlântica Adjacente*, 3rd Meeting on Seismology and Engineering Seismology, IST, Lisbon, Portugal.

DIAS, R. P. (2001), *Neotectónica da Região do Algarve*, Ph.D. Thesis, Lisbon University, 369 pp.

DIAS, R. P. and CABRAL, J. (1995), *Actividade Neotectónica na região do Algarve*, IV National Congress on Geology, Porto University, Portugal, 241–245.

DIAS, R. P. and CABRAL, J. (1997), *Plio-Quaternary Crustal vertical movements in southern Portugal—Algarve*, Cuaternario Ibérico, AEQUA, Huelva (Rodríguez Vital, J., Ed.), 61–68.

DIAS, R. P. and CABRAL, J. (2000), *Deformações Neotectónicas na região do Algarve*, 3rd Symposium on the Atlantic Iberian Margin, Faro, Portugal (Dias, J.A., and Ferreira, O., Coord.), 189–190.

DIAS, R. P., BORGES, J. F., CABRAL, J., FITAS, A., COSTA, P. T., RIBEIRO, A., MATIAS, L., and TERRINHA, P. (1997), *Constrictive Deformation in Algarve Region (Portugal, SW Iberia): Comparison of Neotectonic Data and Earthquake Focal Mechanisms*, IX General Assembly of the European Geophysical Union (EUG), Strasbourg, France, 23–27.

HAVSKOV, J. and OTTEMOLLER, L. (2000), *SEISAN: The Earthquakes Analysis Software Version 7.00*, Institute of Solid Earth Physics, University of Bergen.

KISSLING, E., ELLSWORTH, W., EBERHART-PHILLIPS, D., and KRADOLFER, U. (1994), *Initial Reference Models in Local Earthquake Tomography*, J. Geophys. Research *99*(B10), 19635–19646.

MARTINS, I. and MENDES VICTOR, L. A. (1990), *Contribuição para o estudo da sismicidade de Portugal Continental*. Publ. **18**, Instituto Geofisico do Infante D. Luis, Lisbon University, 67 pp.

MOREIRA, V. (1991), *Historical Seismicity and Seismotectonics of the Area Situated between the Iberian Peninsula, Morocco, Selvagens and Azores Islands*, Seismicity, Seismotectonic and Seismic Risk of the Ibero-Magrebian Region. Publ. IGN, Madrid, Monografia N. 8, pp. 213–225.

REILINGER, R. E., MCCLUSKY, S. C., ORAL, M. B., KING, R. W., TOKSOZ, M. N., BARKA, A. A., KINIK, I., LENK, O., and SANLI, I. (1997), *Global Positioning System Measurements of Present-day Crustal Movements in the Arabia-Africa-Eurasia Plate Collision Zone*, J. Geophys. Res. *102*(B5), 9983–9999.

RIBEIRO, A., CABRAL, J., BAPTISTA, R., and MATIAS, L. (1996), *Stress Pattern in Portugal Mainland and the Adjacent Atlantic Region, West Iberia*, Tectonics *15*(2), 641–659.

RICHTER, C., *Elementary Seismology* (W.H. Freeman and Company, San Francisco 1958).

WALKER, A., and TRANSFRONTIER GROUP (1997), *Technical Report WL/97/21 to CE*, Global Seismology Series, Edinburgh, British Geologycal Survey, 280 pp.

(Received February 11, 2002, revised December 18, 2002, accepted February 5, 2003)

 To access this journal online:
http://www.birkhauser.ch

Pure appl. geophys. 161 (2004) 607–621
0033–4553/04/030607–15
DOI 10.1007/s00024-003-2465-2

❘ Pure and Applied Geophysics

The Ain Temouchent (Algeria) Earthquake of December 22nd, 1999

A. K. Yelles-Chaouche[1], H. Djellit[1], H. Beldjoudi[1],
M. Bezzeghoud[2], and E. Buforn[3]

Abstract — On December 22nd, 1999 an earthquake of Magnitude M_w : 5.7 occurred at Ain Temouchent (northwest Algeria). This moderate seismic event was located in a region characterized by a low seismic activity where few historical events have been observed. The earthquake, with a maximum intensity of VII (MSK scale), caused serious damages to the Ain Temouchent city and its surroundings. In the epicentral area, 25 people died and about 25,000 people were made homeless. Some minor breaks have been observed in several areas in the field. They were mainly related to minor collapses in the landscape or in volcanic cavities. The focal mechanism has been studied by using broadband data at regional and teleseismic distances, and different methods. The fault-plane solution has been estimated from first motions of *P* wave. Depth and source time function have been estimated from the modeling of body waveforms. Scalar seismic moment and source dimension have been obtained from spectral analysis. Results show thrust motion, with a horizontal pressure axis oriented in a NW-SE direction, a depth of 4 km and a simple source time function with time duration of 5 s. Scalar seismic moment estimated from waveform modeling is 4.7×10^{17} Nm, and spectral analysis gives a value of 1.7×10^{17} Nm and a source radius of 7.5 km.

Key words: Ain Temouchent (Algeria), seismic source, damages, focal mechanism, breaks.

Introduction

On December 22nd, 1999 at 17h 37m 30s, Northern Algeria was again struck by a shallow earthquake of Magnitude $M_w = 5.7$, namely the Ain Temouchent earthquake (Yelles *et al.*, 2000). The earthquake occurred in the western part of Algeria, in Ain Temouchent, a town located 70 km southwest of Oran (capital of Oranie, the western region of Algeria) (Fig.1). The earthquake shook the region, causing considerable damages to Ain Temouchent and its surrounding regions and generating great panic among the population of the Oranie region.

The main shock was located by the Algerian seismic network at 35.25°N, 01.30°W, i.e., at Ain Allem (20 km SW of Ain Temouchent). This location was slightly different

[1] Centre de Recherche en Astronomie Astrophysique et Géophysique B.P. 63 Bouzareah, Algiers, Algeria. E-mail: Kyelles@yahoo.fr
[2] Departamento de Física/CGE, Apartado 94, Universidade de Évora, 7002-554 Évora, Portugal.
[3] Departamento de Geofísica y Meteorología, Universidad Complutense, 28040 Madrid, Spain.

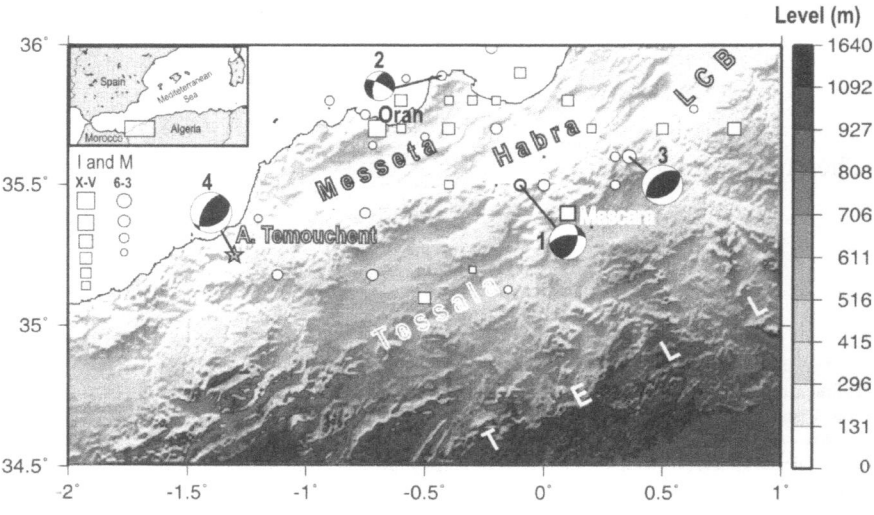

Figure 1

The Ain Temouchent earthquake epicenter location with the focal mechanism (4) determined from waveform analysis in this study. Seismicity for M ≥ 3.0 and I ≥ V for the 1790–1999 period is reported. Other focal mechanisms are also reported (see BEZZEGHOUD and BUFORN, 1999): (1) Mascara earthquakes of August 18, 1994; (2) Oran earthquake of April 19, 1981; and (3) of July 13, 1967. The topography is from digital elevation model (GLOBE TASK TEAM *et al.*, 1999). LCB = Lower Chelif Basin.

than the CSEM and USGS epicenter locations (Table. 1). Investigations carried out rapidly in the region, just after the earthquake resulted in serious damages to the buildings and the socio-economic infrastructures (schools, hospitals, mosques, roads, bridges, water networks). The epicentral area with maximal Intensity of VII (MSK scale) extended to an area of 30 km radius, west of Ain Temouchent (Fig. 2), 25 people died there and about 3000 families were made homeless. In Ain Temouchent, the most important hospital, several schools and public buildings (post offices…) were severely affected; deep cracks on the walls, torsion of the pillars and collapsing of the roof were observed. Some old houses were totally destroyed. It seems that buildings located along the small river which crosses the town suffered more than others because

Table 1

Earthquake location given by different institutions

Origin time UTC (hh.mn:ss)	Lat. N	Lon. W	H (km)	M	Ref.
17:36:53.00	35.25°	01.30°	10	5.7	CRAAG
17:36:56.24	35.32°	01.28°	10	5.7	USGS
17:36:57.00	35.23°	01.39°	10.6	5.7	CSEM

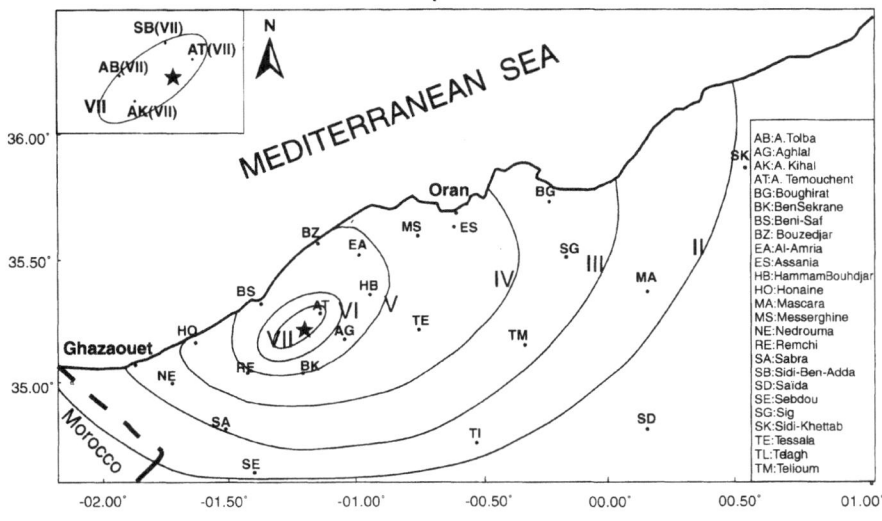

Figure 2

Isoseismal map for the Ain Temouchent earthquake. Black circles show the villages and towns. Isoseismal VII is shown with more detail in the left upper corner.

of the nature of the soil. Besides Ain Temouchent, the villages of Ain Allem, Ain Tolba and Ain Kihal (Fig. 2) were greatly affected. In these villages, the population suffered considerably because the earthquake happened during the winter season. Because of the serious damages caused by the earthquake, the Algerian authorities decided to declare the Wilaya (department or province) of Ain Temouchent a 'disaster area'. An urgency plan was launched to assist the population and to rebuild all socio-economic infrastructures destroyed by the earthquake.

Two days after the main shock, a network of five Spregnether MEQ 800 was deployed in the epicentral area around the Ain Allem-Ain Temouchent axis in order to record the aftershocks sequence. Due to the low number of stations available, it was impossible to carry out a dense coverage of the region. In addition, it was not possible to maintain the seismic monitoring for more than one month. From December 22, 1999 to January 25, 2000, corresponding to the period when the portable network was in activity, 293 events of magnitude ranging between 1.0 and 4.0 were recorded. The aftershock seismic activity lasted about 14 months. During this period three strong aftershocks occurred: on May 27, 2000 at 12h 26mn with a magnitude of 3.5; on July, 30th, 2000 at 02h 25mn with a magnitude of 3.7 causing the death of three people; and on the January 4th 2001 with magnitude of 3.1 followed a few minutes later by a second event of magnitude 2.9.

The area under study (2°W to 0°) is characterized by the occurrence of earthquakes with small magnitude, and consequently no studies of focal mechanism exist for this region (BEZZEGHOUD and BUFORN, 1999). The studied event is the most important earthquake ($M > 5.5$) to strike Ain Temouchent and its vicinity in modern times.

The purpose of this article is to report the seismotectonic effects, and to present a detailed seismological study of the Ain Temouchent earthquake.

Historical Seismicity

According to the historical seismicity of the Oranie region, several moderate to large earthquakes (maximum intensities of IV to X) have occurred. The most important one was the Oran earthquake of October 9th, 1790 with maximum intensity of X that caused the death of about 2000 people. The epicenter located offshore might have been triggered by the offshore extent of the Murdjadjo anticline. Subsequently, another earthquake occurred on March 1819 in the Mascara region. This earthquake of intensity IX destroyed a huge number of farms, wine cellars, etc. Then, on November 29th, 1887 the region was impacted by an earthquake of maximum intensity X, which caused the loss of some 20 people, destroying 331 housing units (BENOUAR *et al.*, 1994).

The Ain Temouchent region, in comparison to the Oran and Mascara regions, is characterized by a low to moderate seismic activity (Fig. 2). Indeed, through history, no important earthquakes have been mentioned by the seismic catalog (ROUSSEL, 1973; BENHALLOU, 1985; MOKRANE *et al.*, 1994; BENOUAR, 1994). Only a few seismic events have been reported in the region during the twentieth century (Table 2). During the last twenty years, we can outline as more important three events that occurred in the vicinity of the epicenter of the Ain Temouchent event. These are the events of 16.01.1980 ($Io = V$, $M = 3.8$), 15.07.1985 ($Io = V$, $M = 4.1$), 17.12.1992 ($Io = V$, $M = 4.8$), all with magnitudes lower than 5.0. We believe that it is difficult to envisage this region with a weak seismicity in the past because it is located in the Eurasiatic-African plate boundary (Fig. 1).

A possible explanation for the low seismic activity of the Ain Temouchent region may be the lack of historical documents, human witnesses, and absence of seismological stations. We recall, also, that a consequence of the earthquake of 1790, which destroyed the city of Oran, was the departure of the Spanish and, probably, a decrease in the population. In the Oranie region, before installation of the Algerian seismological network in 1990, the seismic monitoring was carried out by only one station (BEZZEGHOUD *et al.*, 1996). It was the Tlemcen station (TEC), which operated between 1978 and 1992. Subsequently, there is a major gap in the knowledge of the seismic activity before 1978 and consequently it is not possible to precisely perceive of the real seismic activity of the region.

Table 2

Seismic events reported during the XX century for the studied area

Date	Time	Lat. N	Lon. W	M	I_O	Observations	Ref.
13.05.1964	13 46 21	35.50°	01.50°	5.2	VII	Beni Saf region	MOKRANE *et al.* (1994)
16.08.1967	13 46 09	35.50°	01.30°				MOKRANE *et al.* (1994)
16.01.1980	21 40 00	35.35°	01.03°	3.8	V	Ain Temouchent	MOKRANE *et al.* (1994)
15.07.1985	11 20 39	35.48°	01.22°	3.9	V	Bouzedjar	BENOUAR, 1994
18.07.1985	11 44 00	35.38°	01.20°	3.5	IV	Terga	MOKRANE *et al.* (1994)
17.10.1992	20 43 21	35.18°	01.20°	4.8	V	Ain Temouchent	CRAAG
22.12.1999	17 36 53	35.25°	01.30°	5.7	VII	Ain Temouchent	This study

Io = isoseismal maximum intensity, M = magnitude

The isoseismal map for the Ain Temouchent event (Fig. 2) shows a distribution of intensity in a NE-SW direction, very clear for the VII and VI isoseismal, which covers an elliptical area of approximately 42 × 19 km, with a faster attenuation in a NW-SE direction.

Geological Setting

The Ain Temouchent region is located in the western extremity of the lower Cheliff basin (GUARDIA, 1975; THOMAS, 1985). This was later developed after the major structuring of the Tellian domain, one of the segments of the Alpine chain of Northern Africa (MEGHRAOUI, 1988). The Ain Temouchent region is bordered on the east by the Oran Sebkha (a salt lake) and on the west by the coastal massif of Beni Saf. In the north, the region is limited by the volcanic sedimentary units of Bouzedjar and in the south by the periclinal ending of the Sebah Chioukh Mountains and western Tessala massifs (Fig. 3).

From the geological point of view, two main geological units characterize the region of Ain Temouchent: 1) the metamorphic basement represented by quarzitic units and 2) the volcanic sedimentary cover formed by Plioquaternary rocks. We must notice that in the Oranie region, two volcanic episodes synchronous to the Neogene tectonic phases have been distinguished. The first one dated in the Messinian period corresponding to a calco-alkalin volcanism, and the second one, more recent of basaltic alkaline type and dated in the Quaternary period. In general these volcanic deposits follow the main accidents trending N50°. GUARDIA (1975) and THOMAS (1985) indicate that the Neogene and Quaternary deposits were slightly affected by the NS to NW-SE Quaternary compressionnal phases that affect the Tellian Atlas Mountains. The quasi-tabular aspect of the geological formations confirms this fact. This lack of deformation could explain the rather slow seismic activity level of the region of Ain Temouchent.

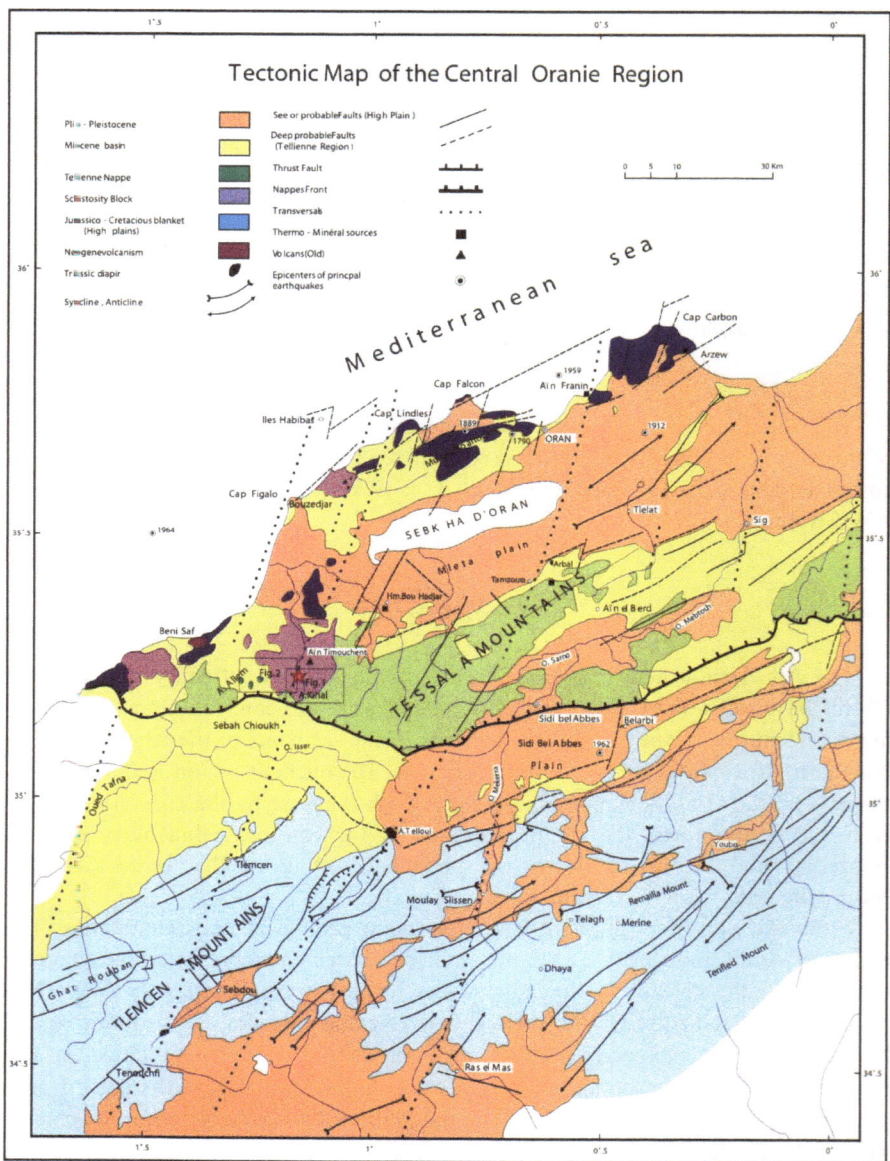

Figure 3
Geological map of the Ain Temouchent region. The legend gives information on the age of the formations and the structures.

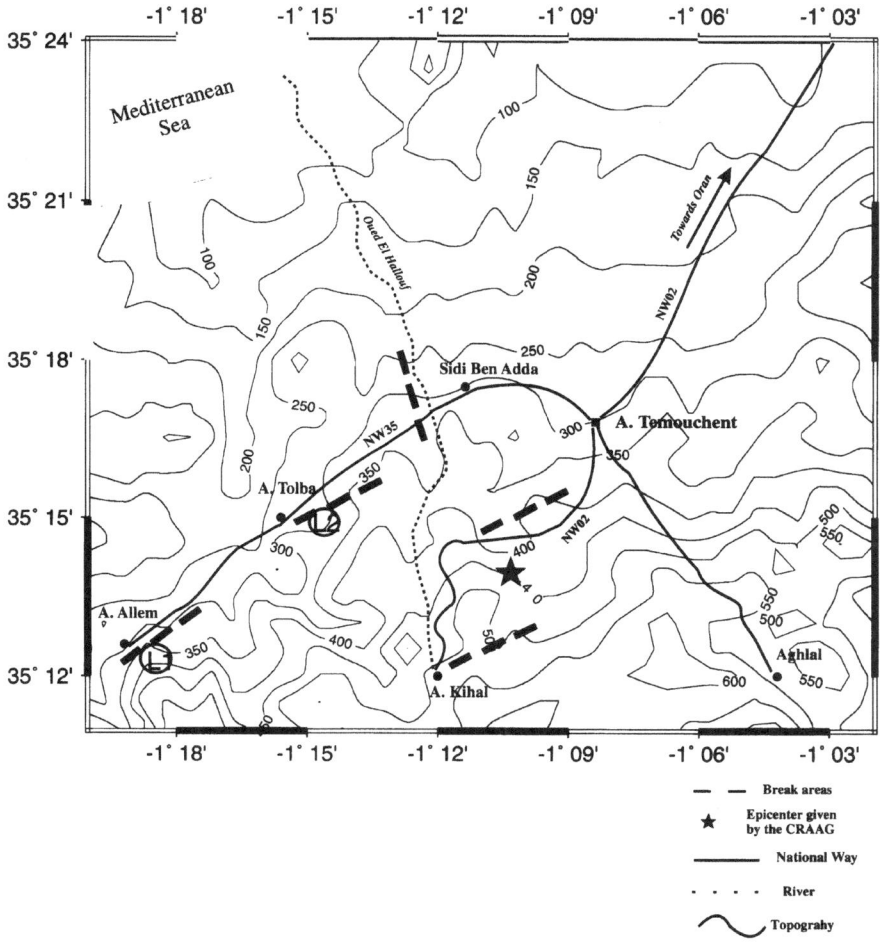

Figure 4
Topographic map of the Ain Temouchent epicentral zone showing break areas. Contour interval is 50 m.
Photographs of Ain Tolba surface breaks (L2) and Ain Allem volcanic cavity (L1) are showed in Figure 5.

Surface Ruptures

Field investigations carried out in the epicentral area just after the occurrence of
the main shock reveal different types of ground effects such as landslides, rock falls,
cavities collapsing and spring water variations (YELLES *et al.*, 2000). However, no
clear set of aligned cracks and fissures has been observed relative to the faulting.
The ground effects were observed in the following zones (Figs. 4 and 5):

Figure 5
a) Photographs of surface breaks in the region of Ain Tolba (L2, Fig. 4) and b) volcanic cavity collapsing in the region of Ain Allem (L1, Fig. 4).

Ain Allem: In this zone the senonian formations located at the entrance of the village displayed a set of NE-SW tensile cracks of many meters in length (Fig. 5a). In the sandstone series a vertical offset of 20 cm was observed, suggesting a minor landslide. In the village, a left-lateral displacement of a water pipe could be observed.

Oued ("river" in Arabic) El Kihal: Four kilometers from the village of Ain Tolba towards the southeast, straight cracks with a total length of 100 m were observed (Fig. 4). They were related to a minor landslide affecting the western flank of the Oued El Kihal. The vertical movement reaches 20 cm. On the bottom of this flank two water sources were created. On the road towards Ain Kihal the earthquake affected the volcanic series where several blocks fell down (Fig. 5b).

Sidi Ben Adda: Five kilometers west of this village, near the Oued El Hallouf, a set of cracks with a N160° direction affects the eastern flank of the river.

Site d'El Baida: South of the Ain Temouchent, along the road leading to Ain Kihal, an NNE-SSW oriented surface breaks crosscut over 200 meters of the senonian formation.

Although no continuous set of cracks affects the neogene series, it is important to take note of the fact that all sparse ground failures observed seem to be distributed along two major directions, NE–SW (N40°–N65°) and NW–SE (N140°/N165°).

Focal Mechanism And Source Parameters

Methodology and Data

The data used for the focal mechanism of the main shock correspond to seismograms recorded at teleseismic distances (30° < 90°) in order to avoid problems with upper mantle wave triplications and diffractions by mantle-core boundary. The fault plane solution has been determined from the first motion (FM) of P-waves by using the algorithm of BRILLINGER et al. (1980). This algorithm determines the maximum likelihood function and it estimates the orientation of the principal stress axes (P and T), nodal planes and their standard errors (UDÍAS and BUFORN, 1988). Take-off angles have been calculated for teleseismic distances from Jeffreys-Bullen tables by using a velocity of 5.8 km/s for shallow focus (depth less than 15 km). Slight changes on the velocity model are not significant in modeling results at teleseismic distances. A total of 27 observations have been used, most of them corresponding to broadband seismograms recorded at teleseismic distances, with the exception of three stations (MELI, SELV and SFUC) located at less than 1000 km. From Figure 6 we observe a good azimuthal coverage.

For the waveform analysis (WA) of P waves, we use the McCaffrey et al. (1991) version of Nabelek's (1984) inversion procedure, which minimizes, in a weighted least square sense, the misfit between observed and synthetic seismograms. Unfortunately, due to the small magnitude of the earthquake, the SH waves are contaminated by noise and it was impossible to use them. Depth, nodal planes, scalar seismic moment and source time function (STF) are inverted simultaneously from broadband teleseismic data. The initial model is taken from the result of the FM method described above. Synthetic seismograms were computed in a homogeneous half space and included P, pP and sP phases; the Q factor was defined with attenuation time constant of $t^* = 1$ second for P waves. The synthetic seismograms are obtained by convolution of Green's function of the propagation with the instrumental response and then by the STF. Unfortunately, for this area no details about the crustal structure are known. For this reason, to generate Green's functions, the crustal model used corresponds to a layer with a thickness of 30 km and P velocity of 6 km/s, over a mantle with P velocity of 8 Km/s. Every record was converted to a seismogram recorded at 40 degrees of epicentral distance (station with same azimuth and common gain).

Finally, the scalar seismic moment and source dimensions were also estimated from spectral analysis (SA) using a total of seven broad band records corresponding to distances between 30° and 80°. The P wave records, with a sample rate of 0.05s, were windowed, detrended, deconvolved from instrumental response and tapered with a cosinus before performing a fast Fourier transformation. Then, the correction of attenuation, the same that we have used on the inversion method and the radiation pattern were applied. The radiation pattern is computed from the fault plane solution determined from FM solution previously described. Scalar seismic moment value obtained from spectral analysis is compared with the value from modeling. We estimated the corner frequency by direct measurement of the intersection of the low and high frequency trends of the amplitude spectra. The STF duration t was computed from the corner frequency using the expression $t = 1/fc$, where fc is the corner frequency. There is a variation of the values (of the order of a factor of 2) of Mo and t from station to station, which may be due to site effects.

Results

The fault-plane solution (FM) for the Ain Temouchent shock was obtained using 27 polarities of P waves at teleseismic distances, and corresponds to a reverse faulting mechanism with planes striking in a NE–SW direction and with horizontal pressure axis oriented in a NW–SE direction (Fig. 6 and Table 3). The plane with 208° of azimuth and dipping 58° is better constrained than the plane dipping 32° to the SE due to the use of four Spanish stations, located at a distance less than 1500 km from the epicenter. Estimation of errors for the trend of the tension axis is high (89°) due to the fact that small changes on the position of this axis over focal sphere correspond to broad variations in the azimuth. However the pressure axis is well constrained with an estimation of standard errors of 15° or lower. Strike-slip mechanisms are generally better constrained than dip-slip solutions. Observations very close to the nodal planes are necessary to obtain a dip-slip solution of good quality.

Table 3

Source parameters for the 22 December 1999 Ain Temouchent earthquake (M_w = 5.7)

	Nodal planes			P axis		T axis		Depth km	Radius km	$M_O \times E$ 17 Nm
	Strike	Dip	Rake	Azim.	Plunge	Azim.	Plunge			
FM	25	32	92	297	13	121		—	—	—
N = 27	(± 63)	(±50)	(±14)	(±15)	(±13)	(±89)	(±13)			
WA	60	36	63	311	11	76	71	4	11	4.1
N = 9								(±1)		(± 0.8)
SA	—	—	—	—	—	—	—	—	7.5	1.7
N = 7									(± 0.1)	(± 0.9)

FM = first motion; WA = waveform inversion; SA = spectral analysis, N = number of data.

Depth, orientation of the nodal planes and source time function have been obtained from the inversion of nine records of *P*-wave broadband data; the starting parameters are taken from the FM focal mechanism (Fig. 6, Table 3) with 4 triangular impulses with a duration of 1.0 second as STF. We inverted the STF, depth and focal parameters. As we have mentioned dip-slip solutions are generally less constrained than strike-slip mechanisms and consequently we tested many models with different source parameters (focal mechanism and depth). The best

The Ain Temouchent (Algeria) earthquake 22-12-99

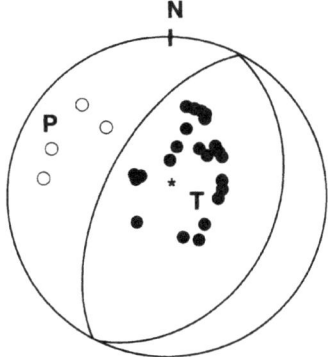

Figure 6
Focal mechanism for the Ain Temouchent earthquake obtained from first motion study (FM). Black circles correspond to compression and white circles to dilatation.

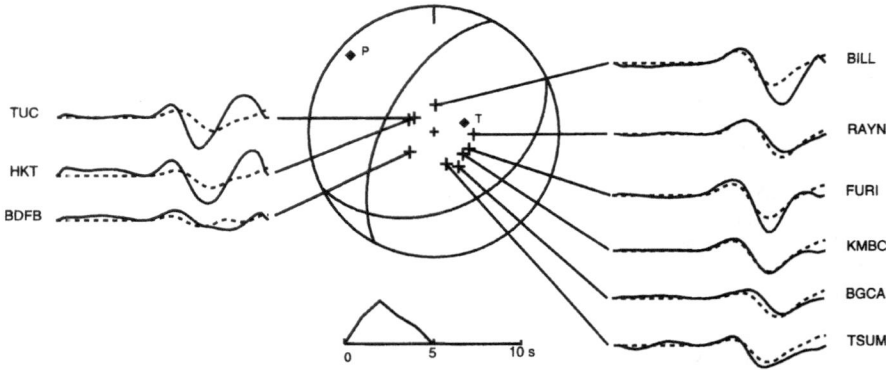

Figure 7
Focal mechanism obtained from inversion of *P* waves for the Ain Temouchent earthquake. Solid line represents the observed seismogram and dashed line the synthetic seismogram. The source time function (STF) is shown below.

solution was the one with the minimum RMS value. The best fitting source model is shown in Figure 7 and the source parameters are listed in Table 3. In this solution, the fit between observed and calculated seismograms is not substantially better than that given by other solutions. However, the chosen solution was preferred because the mechanism is very similar to that obtained from the *P* wave FM solution (Fig. 6, Table 3) and in agreement with the northwestern Algerian tectonic features (BEZZEGHOUD and BUFORN, 1999; BUFORN *et al.*, 2004, this issue). In all cases the focal depth is 4 km, and changing the depth by 1 km produces a noticeable degradation of the fitting of the waveforms. Results of body wave inversion show a simple rupture process: the STF comprises only one event (Mo = $4.1 \times 10e + 17$ Nm) of a duration of 5.0 seconds, with a reverse faulting mechanism striking in a NE–SW direction with horizontal pressure axis trending in a NW–SE direction.

The FM and WA focal mechanisms, determined in this study, are similar to this obtained by Harvard's CMT. However, our focal depth is shallower (4 km) than that (15 km) given by CMT. The CMT resolution is poor for events with depth below 15 km. Scalar seismic moment and dimension obtained from spectral analysis give values of $1.7 \times 10e + 17$ Nm and a radius of 7.5 km respectively. If we compare the values of scalar seismic moment obtained in this study by WA and spectral analysis with the CMT solution ($6.9 \times 10e + 17$ Nm), a factor varying from 1.5 to 4 is noticed. However, the CMT method has a tendency to overestimate scalar seismic moment. A factor of three in the scalar seismic moment values estimated by different authors is common and this may be explained by the frequency content of waveforms (TANIOKA and RUFF, 1997).

Conclusion

Historical seismicity for the Ain Temouchent region indicates a low level of seismic activity, with the occurrence of earthquakes with magnitudes less than 5.5. In the Algerian Maximum Observed Intensity (MOI) map (MOKRANE et al., 1994; BEZZEGHOUD et al., 1996), this region shows a maximum intensity of VI (MM scale). The recent Ain Temouchent earthquake ($M_w = 5.7$) is the largest seismic event which occurred in the Ain Temouchent region with a maximum observed intensity of VII (MSK scale). Therefore, the recent earthquakes which occurred in Western Algeria (Mascara $M_w = 5.7$, 1994; Beni-Ouartilane $M_w = 5.7$, 2000), including the 1999 Ain Temouchent earthquake, should be used to update the Algerian MOI map. The high damages and casualties caused by this shallow seismic event (h = 4 km) are due to a combination of the recent rapid population growth and the fragility of old traditional houses or modern constructions. In the last twenty years (1980–2000), the buildings in Algeria, and in particular in rural regions, have shown a low strength and high vulnerability to the recurrence of destructive earthquakes, in spite of the large El Asnam 1980 earthquake.

From a seismotectonic point of view, the Ain Temouchent earthquake allows us to ascertain with more accuracy the stress pattern in the western part of Algeria. The deduction of regional stress from fault-plane solutions is not exempt from ambiguity. The maximum compressive stress may have an orientation anywhere within the dilatational quadrant and not necessarily at 45 degrees of the fault plane. However, the stress axes derived from fault plane solutions or inversion methods for earthquakes with a magnitude larger than 5.5 may serve as an indication of their general trend. BEZZEGHOUD and BUFORN (1999) demonstrate that the reverse mechanisms are predominant in the region located between 3° E and 0° (Tell Atlas), whereas between 0° and 6°W (Betic-Rif mountains and Alboran Sea), strike-slip and oblique mechanisms with normal component are predominant. The intermediate region, between 2° W and 0°, where the Ain Temouchent earthquake occurred, was difficult to classify due to the absence of focal mechanisms for earthquakes in this area. The occurrence of the Ain Temouchent has made it possible to estimate its focal mechanism that corresponds to a rupture in reverse-faulting with planes oriented in the NE–SW direction, and horizontal P axis trending in a NW–SE direction. The character of this solution agrees with the focal mechanisms of the seismic event of the Tell Atlas region, and particularly with the 1994 Mascara earthquake (Fig. 1). Therefore, the recent Ain Temouchent earthquake, together with the other focal mechanisms of the western part of Algeria (BEZZEGHOUD and BUFORN, 1999), allow us to confirm that the regional stress regime corresponds to horizontal compression in a NW-SE direction, associated with the convergence between Eurasia and Africa.

In this study the final results obtained for the 22 December 1999 Ain Temouchent earthquake can be summarized as follows: maximum intensity I_o = VII (MSK);

motion of thrust faulting with horizontal pressure axis oriented in a NW-SE direction; a shallow depth of 4 km; a single source time function with 5 s duration; scalar seismic moment from body wave inversion and spectral analysis are $4.1 \times 10e + 17$ Nm and $1.7 \times 10e + 17$ Nm respectively; $M_w = 5.7$.

Acknowledgments

We are grateful to the entire CRAAG team (researchers and technicians) of the seismological service for their valuable contribution. We are indebted to Mr. Deramchi and Ferkoul for their contributions to the macroseismic map. The local authorities of Ain Temouchent are gratefully acknowledged for the logistic support. Part of this work has been supported by Ministerio de Ciencia y Tecnología (Spain), project REN2000-0777/C02-01. The fourth author (M.B.) appreciatively acknowledges the support from CGE through Professors Ana Maria Silva and Augusto Fitas of the University of Évora.

REFERENCES

BENHALLOU, H. (1985), *Les catastrophes sismiques de la région d'Echelif dans le contexte de la seismicité historique de l'Algèrie*, Ph.D. Thesis, USTHB, Alger, 294 pp.

BENOUAR, D., AOUDIA, A., MAOUCHE, S., and Meghraoui, M. (1994), *The 18 August 1994 Mascara (Algeria) Earthquake: A Quick Look Report*, Terra Nova 6, 634–637.

BENOUAR, D. (1994), *The Seismicity of Algeria and the Maghreb during the Twentieth Century*, Ph.D. Thesis, Imperial College London, U.K.

BEZZEGHOUD, M., AYADI, A., SEBAI, A., AIT MESSAOUD, M., MOKRANE, A., and BENHALLOU, H. (1996), *Seismicity of Algeria between 1365 and 1989 : Map of Maximum Observed Intensities (MOI)*, Advances en Geofisica y Geodesia, *I*, Pub. de IGN, (Madrid), 107–114.

BEZZEGHOUD, M. and BUFORN, E. (1999), *Source parameters of the 1992 Melilla (Spain, Mw = 4.8), 1994 Alhoceima (Morocco, Mw = 5.8) and 1994 (Algeria, Mw = 5.7) Earthquakes and Seismotectonic Implications*, Bull. Seismol. Soc. Am. 89, 2, 359–372.

BRILLINGER, D.R., UDIAS, A., and BOLT, A. (1980), *A probability model for regional focal mechanism solutions*, Bull. Seism. Soc. Am. 70, 149–170.

BUFORN E., SANZ DE GALDEANO C., and UDIAS, A. (1995), *Seismotectonics of the Ibero-Maghrebian Region*, Tectonophysics 248, 247–261.

BUFORN, E., BEZZEGHOUD, M., UDIAS, A., and PRO, C. (2004), *Seismic sources on the Iberia-African plate boundary and their tectonic implications*. Pure. Appl. Geophys. this issue.

GLOBE Task Team and others (Hastings, David A., Paula K. Dunbar, Gerald M. Elphingstone, Mark Bootz, Hiroshi Murakami, Hiroshi Maruyama, Peter Masaharu, Peter Holland, John Payne, Nevin A. Bryant, Thomas L. Logan, J.-P. Muller, Gunter Schreier, and John S. MacDonald), eds. (1999), *The Global Land One-kilometer Base Elevation (GLOBE) Digital Elevation Model, Version 1.0*, National Oceanic and Atmospheric Administration, National Geophysical Data Center, 325 Broadway, Boulder, Colorado 80303, U.S.A. Digital Data Base on the World Wide Web (URL: HYPERLINK http://www.ngdc.noaa.gov/seg/topo/globe.shtml) and CD-ROMs.

GUARD A P. (1975), *Géodynamique de la marge alpine du continent africain d'après l'étude de l'Oranie Nord – Occidentale*, Ph.D. Thesis, Univiversité de Nice, France, 286 pp.

McCAFFREY, R., ABERS, G., and ZWICK, P. (1991), *Inversion of Teleseismic Body Waves*, IASPEI, 81–166.

MEGHRAOUI, M. (1988), *Géologie des zones sismiques du nord de l'Algèrie: Paléosismologie, Tectonique active et Synthèse Sismotectonique*, Ph.D. Thesis, Univiversité de Paris XI, France, 356 pp.

MOKRANE, A., AIT MESSAOUD, A., SEBAI, A. Menia., N. Ayadi., A. and BEZZEGHOUD, M. (1994), *Les séismes en Algérie de 1365 à 1992*. Sous la direction de Bezzeghoud M. et Benhallou H. Publication du CRAAG, Alger-Bouzaréah, 277pp.

NABELEK, J.-L. (1984), *Determinations of Earthquake Source Parameters from Inversion of Body Waves*. Ph.D. Thesis, MIT, Cambridge, USA.

PHILLIP, H. (1987), *Plioquaternary Evolution of the Stress Field in the Mediterranean Zones of Subduction and Collision*, Annales Geophysicae. *5B*, 301–320.

ROUSSEL, J. (1973), *Les zones actives et la fréquence des Seismes en Algerie (1716–1970)*, Bull. Soc. Hist. Nat. Afr. Nord, Alger, t. 64. Fasc. 3 et 4, 211p.

TANIOKA, Y., and RUFF, L. (1997), *Source time functions*, Seism. Res. Lett. *68*, 386–397.

THOMAS, G. (1985), *Géodynamique d'un bassin intramontagneux, ; le bassin du bas Cheliff occidental (Algérie) durant le Mioplioquaternaire*, Ph.D. Thesis, Université de Pau et Pays de l'Adour, France.

UDIAS, A. and BUFORN, E. (1988), *Single and joint fault-plane solutions from first data in Seismological Algorithms*, D. Doornbos (Editor) Academic, London, 443–453.

YELLES-CHAOUCHE, A.K., DJELLIT, H., DERDER, M., ABTOUT, A., and BELDJOUDI, H., (2000), *The Ain Temouchent Earthquake of December 22 th,1999: Preliminary Investigations*. XXVII General Assembly of the European Seismological Commission ESC. Lisbon, September.

YELLES-CHAOUCHE, A.K. (2001), *Recent Seismic Activity in Algeria*, Workshop on the Geodynamics of the Western Part of the Eurasia-Africa Plate Boundary (Azores-Tunisia) San Fernando (Cadiz, Spain), 31 May–2 June.

YIELDING, G., OUYED, M., KING, G. C. P., and HATZFELD, D. (1989), *Active Tectonics of the Algerian Atlas Mountains: Evidence from Aftershocks of the 1980 El-Asnam Earthquake*, Geophys. J. Int. *99*, 761–788.

(Received April 30, 2002, revised December 9, 2002, accepted January 15, 2003)

 To access this journal online:
http://www.birkhauser.ch

Pure appl. geophys. 161 (2004) 623–646
0033–4553/04/030623–24
DOI 10.1007/s00024-003-2466-1

❘ Pure and Applied Geophysics

Seismic Sources on the Iberia-African Plate Boundary and their Tectonic Implications

E. Buforn[1], M. Bezzeghoud[2], A. Udías[1], and C. Pro[3]

Abstract — The plate boundary between Iberia and Africa has been studied using data on seismicity and focal mechanisms. The region has been divided into three areas: A; the Gulf of Cadiz; B, the Betics, Alboran Sea and northern Morocco; and C, Algeria. Seismicity shows a complex behavior, large shallow earthquakes (h < 30 km) occur in areas A and C and moderate shocks in area B; intermediate-depth activity (30 < h < 150 km) is located in area B; the depth earthquakes (h ≈ 650 km) are located to the south of Granada. Moment rate, slip velocity and b values have been estimated for shallow shocks, and show similar characteristics for the Gulf of Cadiz and Algeria, and quite different ones for the central region. Focal mechanisms of 80 selected shallow earthquakes (8 ≥ m_b ≥ 4) show thrust faulting in the Gulf of Cadiz and Algeria with horizontal NNW-SSE compression, and normal faulting in the Alboran Sea with E-W extension. Focal mechanisms of 26 intermediate-depth earthquakes in the Alboran Sea display vertical motions, with a predominant plane trending E-W. Solutions for very deep shocks correspond to vertical dip-slip along N-S trends. Frohlich diagrams and seismic moment tensors show different behavior in the Gulf of Cadiz, Betic-Alboran Sea and northern Morocco, and northern Algeria for shallow events. The stress pattern of intermediate-depth and very deep earthquakes has different directions: vertical extension in the NW-SE direction for intermediate depth earthquakes, and tension and pressure axes dipping about 45 ° for very deep earthquakes. Regional stress pattern may result from the collision between the African plate and Iberia, with extension and subduction of lithospheric material in the Alboran Sea at intermediate depth. The very deep seismicity may be correlated with older subduction processes.

Key words: Plate boundary, Iberia-African, seismicity, focal mechanisms, intermediate and deep depth earthquakes, subduction.

Introduction

The interaction between Iberia and Africa results in a complex region located in the western part of the Eurasian-African plate boundary. This region corresponds to the transition from an oceanic boundary (between the Azores and the Gorringe Bank), to a continental boundary where Iberia and Africa meet. The plate boundary is very well delimited in the oceanic part, from the Azores Islands along the Azores-Gibraltar fault to approximately 12°W (west of the Strait of Gibraltar). From 12°W

---

[1] Dpt. de Geofísica, Universidad Complutense, Madrid, Spain.
[2] Dpt. de Física, Universidade de Evora and CGE, Evora, Portugal.
[3] Dpt. de Física, Universidad de Extremadura, Spain.

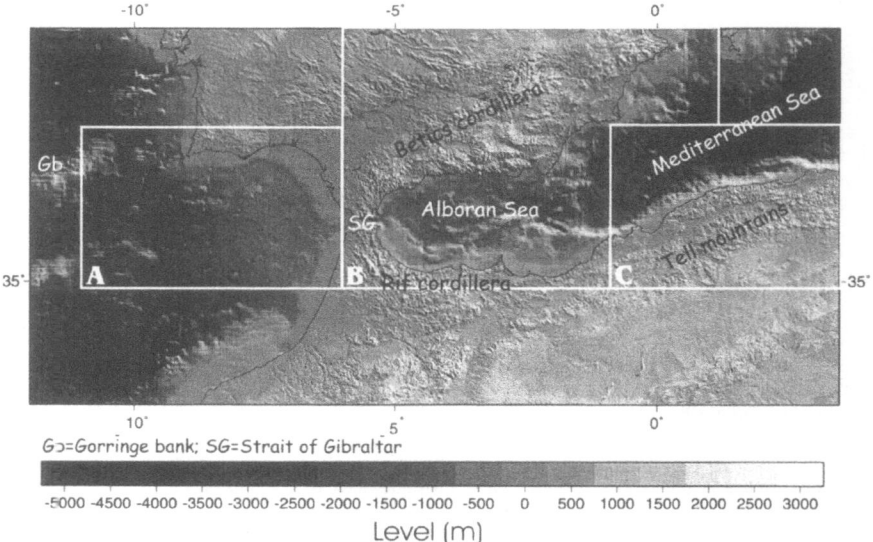

Figure 1
Topography and bathymetry of the Iberia-African region.

to 3.5°E, including the Iberia-African region and extending to the western part of Algeria, the boundary is more diffuse and forms a wider area of deformation (BUFCRN *et al.*, 1988a; MOREL and MEGHRAOUI, 1996; HAYWARD *et al.*, 1999).

The complexity of the region is reflected in its bathymetry, seismicity, stress regime and tectonics. In Figure 1 bathymetry and topography for this region are shown. On land, the main geological features are the Betics, the Rif Cordilleras and the Tell Mountains, formed basically by the Alpine domain, a consequence of the collision between Eurasia and Africa. The bathymetry shows, as main features, the Gorringe Bank region, located west of the Strait of Gibraltar, with a number of seamounts, banks and submarine ridges, and the Alboran Sea east of the Strait of Gibraltar, with important regional crustal thickness variations (TORNÉ *et al.*, 2000).

Plate kinematic models for the region have estimated convergence rates of 4mm/ yr at the Gorringe Bank and Gibraltar, and 6 mm/yr in the Algerian region (ARGUS *et al.*, 1989; DEMETS *et al.*, 1990), and 7.6 mm/yr in northwestern Algeria from seismic slip rate (LAMMALI *et al.*, 1997). From west to east along the plate boundary between Eurasia and Africa, the general stress pattern corresponds to extension in the Azores region, right-lateral strike-slip motion in the central part and compression in the eastern region, from Gorringe Bank to the Strait of Gibraltar and Algeria (MCKENZIE, 1972; UDÍAS *et al.*, 1976; GRIMISON and CHEN, 1986; BUFORN *et al.*, 1988a; JIMENEZ-MUNT *et al.*, 2001; NEGREDO *et al.*, 2002).

In this paper we have divided the Iberia-African plate boundary region into three areas: A (Gulf of Cadiz to Gorringe Bank), B (central part including Betic and Rif Cordilleras and the Alboran Sea) and C (northwest Algeria and Tell Mountains). We will examine the different characteristics of these three areas using observations of seismicity and focal mechanisms.

Shallow Earthquakes

Seismicity of the Iberia-African region is characterized by the occurrence of earthquakes of moderate magnitude, most of them with focus at shallow depth (0 < h < 40 km). The distribution of epicenters corresponding to magnitude $m_b \geq 3.5$, taken from the Instituto Geográfico Nacional Data File (IGN Data File) for the period 1980–1999, is shown in Figure 2. Magnitudes in the IGN Data File have been determined using Lg waves adjusted to values of m_b (BOLETÍN DE SISMOS PRÓXIMOS, 1989). Most shocks correspond to shallow events with magnitudes, in general, less than 5.5. For the 1980–1999 time period, only one earthquake has magnitude greater than 6.0 (El Asnam, Algeria, earthquake 10.10.1980; $M_s = 7.3$). West of Gibraltar, from the Gulf of Cadiz to the Gorringe bank region (area A), epicenters are distributed in an E-W direction, across a band about 100-km wide, with foci at shallow and intermediate depth (Fig. 2). Before the period represented in Figure 2, two large earthquakes had occurred in this region, one in the Gulf of Cádiz (15.3.1964; $M_s = 6.4$) and another west of the San Vicente Cape (29.2.1969; $M_s = 8$)

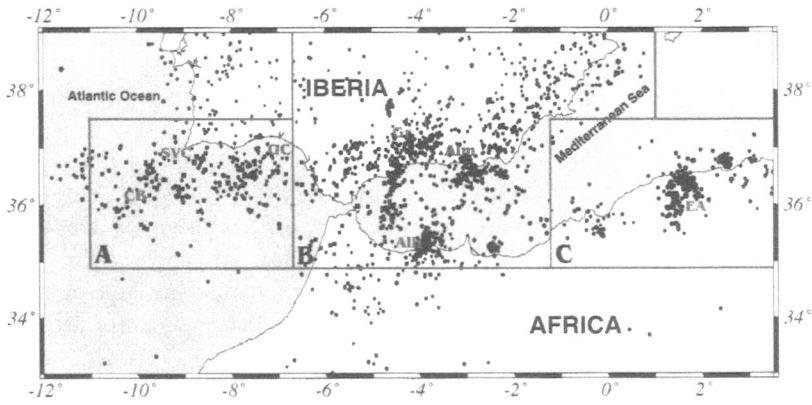

Figure 2

Distribution of epicenters for the period 1980–1999 ($m_b \geq 3.5$) taken from the IGN Data File. Circles correspond to shallow earthquakes (h < 40 km), squares to intermediate depth (40 < h < 150 km) and triangles to very deep (h > 600 km). GB = Gorringe Bank, SVC = San Vicente Cape, GC = Gulf of Cádiz, Gr = Granada, Alm = Almería, ALH = Alhoceima; EA = El Asnam.

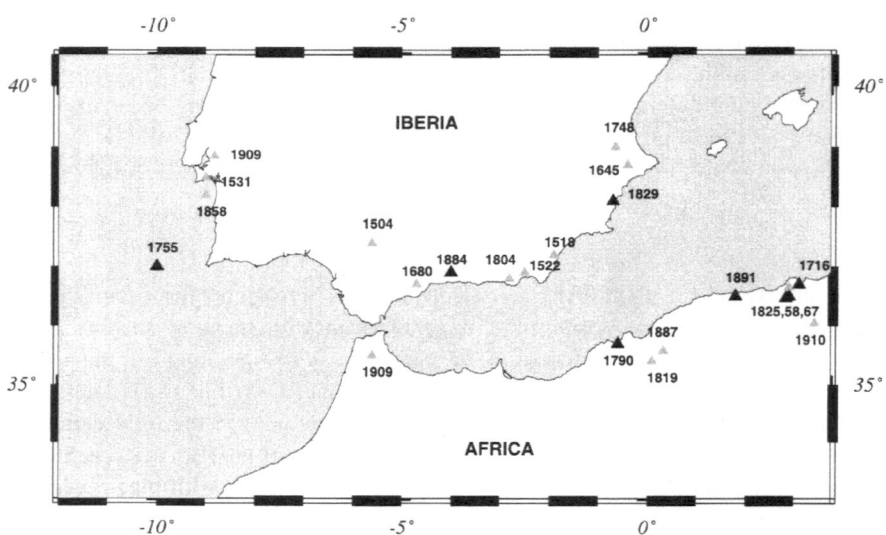

Figure 3
Historical seismicity for the studied region. In black earthquakes with maximum intensity = X, in grey maximum intensity = IX.

(UDíAS *et al.*, 1976; BUFORN *et al.*, 1988b). In area B there are three main concentrations of epicenters, at 4°W (the Alhoceima region, northern Morocco), where in 1994 an earthquake of magnitude $M_w = 5.8$ occurred (CALVERT *et al.*, 1997; BEZZEGHOUD and BUFORN, 1999); at 2.5°W in the Almeria region (SE Spain) where a swarm occurred in 1993–1994 with two shocks of magnitudes 5.0 (RUEDA *et al.*, 1996), and at 4.5°W in the Granada Basin with frequent shocks of m_b about 3. In area C shocks are concentrated in the El Asnam region, Algeria, at 2°E, where earthquakes with $M_s > 6$ occurred in 1954 ($M_s = 6.5$) and 1980 ($M_s = 7.3$).

Historical seismicity (shocks with maximum intensity of IX or X, occurring between 1500 and 1910 for the region are represented in Figure 3 (MÉZCUA and MARTINEZ SOLARES, 1983; MUÑOZ and UDíAS, 1988; MOKRANE *et al.*, 1994). Larger earthquakes (Io = X) are located west of S. Vicente Cape (Lisbon earthquake, 1755), southern Iberia (1829 and 1884) and northeastern Algeria (1716, 1790, 1825, 1858 and 1891). In northern Morocco only one large earthquake has occurred with maximum intensity IX in 1909. In Spain, except for the earthquake of 1504 (Carmona, Sevilla), shocks are located very near the coast in the south and southeast. Another area of high activity is the Lower Tajo Valley, near Lisbon, Portugal.

A selection of focal mechanisms for shallow earthquakes in this region taken from previous studies, together with four new solutions estimated in this study (Annex 1), are shown in Figure 4 and Table 1. Criteria used for the selection of the

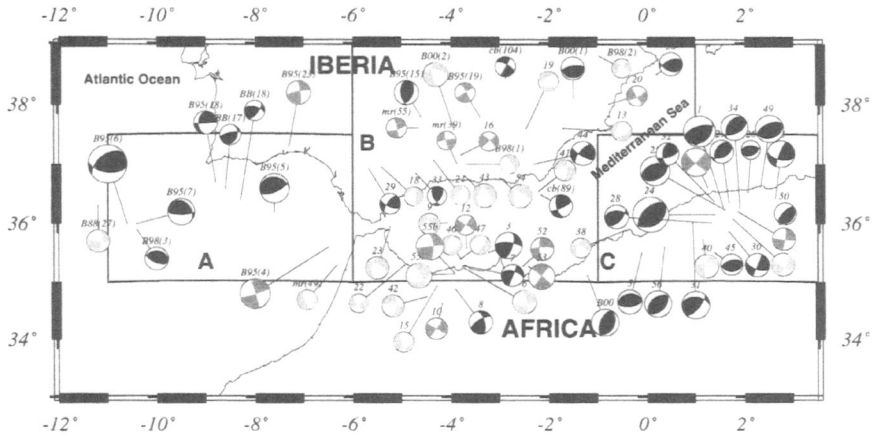

Figure 4
Focal mechanisms for shallow earthquakes (h < 40 km) and $m_b \geq 5.0$, before 1975 and $m_b \geq 4.0$ after 1975. In black thrusting solutions, in dark grey strike-slip and in grey normal solutions. Size is proportional to magnitude. Numbers correspond to Table 1.

solutions are: $m_b \geq 5$ for shocks that occurred before 1975, and $m_b \geq 4$ for shocks after 1975. The new solutions are based on more than 15 polarity data, and nodal planes are well defined. Six solutions were obtained from modelling or inversion of body waves and the rest from polarities of P waves. This selection of focal mechanisms has been chosen so that the solutions represent the regional stress pattern, and the local effects present in smaller earthquakes are avoided.

From Figure 4 we observe that fault plane solutions show predominantly thrusting motion in areas A and C (Gulf of Cadiz and Algeria regions), with an average horizontal compression in the NNW-SSE direction. There exist some strike-slip mechanisms in A (events B95 (4), B95 (23)) and in B (events 2, 51, 40), which are also compatible with a horizontal compression in the NNW-SSE direction. For these earthquakes tension axes are also horizontal in the NNE-SSW direction. Strike-slip solutions have a predominantly E-W plane with right-lateral motion, the northern block moving east. In area B (central region) focal mechanisms show a greater variety of solutions. Solutions correspond to normal faulting (5), strike-slip mechanisms with a large component of normal motion (10), reverse faulting (4), strike-slip with a large component of reverse motion (7), and pure strike-slip faulting (15). Thus a normal component of motion is present in 15 shocks. In this respect area B differs from A and C, where the predominant motion is of reverse character. Most solutions in area B are compatible with a stress pattern consisting of a general horizontal tension axis in the E-W direction and a horizontal pressure axis in the NW-SE direction. Solutions with a large normal component and vertical P axis are more frequent on the south coast of Spain and in the Alboran Sea.

Table 1

Solutions for focal mechanisms of shallow earthquakes represented in Figure 4. φ: strike, δ: dip, λ: slip. NF is the reference plotted with each solution

Date	Lat. N	Long.E	Depth	M	φ	δ	λ	NF	Ref.
190551	37.58	−3.93	30	5.1	169	69	−35	B00(2)	9
090954	36.28	1.57	10	6.5	253	61	104	1	1
100954	36.62	1.24	30	6.0	44	90	−8	2	1
230859	35.51	−3.23	20	5.5	276	70	153	3	1
051260	35.60	−6.50	15	6.2	73	86	−178	B95(4)	2
150364	36.2	−7.60	12	6.1	*276	24	117	B95(5)	2
130767	35.50	−0.10	5	5.1	260	30	87	5	1
170468	35.24	−3.73	22	5.0	83	70	−162	6	1
301068	35.28	−3.76	5	4.6	286	55	145	7	1
280269	36.10	−10.60	22	8.0	231	47	54	B95(6)	2
050569	36.00	−10.40	29	5.5	324	24	142	B95(7)	2
070470	34.87	−3.90	5	4.8	244	64	151	8	1
180472	36.30	−11.20	15	4.7	8	65	−2	B88(27)	7
221172	36.02	−4.07	5	4.4	234	50	−15	9	1
290473	34.63	−4.17	10	4.5	212	90	1	10	1
140774	35.58	−3.68	5	4.3	305	90	180	12	1
060677	37.60	−1.70	10	4.2	208	45	−121	13	1
150777	35.17	−3.73	13	4.0	211	70	−25	14	1
240279	34.93	−4.28	5	4.3	51	40	−23	15	1
200379	37.16	−3.79	5	4.1	316	78	−179	16	1
210479	35.03	−4.00	5	4.0	173	71	148	17	1
010579	36.95	−5.42	24	4.0	249	35	−24	18	1
140579	37.70	−2.46	5	4.2	107	49	−40	19	1
251079	38.01	−0.77	20	4.3	59	81	−7	20	1
221279	37.06	−4.34	40	4.0	210	64	−86	21	1
100280	35.29	−4.94	5	4.0	55	85	−18	22	1
220680	35.96	−5.93	30	4.7	304	66	−135	23	1
101080	36.16	1.39	5	7.3	*225	54	83	24	1
101080	36.24	1.59	10	6.1	58	43	81	25	1
131080	36.53	2.07	5	4.0	63	42	69	26	1
301080	36.26	1.68	5	4.8	210	46	64	27	1
081180	36.02	1.32	5	5.0	270	45	126	28	1
031280	36.92	−5.67	27	4.3	114	68	155	29	1
051280	35.87	1.68	5	5.0	112	61	−179	30	1
071280	36.02	0.94	5	5.8	277	40	140	31	1
150181	36.38	1.38	8	4.7	181	53	29	32	1
210181	36.85	−4.71	5	4.0	153	56	46	33	1
010281	36.27	1.90	11	5.5	210	43	64	34	1
140281	36.08	1.76	26	4.9	26	67	−18	35	1
050381	38.50	0.20	5	4.9	113	42	128	36	1
200381	35.13	−3.90	5	4.0	164	89	133	37	1
190481	35.89	−0.43	16	4.2	198	57	−16	38	1
070481	35.12	−3.98		4.0	182	75	132	39	1
151182	35.73	1.15	7	5.0	274	70	−169	40	1
060183	36.49	−2.15	12	4.7	163	58	14	cb(89)	3
200383	36.55	−2.20	6	4.4	266	62	−18	41	1
241183	34.74	−4.49	40	4.6	272	74	−23	42	1
240684	36.80	−3.70	5	5.0	201	48	−46	43	1

Table 1 (contd.)

Date	Lat. N	Long.E	Depth	M	φ	δ	λ	NF	Ref.
130984	37.00	−2.30	9	5.1	121	73	156	44	1
100485	38.43	−2.88	5	4.2	298	67	−3	cb(104)	3
030585	35.50	1.40		4.5	225	54	83	45	1
260585	37.80	−4.60	5	5.1	174	51	70	B95(15)	2
201086	36.70	−8.80	37	4.8	180	37	3	B95(18)	2
110387	37.80	−3.40	7	4.2	329	80	2	B95(19)	2
091287	35.40	−3.82	14	4.2	54	49	−58	46	1
051088	35.40	−3.80	11	4.2	248	26	−58	47	1
311088	36.44	2.63	13	5.7	103	55	167	48	1
051288	37.01	−3.88	5	4.0	169	82	73	mr(39)	6
291089	36.61	2.33	31	5.8	242	55	87	49	1
201289	37.30	−7.30	23	5.0	351	77	10	B95(23)	2
090290	36.26	2.83	18	4.5	49	18	95	50	1
071190	37.00	−3.68	2	4.0	165	16	−74	B98(1)	9
150691	35.90	−10.40	6	4.8	273	30	73	B98(3)	9
140891	38.80	−0.96	4	4.1	314	72	−164	B98(2)	9
190192	36.21	1.86	4	4.7	277	85	−169	51	1
120392	35.27	−2.53	8	4.8	268	76	−161	52	1
160293	36.60	−8.60	26	4.3	17	33	34	BB(17)	5
010593	35.29	−6.33	30	4.2	15	25	−60	mr(49)	6
230593	35.27	−2.42	6	5.4	308	86	4	53	1
220693	36.40	−8.30	15	4.3	37	62	40	BB(18)	5
231293	36.77	−2.99	8	4.9	300	70	−130	54	1
260594	35.14	−3.92	7	5.3	*330	77	−45	55a	1
260594	35.16	−3.92	8	5.7	*355	79	2	55b	1
180894	35.60	0.36	4	5.7	*58	45	95	56	1
160496	37.61	−4.66	8	4.3	75	76	−179	mr(55)	6
020299	38.11	−1.49	5	4.8	260	67	89	B00(1)	4
221299	35.26	−1.45	6	5.6	*25	31	92	B00	8

1. BEZZEGHOUD and BUFORN. (1999); 2. BUFORN et al. (1995); 3. COCA and BUFORN (1994); 4. BUFORN and SANZ DE GALDEANO (2001); 5. BORGES et al. (2001); 6. MEZCUA and RUEDA (1997); 7. BUFORN et al. (1988a); 8. YELLES-CHAOUCHE et al., This issue; 9. This paper; *Modeling or inversion of body waves.

Intermediate Depth Earthquakes

Important seismic activity at intermediate depth (150 > h > 40 km) is present in parts A and B, as can be seen in Figure 2 (MUNUERA, 1963; HATZFELD, 1978; GRIMISON and CHENG, 1986; BUFORN et al., 1988b, 1991a, b, 1997; SEBER et al., 1996; SERRANO et al., 1998). This activity is shown in more detail for the period 1980–1999 and $m_b \geq 3.5$, in Figure 5 (I.G.N. Data File). The reliability of depth determinations for this time period was improved by the installation of the network of the IGN in Spain (Tejedor and García, 1993). Magnitudes m_b of earthquakes in this figure are between 3.5 and 5. In part A, intermediate-depth shocks are spread over a band approximately 100-km wide, (between 36°N and 37°N), extending E-W from 8°W to 11°W. The most important concentration of foci at intermediate depth

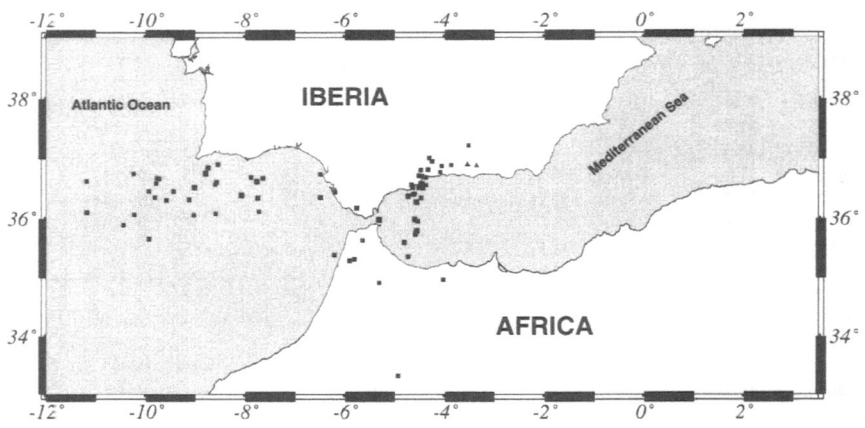

Figure 5
Distribution of epicenters at intermediate depth (40 < h < 150 km, squares) and very deep earthquakes (h > 600 km, triangles). The period represented is 1980–1999 (Instituto Geográfico Nacional, Data File).

is in part B, at the eastern side of the Strait of Gibraltar in a narrow N-S trending band less than 50-km wide, centered at 4.5°W and extending from 35°N to 37°N. Thus the distribution of intermediate-depth shocks differs in area A (E-W trend) from that of area B (N-S trend). No intermediate depth seismic activity is observed east of 3°W in Spain, Morocco or Algeria.

A vertical E-W cross section of the seismicity, showing the distribution of hypocenters along a band between 12°W and 2°W and centered at 36°N, is shown in Figure 6a. From this figure we observe that most earthquakes in part A have depths less than about 60 km. In part B, at about 4.5°W there is a concentration of shocks to depths of 100 km in a very narrow band of less than 50 km. To the east of this area depths decrease rapidly with maximum depths less than 40 km to the east of 3°W (to Algeria). No intermediate-depth activity is found in part C.

Figures 6b and 6c show N-S vertical cross sections along bands centered at longitudes 9°W (Gulf of Cadiz) and 4.5°W (western part of the Alboran Sea). For the Gulf of Cadiz (Fig. 6b), most earthquakes occur at depths less than 60 km, with most foci concentrated between latitudes 36°N and 37°N. In the Alboran Sea (Fig. 6c) depths increase to values around 100 km, with an important concentration of hypocenters between 60 and 100 km occurring between latitudes 35°N and 37°N. These shocks correspond to the narrow band in Figure 6a. There is an apparent gap between 30 and 50 km for the same latitudes however this may be due to the crustal model used in the hypocentral determinations.

Focal mechanisms for 27 selected intermediate-depth events ($m_b \geq 3.5$), from previous studies, together with eight solutions determined for this study (Annex 1), are shown in Figure 7 and listed in Table 2. Most solutions correspond to

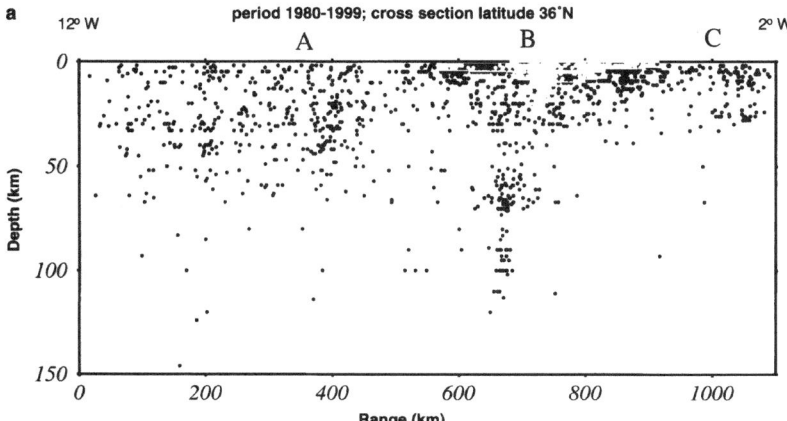

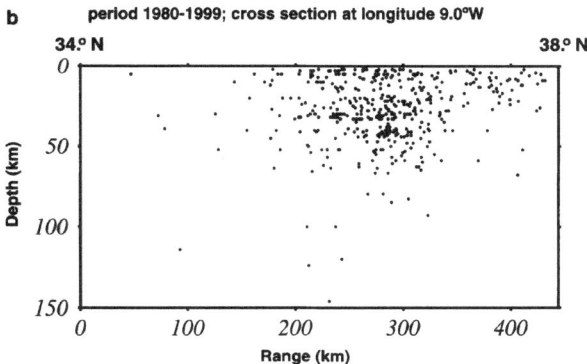

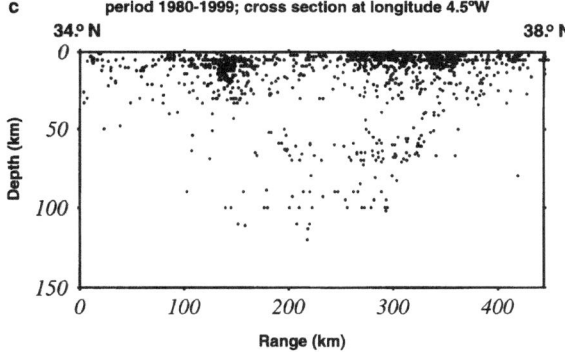

Figure 6

Vertical cross sections for the period 1980–1999 and magnitude greater than or equal to 3.5, (a) E-W section centered along 36°N, (b) N-S centered along 9°W and (c) NS centered along 4.5°W.

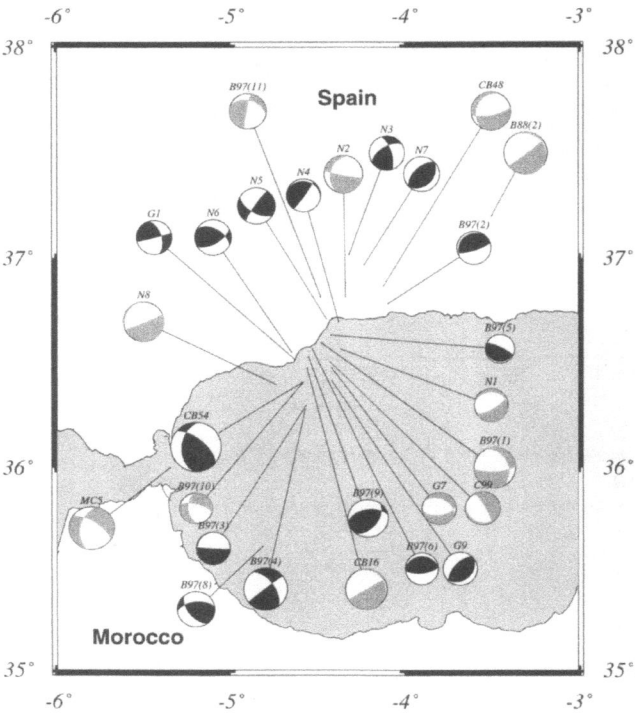

Figure 7
Focal mechanisms for intermediate depth earthquakes (40 < h < 150 km). Thrusting solutions in black and normal solutions in grey. Numbers correspond to Table 2.

earthquakes located very near the Spanish coast, where the concentration of epicenters is greater. Twelve of these solutions were obtained by modelling and inversion of body waves (Table 2) and the rest from first motion of *P* waves. Most solutions correspond to dip-slip motion, 7 to normal component and 9 to reverse component of motion. In most solutions the nearly vertical nodal plane has a mean orientation NE-SW. Only four shocks have an appreciable strike-slip component of motion. A nearly vertical tension axis has been obtained for 15 earthquakes (7 of them estimated from modelling or inversion of body waves).

Deep Earthquakes

An important feature of the seismicity of this region is the occurrence of very deep earthquakes at a depth of about 630 km. The largest of these earthquakes (M = 7) took place in 1954. These are the deepest earthquakes in the Mediterranean

Table 2

Solutions for focal mechanisms of intermediate depth earthquakes represented in Figure 7. φ: strike, :δ dip, λ: slip. NF is the reference plotted with each solution

Date	Lat. N.	Long.E	Depth	M	φ	δ	λ	NF	Ref.
130268	36.48	−4.56	91	4.3	334	10	5	CD16	3
130674	36.87	−4.12	60	4.1	78	72	−69	CB48	3
070875	36.41	−4.59	105	5.2	186	42	138	CB54	3
200679	37.20	−3.50	60	4.5	52	87	−102	B88(2)	4
220680	35.96	−5.23	81	4.7	304	76	−135	MC5	5
130586	36.60	−4.48	90	4.3	*87	74	−123	B97(1)	1
270387	36.79	−4.10	79	3.5	*69	72	76	B97(2)	1
300588	36.52	−4.63	80	3.6	75	88	35	G1	6
281188	36.30	−4.57	100	3.5	*93	88	−85	B97(3)	1
121288	36.28	−4.57	95	4.5	*232	87	146	B97(4)	1
190789	36.64	−4.43	95	3.0	*296	79	94	B97(5)	1
060290	36.57	−4.53	68	3.4	*270	23	96	B97(6)	1
130490	35.61	−4.82	89	3.9	*263	53	45	B97(8)	1
020590	36.53	−4.55	95	4.2	*36	49	57	B97(9)	1
181190	36.41	−4.59	85	3.4	*175	51	−30	B97(10)	1
250891	36.82	−4.48	58	3.8	*286	39	−173	B97(11)	1
140392	36.51	−4.43	64	3.6	*118	14	−123	C99	2
030992	36.48	−4.42	86	3.5	298	41	−61	G7	6
091193	36.42	−4.42	70	3.5	223	60	86	G9	6
010194	36.57	−4.37	68	3.5	60	71	−103	N1	7
170395	36.82	−4.34	56	4.0	100	85	−56	N2	7
181195	37.02	−4.32	52	3.6	238	59	154	N3	7
281195	36.70	−4.38	68	3.5	35	84	76	N4	7
220696	36.71	−4.45	68	3.9	120	58	172	N5	7
271296	36.56	−4.65	59	3.8	60	60	49	N6	7
180397	36.96	−4.23	56	3.7	43	34	87	N7	7
200897	36.40	−4.65	68	4.2	67	86	−63	N8	7

1. BUFORN et al. (1997); 2. COCA (1999); 3. COCA and BUFORN (1994); 4. BUFORN et al. (1988b); 5. MEDINA and CHERKAOUI (1992); 6. MORALES et al. (1999); 7. This paper; *wave-form modelling or inverison.

region and their origin is still an undecided question. During the period represented in Figure 2 very deep events occurred in 1990 and 1993 (Table 3). The 1990 event ($M_w = 4.8$) has been studied previously (BUFORN et al., 1991b, 1997) and the 1993 earthquake, with a lower magnitude ($M_w = 4.4$), is studied in this paper (Annex 1). Their foci are located to the east of the concentration of the intermediate depth earthquakes at about the same location as that of the 1954 and 1973 events (CHUNG and KANAMORI, 1976; BUFORN et al., 1991a, 1997). This occurrence of deep seismicity is concentrated in a very small area inland south of Granada, the east of the intermediate-depth seismicity that has most foci in the Alboran Sea. Focal mechanism solutions for deep earthquakes are all very similar (Fig. 8 and Table 3). Solutions have a vertical plane oriented N-S and the other nearly horizontal. Pressure and tension axes dip about 45°, with the pressure axis dipping to the east and the tension axis to the west. In this regard solutions differ from those for the

Table 3

Solution for focal mechanisms of deep depth earthquakes represented in Figure 8. ϕ: strike, δ: dip, λ: slip. NF is the reference plotted with each solutions

Date	Lat. N.	Long.E	Depth	M	ϕ	δ	λ	NF	Ref.
210354	37.00	−3.70	640	7.0	179	88	−122	B91(1)	1
300173	36.90	−3.70	660	4.8	191	74	−56	B91(2)	1
080390	37.00	−3.60	637	4.8	177	62	−91	B91(8)	1
310793	36.80	−3.43	663	4.4	177	60	−91	B02	2

1. Buforn *et al.* 1991a; 2. This paper

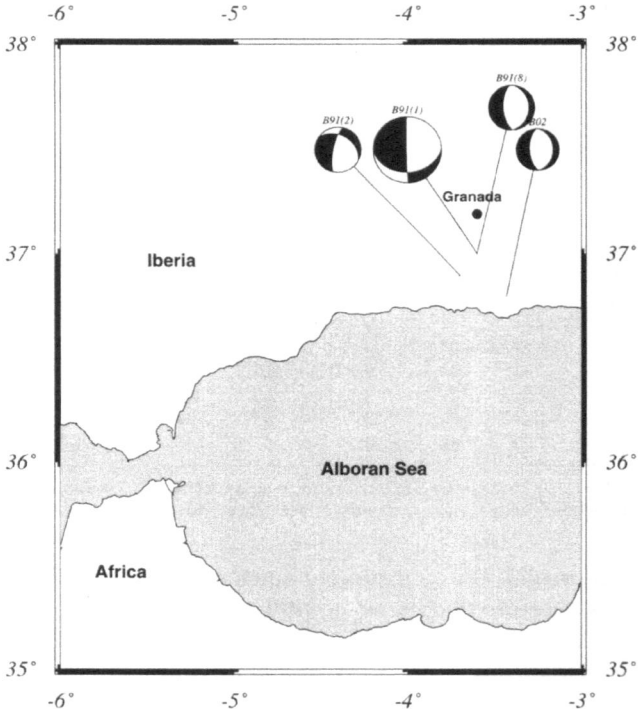

Figure 8

Focal mechanisms for very deep depth earthquakes (h > 600 km). Numbers correspond to Table 3

intermediate-depth earthquakes in which pressure axes dip to the south or southeast. From Figures 7 and 8 we observe, as a common characteristic of the focal mechanisms for intermediate-depth and very deep earthquakes, a nearly vertical plane. However the orientation of this plane is approximately E-W for intermediate shocks and N-S for the very deep ones.

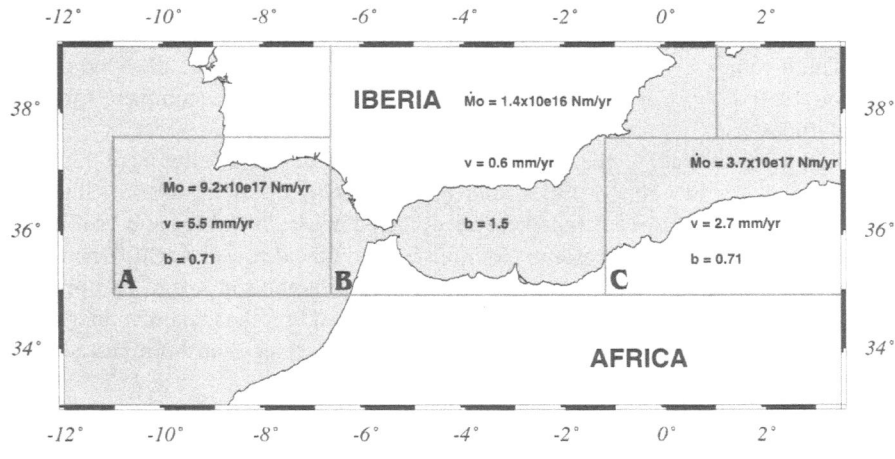

Figure 9
Moment rate, slip velocity and b values for the three studied areas (A, B and C).

Moment Rates, Slip Velocities and b Values

From the seismicity and focal mechanisms of shallow and intermediate-depth events, different behavior of the plate boundary in the three areas A, B and C from the Gulf of Cadiz to Algeria may be deduced. In order to quantify the characteristics of the seismicity in these three areas, the seismic moment rate, slip velocity and b value have been estimated for each of them. Results are shown in Figure 9. The moment rate was estimated from shallow earthquakes occurring in each area during the period 1900–1999 with magnitude $m_b \geq 5$. We used an empirical relation between magnitude m_b and scalar seismic moment M_o obtained from events with M_o values determined from spectra of body waves. The relation is similar to that obtained by other authors (BUFORN *et al.*, 1988a; EKSTRÖM and DZIEWONSKI, 1988; BADAL *et al.*, 2000).

$$\log \mathbf{M}_0 = 1.54\mathbf{m}_b + 8.7. \tag{1}$$

The moment rate M_0 was estimated by dividing the sum of the scalar seismic moment in each of the three areas by the time period (100 years).

Average slip velocity was estimated from the moment rate according to the expression

$$\Delta \dot{u} = \dot{M}_0/\mu S, \tag{2}$$

where μ is the rigidity coefficient and S the fault area. As an estimate of the slip velocity we used $\mu = 3 \times 10^4$ MPa, and for S we took the area of a vertical fault with a length of 550 km for parts A and B and 440 km for part C, and a width of 10 km (the average depth for shallow shocks). This is equivalent to considering a fault with these dimensions as the origin of the shallow seismicity (Fig. 2).

The *b* values were obtained using earthquakes with magnitude $m_b \geq 3.0$ for the period 1950–1999. For this period the catalogue is considered complete for these magnitudes.

For areas A and C the same order of magnitude for the moment rate was obtained: 9.2×10^{17} Nm/yr for A and 3.7×10^{17} Nm/yr for C. These values are conditioned by the large earthquakes that occurred in area A in 1964 and 1969 and in area C in 1954 and 1980. However, for area B the moment rate (1.4×10^{16} Nm/yr) is nearly two orders of magnitude lower than that obtained for A and one order of magnitude lower than that for C. This is due to the fact that the maximum magnitude of earthquakes during the 20th century in area B is lower than 6 (largest shallow earthquake was $M_s = 5.1$ in 1951).

Values of slip velocity show similar results as those of moment rate: 5.5 mm/yr and 2.7 mm/yr for areas A and C and 0.6 mm/yr for area B. We have compared these results with the relative motions of Africa with relation to Eurasia predicted by the models NUVEL-1a (DeMets *et al.*, 1990) and DEOSK2 (Rui Fernandes, personal communication) (Table 4). These values have been estimated at the following points: 36.25°N, -8.5°W for area A; 37.0°N, −2.5°W for area B and 36.25°N, 1.25°E for area C. For area A, similar values are obtained in our study and by NUVEL-1a and DEOSK2 (Table 4), about 5 mm/yr in each case. For area C our velocity of 2.7 mm/yr is approximately 50% of the value obtained using NUVEL-1a and DEOSK2 models. However, the largest difference occurs in area B, where the velocity predicted by the models is very similar (5.2 mm/yr and 5.4 mm/yr, respectively) while our estimations give only 0.6 mm/yr, that is only about 10% of the modeled values. This may indicate that only a small fraction of the deformation was released seismically during the 20th century.

The *b* value for each area is as follows: in area B it is 1.5, and in both areas A and C it is 0.71. The difference in *b* values in areas A and C and area B corresponds to the different behavior of the seismic activity, with a large number of small to moderate earthquakes and an absence of large shocks in the last 50 years in area B, and larger earthquakes and a lower number of small ones in areas A and C.

Table 4

Values of slip velocity obtained from NUVEL 1a, DEOSK2 models and in this study

	NUVEL 1aa (mm/yr)	DEOSK2 (mm/yr)	This study
Gulf of Cadiz	4.3	5.0	5.5
Central region	5.2	5.3	0.6
Algeria	5.6	5.4	2.7

For intermediate-depth and deep earthquakes the moment rate, slip velocity and *b* values have not been estimated due to the low number of these earthquakes, for which depth determinations are reliable corresponding to recent years only.

In consequence, from Figure 9 we can conclude that in the Gulf of Cadiz and Algeria (areas A and C) the plate boundary between Eurasia and Africa corresponds to an area where the material is relatively rigid and the stresses are released by larger earthquakes. In the Betics, Alboran and Rif regions (area B) the material is more fragmented, with a large number of small faults, and consequently the stresses are released by frequent small to moderate earthquakes. As a consequence, the plate boundary is not well defined in area B and it corresponds to a wide area where deformation is manifested by the continuous occurrence of small earthquakes, and only occasionally, some moderate events occur. However, in the past, large events also have occurred in this area as shown by the historical seismicity (Fig. 3) and the lack of large earthquakes in the 1900–1999 period and the consequent low values of seismic moment rate and slip velocity may be due only to an anomalous period of seismic quiescence during the last century. In the 19th century at least two earthquakes (1829 and 1884) took place in southern Spain with magnitude greater than 6. For this region the time period selected (1900–1999) does not adequately represent the long range seismic activity of the region.

Frohlich Diagrams and Total Moment Tensor

From the results of focal mechanisms for shallow earthquakes shown in Figure 4 and Table 1, the Frohlich diagrams (FROHLICH and APPERSON, 1992) and the total seismic moment tensor for the three areas have been estimated (Fig. 10). The Frohlich diagrams for areas A and C show that most solutions correspond to reverse faulting, with only three mechanisms of pure strike-slip in area A and five in area C. A single solution corresponding to a normal motion is found in area A and there is a total absence of this type of faulting in area C. In area B the stress regime is more complex, strike-slip is the predominant motion and with a large component of normal and reverse faulting. A normal component is most frequent; it is present in 18 cases while reverse component is only present in 10 cases. Five mechanisms show pure normal faulting and four pure reverse faulting.

A problem in the use of the Frohlich diagrams is that in this representation all the earthquakes have the same weight, independent of their magnitude, and in consequence it is difficult to quantify the stress regime in an area. This may be solved with the use of the total seismic moment tensor defined as the sum of the moment tensors calculated from individual solutions:

$$M_{ij}^{\text{total}} = \sum_{k=1}^{N} M_0^k m_{ij}^k, \tag{3}$$

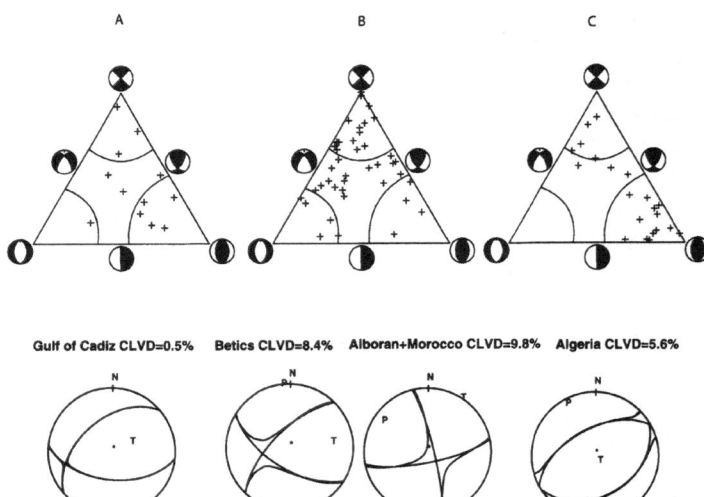

Figure 10
Frohlich diagrams and total seismic moment tensor for shallow earthquakes in the three studied regions.
The CLVD component for each region is indicated.

where k is the number of earthquakes, M_0 the scalar seismic moment of each event and m_{ij} the seismic moment tensor components. Larger earthquakes (with high values of M_0) make a larger contribution in the estimation of the total seismic moment tensor. Equation (3) was used to estimate the components of total M_{ij} for the three regions using the solutions of Table 1. The total seismic moment tensor was separated into a DC and a non-DC (CLVD) component (DZIEWONSKI and WOODHOUSE, 1983). Results are shown in Figure 10. For area A (Fig. 10) the seismic moment tensor obtained corresponds to thrusting motion. This result is due to the solution of the 1969 earthquake. A similar result has been obtained for area C; the total seismic moment tensor shows thrusting motion due to the solutions of the earthquakes in the El Asnam region, especially that of the 1980 earthquake. The small non-DC components obtained (0.5% and 5.6%, respectively) confirm that solutions obtained for both regions are very similar and that large earthquakes control the stress regime in areas A and C.

In order to obtain coherent results, for area B it was necessary to subdivide the region into two: one part corresponds to the Betics and the other to the Alboran Sea and Morocco. For the Betics the stress regime corresponds to strike-slip faulting with a component of reverse motion, with the pressure axis nearly horizontal and oriented in N-S direction. For Alboran and Morocco the total moment tensor shows strike-slip motion with a small component of normal motion and horizontal tension axis oriented in a NE-SW direction. The amount of non-DC component is 8.4% and

9.8% for the Betics and Alboran and Morocco, respectively. The non-DC values obtained for both parts of area B, each less than 15%, indicates that for these regions the total seismic moment tensor obtained can be considered to represent the stress regime in the area. Thus in the three areas (A, B and C) there is a common orientation of the pressure axis which is horizontal and trending N-S to NW-SE. The tension axis is nearly vertical in areas A and C, and nearly horizontal in B trending E-W to NE-SW.

A similar study has been carried out for the intermediate-depth events (Fig. 11). Frohlich's diagram shows that most of the mechanisms correspond to dip-slip solutions, with a greater component of reverse motions. Only three solutions correspond to pure normal motion while 10 correspond to reverse motion. About 12 solutions correspond to motion on a nearly vertical or horizontal plane. From the total seismic moment tensor, a solution is obtained with a steeply dipping plane oriented NW-SE and a near horizontal plane. The pressure axis is horizontal and trending to the NE and the tension axis is almost vertical, with a small dip to the SE. The amount of non-DC component (0.8%) indicates that the stress regime resulting from the focal mechanism solutions for the intermediate-depth events is fairly uniform and may be represented by the solution shown in Figure 11.

For very deep earthquakes Frohlichs diagrams and the total seismic moment tensor have not been estimated. The reason is the small number (only four earthquakes) of similar solutions of focal mechanism and the large magnitude of the

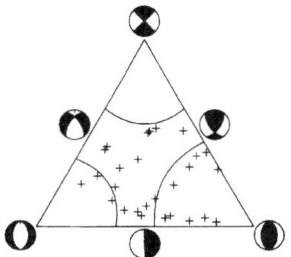

Alboran intermediate depth CLVD=0.8%

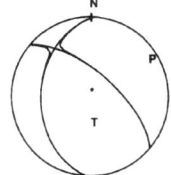

Figure 11
Frohlich diagrams and total seismic moment tensor for intermediate depth earthquakes in the Alboran region, the CLVD component is indicated.

1954 earthquake. In this case the total seismic moment tensor is controlled by the largest event, with $M = 7$, versus values of less than 5 for the other three earthquakes.

Seismotectonic Interpretation

Figure 12 depicts a proposed seismotectonic scheme for the studied region. The area delimited by the epicenters of shallow earthquakes (enclosed by the dashed lines) represents the surface expression of the plate boundary. In areas A and C (Gulf of Cadiz and Algeria regions) the plate boundary corresponds to a narrow band well defined by the seismicity, where large earthquakes ($M > 7$) occur in association with horizontal compression N-S to NNW-SSE due to the convergence of Eurasia and Africa. The intermediate-depth earthquakes, with a distribution in the E-W direction and delimited by a narrow band less than 20-km wide that broadens as we move to the Strait of Gibraltar, may also be associated with the convergence process of the

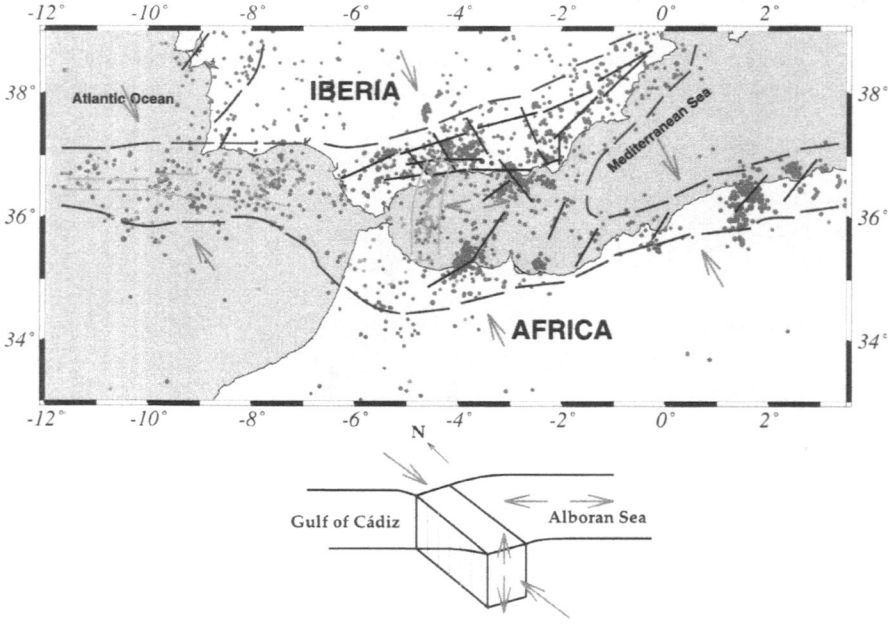

Figure 12

Seismotectonic scheme for the Iberia-Africa region. Shallow earthquakes are plotted in dark, intermediate depth events in grey. Arrows indicate the surface stress regime obtained from focal mechanisms of earthquakes. At bottom the seismogenic block proposed as origin of the intermediate depth seismicity at the western part of Alboran Sea.

Eurasia-Africa plates. No intermediate-depth earthquakes have been observed in the eastern part of the region (area C).

In area B the plate boundary is more diffuse and corresponds to a wider area that includes the Betics, the Alboran Sea and the Rif. It is difficult in this case to identify a simple line that corresponds to the plate boundary. In area B results of moment rate, slip velocity and b values indicate that the strain accumulated in the region is released partly in a continuous seismic activity of moderate magnitude over the whole area. Earthquakes with magnitudes larger than 6 occur at prolonged intervals. From historical seismicity it may be concluded that the 20th century was a period of anomalously low levels of seismic activity in this area. The stress regime obtained from the focal mechanisms of shallow events (Frohlich diagrams and total moment tensor) is compatible with horizontal N-S to NW-SE convergence of Eurasia and Africa. However in the Betics-Alboran area there is also a horizontal extension in an approximately E-W direction.

The existence of important seismic activity at intermediate depth (60 to 150 km) extending in a very narrow vertical band 50-km wide in N-S direction may be explained by the existence of a seismogenic block, of approximate dimensions 200-km long, 150-km deep and 50-km wide, on the eastern side of the Strait of Gibraltar. Inside this block the stress regime, deduced from focal mechanisms of earthquakes, corresponds to nearly vertical tension dipping to the SE. Different tectonic models have been proposed for this region, such as some kind of subduction process (BUFORN et al., 1988b, 1991a; MORALES et al., 1999), extensional collapse of thickened continental lithosphere (PLATT and VISSERS, 1989; HOUSEMAN, 1996), continental lithospheric delamination (DOCHERTY and BANDA, 1995; MÉZCUA and RUEDA, 1996; SEBER et al., 1996; CALVERT et al., 2000), backarc extension caused by subduction rollback (MORLEY, 1993; LONERGAN and WHITE, 1997; MICHARD et al., 2002), convective thinning (HOUSEMAN, 1996) or subduction and breaking of a slab of material (ZECK, 1996). Some of these models, such as continental lithospheric delamination, are not compatible with the presence of the intermediate-depth earthquakes and their focal mechanisms. The results presented here are consistent with the model presented by BUFORN et al. (1997) of an almost vertical slab of material with strike N-S driven by the extensional E-W forces present on the Alboran Sea, and under NW-SE compressive forces. The slab is being stretched downward, possibly by gravitational instability processes. Models which propose very low angle subduction or delamination (CALVERT et al., 2000) are not compatible with the results presented in this paper, due to the vertical distribution of hypocenters. Whatever explanation is given for the tectonics of this region, it must satisfy the geometry of the location of hypocenters and their focal mechanisms. Tomographic studies in this area show the existence of low velocity in the upper mantle between the Betic Cordillera and the Alboran Sea (SERRANO et al., 1998; MORALES et al., 1999). This anomaly is located in the western part of the Alboran Sea and it extends to 100-km depth, in the same region where the intermediate-depth events occur.

The presence of the very deep earthquakes (650 km) under southern Spain is a further sign of the complexity of area B. Their focal mechanisms correspond to pressure and tension axes trending E-W and dipping about 45°. Tomographic studies of the area show the existence of an anomalous high velocity region extending in depth from 200 km to 700 km (BLANCO and SPAKMAN, 1993). Only a very small part of the area, a volume where the foci are located, is seismically active. The relation of this deep activity with that of intermediate depth is not clear, but results of focal mechanisms and tomographic studies suggest different origins for them. In both cases these may be related to subduction processes, more recent for intermediate-depth shocks and older for very deep activity.

Acknowledgements

The authors wish to thank the Instituto Geográfico Nacional (Madrid, Spain) for providing part of the data. The authors also appreciate valuable comments and discussion given by Dr. D. McKnight. This work has been supported in part by the Ministerio de Ciencia y Tecnología (Spain), project REN2000-0777-C02-01.

REFERENCES

ARGUS, D., GORDON, R., DEMETS, C., and STEIN, S. (1989), *Closure of the Africa-Eurasia-North America Plate Motion Circuit and Tectonics of the Gloria Fault*, J. Geophys. Res. *94*, 5585–5602.

BLANCO, M.J. and SPAKMAN, W. (1993), *The P Velocity Structure of the Mantle below the Iberian Peninsula: Evidence for Subducted Lithosphere below South Spain*, Tectonophysics *221*, 13–43.

BADAL, J., SAMARDJIEVA, E., and PAYO, G. (2000), *Moment Magnitudes for Early (1923–1961) Instrumental Iberian Earthquakes*, Bull. Seismol. Soc. Am *90*,1161–1173.

BEZZEGHOUD, M. and BUFORN, E. (1999). *Source Parameters of the 1992 Melilla (Spain, $M_w = 4.8$), 1994 Alhoceima (Morocco, $M_w = 5.8$) and 1994 Mascara (Algeria, $M_w = 5.7$) Earthquakes and Seismotectonic Implications*, Bull. Seismol. Soc. Am. *89*, 359–372.

BOLETÍN DE SISMOS PRÓXIMOS (1989), Instituto Geográfico Nacional, Madrid.

BORGES, J.F., FITAS, A., BEZZEGHOUD, M., and TEVES-COSTA, P. (2001), *Seismotectonics of Portugal and its adjacent area*, Tectonophysics *337*, 373–387.

BUFORN, E., UDÍAS, A., and COLOMBÁS, M.A. (1988a), *Seismicity, Source Mechanisms and Seismotectonics of the Azores-Gibraltar Plate Boundary*, Tectonophysics *152*, 89–118.

BUFORN, E., UDÍAS, A., and MÉZCUA, J. (1988b), *Seismicity and Focal Mechanisms in South Spain*, Bull. Seismol. Soc. Am. *78*, 2008–2224.

BUFORN, E., UDÍAS, A., and MADARIAGA, R. (1991a), *Intermediate and Deep Earthquakes in Spain*, Pure Appl Geophys. *136*, 375–393.

BUFORN, E., UDÍAS, A., MÉZCUA, J., and MADARIAGA, R. (1991b), *A deep Earthquake under South Spain, 8 March 1990*, Bull. Seismol. Soc. Am. *81*, 1403–1407.

BUFORN, E., SANZ DE GALDEANO, C., and UDÍAS, A. (1995), *Seismotectonics of the Ibero-Maghrebian Region*, Tectonophysics *248*, 247–261.

BUFORN, E., COCA, P., UDÍAS, A., and LASA, C. (1997), *Source Mechanism of Intermediate and Deep Earthquakes in Southern Spain*, J. Seism. *1*, 113–130.

BUFORN, E. and SANZ DE GALDEANO, C. (2001), *Focal Mechanism of Mula (Murcia, Spain) Earthquake of February 2, 1999*, J. Seism. *5*, 277–280.

CALVERT, A., GOMEZ, F., SEBER, D., BARAZANGI, M., JABOUR, N., IBENBRAHIM, A., and DEMNATI (1997), *An Integrate Geophysical Investigation of Recent Seismicity in the Al-Hoceima Region of North Morocco*, Bull. Seismol. Soc. Am. *87*, 637–651.

CHUNG, W. and KANAMORI, H. (1976), *Source Process and Tectonic Implications of the Spanish Deep-focus Earthquake of March 29, 1954*, Phys. Earth and Plan. Int. *13*, 85–96.

COCA, P. and BUFORN, E. (1994), *Mecanismos focales en el sur de España: periodo 1965–1985*, Estudios Geológicos, Madrid *50*, 1–2, 33–45.

COCA, P. (1999), *Métodos para la inversión del tensor momento sísmico. Terremotos del Sur de España*, Ph.D. Thesis, Universidad Complutense, Madrid, 300 pp.

DEMETS, C., GORDON, R., ARGUS, D., and STEIN, S. (1990), *Current Plate Motions*, Geophys. J. Int. *101*, 425–478.

DOCHERTY, C. and BANDA, E. (1995), *Evidence of Eastward Migration of the Alboran Sea Based on Regional Subsidence Analysis: A Case for Basin Formation by Delamination of Subcrustal Lithosphere?* Tectonics *14*, 430–433.

DZIEWONSKI, A. and WOODHOUSE, J. (1983), *An Experiment in Systematic Study of Global Seismicity: Centroid Moment Tensor Solutions*, J. Geophys. Res. *84*, 3247–3271.

EKSTRÖM, G. and DZIEWONSKI, A. (1988), *Evidence in on Bias in Estimations of Earthquakes Size*, Nature *332*, 319–323.

FROHLICH, C. and APPERSON, K.D. (1992), *Earthquake Focal Mechanisms, Moment Tensors and Consistency of Seismic Activity near Plate Boundaries*, Tectonics *11*, 279–296.

GRIMISON, N. and CHENG, W. (1986), *The Azores-Gibraltar Plate Boundary: Focal Mechanisms, Depths of Earthquakes and their Tectonic Implications*, J. Geophys. Res. *91*, 2029–2047.

HATZFELD, D. (1978), *Etude sismotectonique de la zone de collision Ibero-Maghrébine*, Ph.D. Thesis, Grenoble (France), 281 pp.

HAYWARD, N., WATTS, A.B., WESTBROOK, G.K. and COLLIER, J.S. (1999), *A Seismic Reflection and GLORIA Study of Compressional Deformation in the Gorringe Bank Region, Eastern North Atlantic*, Geophys. J. Int. *138*, 831–850.

HOUSEMAN, G. (1996), *From Mountains to Basin*, Nature *379*, 771–772.

JIMENEZ-MUNT, I., BIRD, P., and FERNANDEZ, M. (2001), *Thin-shell Modeling of Neotectonics in the Azores-Gibraltar Region*, Geophys. Res. Lett. *28*, 6, 1083–1086.

LAMMALI, K., BEZZEGHOUD, M., OUSSADOU, F., DIMITROV, D., and BENHALLOU, H. (1997), *Postseismic Deformation at El Asnam (Algeria) in the Seismotectonic Context of Northwestern Algeria*, Geophys. J. Int. *129*, 597–612.

LONERGAN, L. and WHITE, N. (1997), *Origin of the Betic-Rif Mountain Belt*, Tectonics *16*, 3, 504–522.

MCKENZIE, D. (1972), *Active Tectonics of the Mediterranean Region*, Geophys. J. R. Astron. Soc. *30*, 109–185.

MEDINA, F. and CHERKAOUI, T.E. (1992). *Mechanismes au foyer des seismes de Maroc et des régions voisines (1959–1986), Conséquences tectoniques*, Eclogae Geol. Helv. *85*, 433–457.

MÉZCUA, J. and MARTÍNEZ SOLARES, J.M. (1983), *Sismicidad del área Ibero-Mogrebí*, Instituto Geográfico Nacional Madrid.

MÉZCUA, J. and RUEDA, J. (1997), *Seismological Evidence for a Delamination Process in the Lithosphere under Alborán Sea*, Geophys. J. Int. *129*, F1–F8.

MICHARD, A., CHALOUAN, A., FEINBERG, H., GOFFÉ, B., and MONTIGNY, R. (2002), *How Does the Alpine Belt End between Spain and Morocco?* Bull. Soc. Geol. France *173*, 3–15.

MOKRANE, A., AIT MESSAOUD, A., SEBAI, A., MENIA, N., AYADI, A., BEZZEGHOUD, M. (1994), *Les séismes en Algérie de 1365 à 1992*, Bezzeghoud, M. and Benhallou, H. (eds). Publication du CRAAG, Alger-Bouzaréah, 277 pp.

MORALES, J., SERRANO, I., JABALOY, A., GALINDO-ZALDIVAR, J., ZHAO, D., TORCAL, F., VIDAL, F., and GONZALEZ-LODEIRO, F. (1999), *Active Continental Subduction beneath the Betic Cordillera and the Alboran Sea*, Geology *27*, 735–738.

MOREL, J. and MEGHRAOUI, M. (1996), *Gorringe-Alboran-Tell Tectonic Zone: A Transpressio System along the Africa-Eurasia Plate Boundary*, Geology *24*, 755–758.

MORLEY, C. (1993), *Discussion of the Origins of Hinterland Basins to the Rif-Betic Cordillera and Carpathians*, Tectonophysics *226*, 359–376.

MUNUERA, J.M. (1963), *Datos básicos para un estudio de sismicidad en la región de la Península Ibérica.* Mem Inst. Geog. Cat., Madrid, 32, 93 pp.

MUÑOZ D. and UDÍAS, A. (1988), *Evaluation of damage and source parameters of the Málaga earthquake of 9 October 1680.* In W.H.K. Lee, H. Meyer and K. Shimazaki, (eds), *Historical Seismograms and Earthquakes of the World* (Academic Press, San Diego 1988), pp. 208–221.

NEGREDO, A., BIRD, P., SANZ DE GALDEANO, C., and BUFORN, E. (2002), *Neotectonic Modeling of the Ibero-Maghrebian Region*, J. Geophys. Res., in press.

PLATT, J.P. and VISSERS, R. (1989), *Extensional Collapse of Thickened Continental Lithosphere: A Working Hypothesis for the Alboran Sea and Gibraltar Arc*, Geology *17*, 540–543.

RUEDA, J., MÉZCUA, J., and SANCHEZ RAMOS, M. (1996), *La serie sísmica de Adra (Almeria) de 1993–1994 y sus principales consecuencias sismotectónicas*, Avances en Geofísica y Geodesia, Instituto Geográfico Nacional, Madrid, 91–98.

SEBER, D., BARAZANGI, M., IBENBRAHIM, and DEMNATI, A. (1996), *Geophysical Evidence for Lithospheric Delamination beneath the Alboran Sea and Rif-Betic Mountains*, Nature *379*, 785–790.

SERRANO, I., MORALES, J., ZHAO, D., TORCAL, F., and VIDAL, F. (1998), *P-wave Tomographic Images in the Central Betics-Alboran Sea (South Spain) Using Local Earthquakes: Contribution for a Continental Collision*, Geophys. Res. Lett. *25*, 4031–4034.

TEJEDOR, J.M. and GARCÍA , O. (1993), *Funciones de transferencia de las estaciones de la red Sísmica Nacional.* Instituto Geográfico Nacional, Madrid, 82 pp.

TORNÉ, M., FERNANDEZ, M., COMAS, M.C., and SOTO, J.I. (2000), *Lithospheric Structure beneath the Alboran Basin: Results from 3D Gravity Modeling and Tectonic Revelance*, J. Geophys. Res. *105*, 3209–3228.

UDÍAS, A., LÓPEZ ARROYO, A., and MÉZCUA, J. (1976), *Seismotectonics of the Azores-Alboran Region*, Tectonophysics *31*, 259–289.

YELLES-CHAOUCHE, A.K., DJELLIT, H., BELDJOUDI, H., BEZZEGHOUD, M. and BUFORN, E, *The Ain Temouchent Earthquake of December 22th, 1999*, Pure Appl. Geophys., this issue.

ZECK, H. (1996), *Betic-Rif Orogeny: subduction of Mesozoic Tethys under E-ward Drifting Iberia, Slab Detachment Shortly before 22 Ma and Subsequent Uplift and Extensional Tectonics*, Tectonophysics *254*, 1–16.

(Received May 3, 2002, revised November 5, 2002, accepted December 12, 2002)

To access this journal online:
http://www.birkhauser.ch

Annex 1

Focal mechanisms of earthquakes studied in this paper obtained from polarities of P-waves. At top date of the shock, black circles correspond to compressions and white circles to dilatations. T and P indicate the tension and pressure axis.

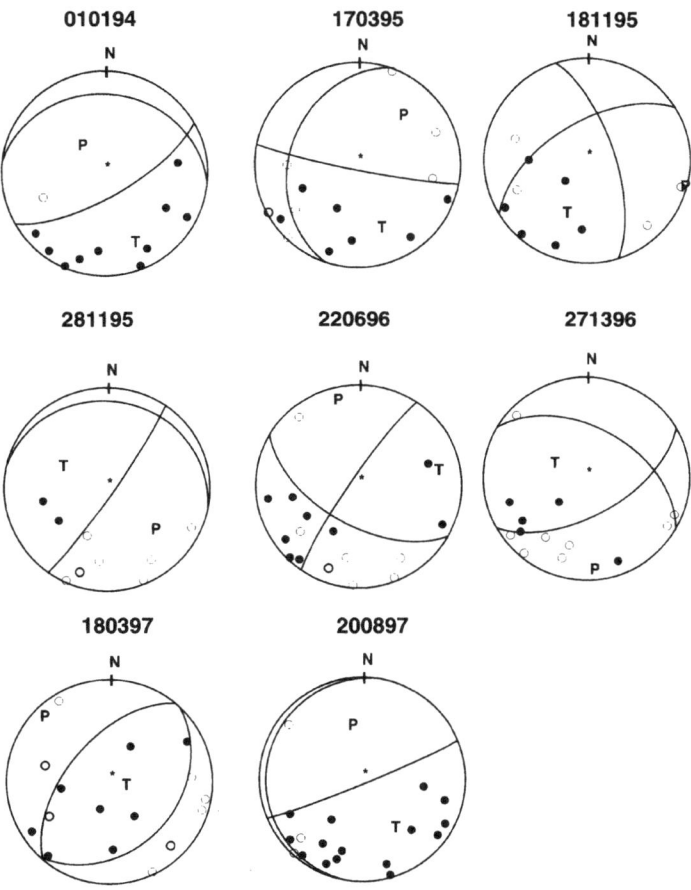

E. Buforn *et al.*   Pure appl. geophys.,

Annex 2

190551

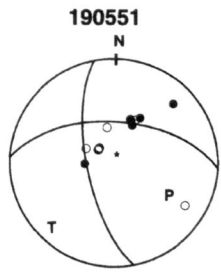

071190

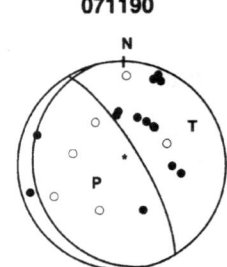

150691

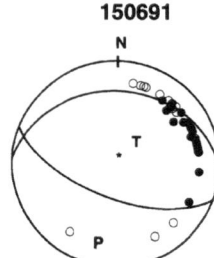

140891

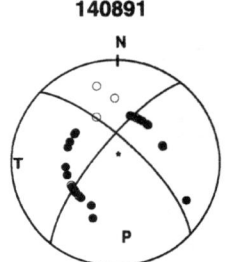

310793

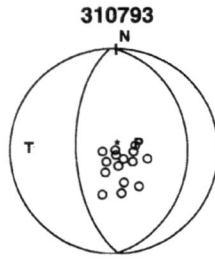

Pure appl. geophys. 161 (2004) 647–659
0033–4553/04/030647–13
DOI 10.1007/s00024-003-2467-0

❙ Pure and Applied Geophysics

Calibration of Local Magnitude M_L in the Azores Archipelago Based on Recent Digital Recordings

Eva Góngora[1], Fernando Carrilho[2], and Carlos S. Oliveira[3]

Abstract—The development of the digital seismic network in the Azores Archipelago during recent years made it possible to obtain the amplitudes (waveform) of recorded motion in a large set of stations. With this new data, maximum amplitudes of the Wood Anderson seismograph are computed, for each station/component, which, together with epicentral distances, allows for the estimation of local magnitude M_L.

We used data recorded in 8 digital permanent three-component stations, with inter-station's distances up to 300 km, in the period June 1998 – June 2000, corresponding to a set of 1315 events with magnitude (M_L or M_D) $2 < M < 5.8$ and epicenters located in the Azores region, to estimate the coefficients of the equation to compute M_L, as well as to determine the corrections to be applied to each station. The new set of parameters, formed by attenuation coefficients and station corrections, were introduced in the calculations of the M_L, leading to smaller dispersions in the analyzed dataset. We also conclude that the attenuation in the first 150 km is similar to the California values, although higher for longer distances.

Key words: Azores, amplitudes, attenuation, calibration of Local Magnitude.

1. Introduction

The Azores Archipelago is located in the Northern Atlantic Ocean, near the triple junction where the American, African and Euro-Asiatic plates converge. As a consequence, the area is characterized by a high seismicity along the arch where the islands are located.

With the recent development of the digital seismic network in the Azores Archipelago in the last few years it became possible to increase quite significantly the quality of the earthquake parameters and the perceptibility of the occurrences. Data are collected in a database from which the Seismological Preliminary Bulletin of Azores (*Boletim Sismológico Preliminar dos Açores*, 1998–2000) is published by SIVISA (Meteorological Institute/University of Azores) on a monthly basis. The activity since 1998 has been dominated by the aftershock sequence of the July 9, 1998

---

[1] Delegação Regional dos Açores do Instituto de Meteorologia, P. Delgada, Açores, Portugal. E-mail: eva@alf.uac.pt

[2] Instituto de Meteorologia, Lisboa, Portugal. E-mail: fernando.carrilho@meteo.pt

[3] Instituto Superior Técnico,DECivil/ICIST, Lisboa, Portugal. E-mail: csoliv@civil.ist.utl.pt

Faial/Pico earthquake, magnitude (M_D) 5.8, which consisted of numerous seismic events. Special attention is focused on the attenuation of seismic waves and to the calibration of M_L determinations.

The concept of local magnitude (M_L) was first introduced by RICHTER (1935) in California and, with the advent of modern digital seismology, several authors revisited the problem of the local magnitude estimation. BAKUN and JOYNER, (1984) and HUTTON and BOORE, (1987) looked to updated data in California, KIM (1998) analyzed data from the NE USA, SECANELL *et al.* (1996), GONZÁLEZ (1999) and GONZÁLEZ *et al.* (2000) data from Catalunya, Spain, and KANG *et al.* (2000) data from the Republic of Korea. More work is being developed presently, BAUMBACH *et al.* (2002), SPALLAROSSA *et al.* (2002), FERRETTI *et al.* (2002), demonstrating the great importance of this topic in modern seismology.

Since the new digital seismic network was installed in the Azores the amplitudes referred to the Wood-Anderson seismograph were determined for all stations and components. These amplitudes are collected in a mentioned database. To date, the local magnitude (M_L) for each event, which is also contained in the Bulletin, was estimated based on the algorithm (see section 4) with constants taken from the southern California observations.

2. The Digital Network and Earthquake Database

The installation of digital seismological stations from SIVISA in the Azores was initiated in June 1998, and 8 stations have been in place since October 1999 (Fig. 1). They are all three-component Lennartz MARS88 with 20-bit, some equipped with sensors with constant velocity response in the frequency band 1 Hz – 80 Hz, and others within 0.2 Hz – 40 Hz. Presently, the digital network comprises more stations. Sampling rate is 62.5 per second.

In this study we selected all events recorded in the period from June 1998 to June 2000. These include the events with $5.8 > M > 2.0$ within the geographic region of the Archipelago (with M the maximum from M_L and M_D; the value of magnitude M_L used for the selection was taken from the Bulletin — coefficients taken from California; M_D also reported on the Bulletin). The digital stations considered for the analysis, totaling 8, are presented in Table 1 and Figure 1.

Besides the epicentral location, depth, M_L (computed as referred before with the coefficients taken from California), errors in the determinations and arrival times for the identified phases, the database of SIVISA contains, station by station, information on the amplitudes in nanometers (nm) of each component correspondent to the maximum amplitude of the Wood-Anderson response. The determination of these amplitudes was obtained from the original record in velocity, adequately transformed through an algorithm of SEISAN® (HAVSKOV and OTTEMÖLLER, 1999) to reproduce the response of the Wood-Anderson

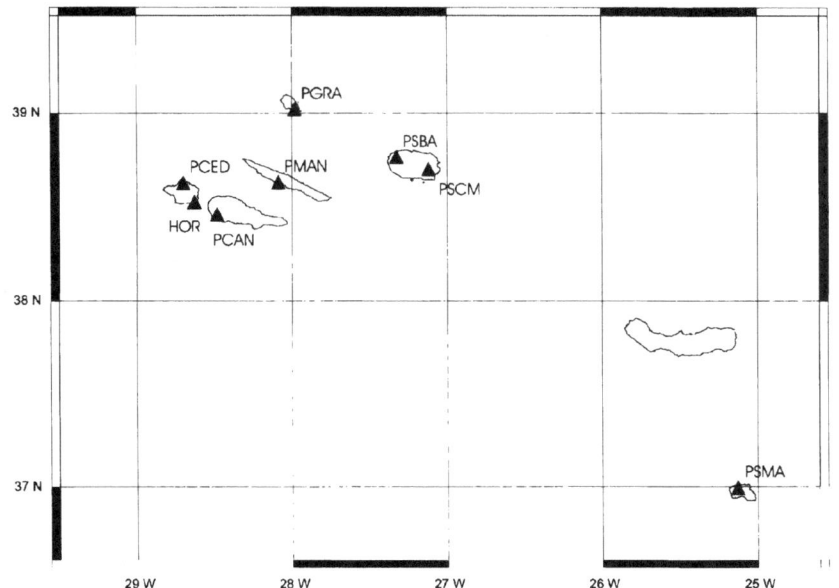

Figure 1
Localization of the digital stations used in this work.

seismometer (period $T_0 = 0.8$ sec, damping ratio $\xi = 80\%$ and a nominal amplification equal to 2800).

The magnitude M_L is then obtained from the average of the individual values of
horizontal components by application of eqs. (1 and 2) given in section 4.

In the computation of regression coefficients, we used two numerical algorithms.
One based on SOLVER (Excel ®) and the other from the direct inversion of the
equation system.

Table 1

Identification of stations

Code	Name	Island
HOR	Horta	Faial
PCED	Cedros	Faial
PCAN	Candelária	Pico
PMAN	Manadas	S. Jorge
PGRA	Caldeira	Graciosa
PSCM	Serra do Cume	Terceira
PSBA	Santa Bárbara	Terceira
PSMA	Faneca	Santa Maria

We also investigated the causes for the observed dispersion on the data, and for the existence of systematic deviations from the mean values in a few stations/components. Such deviations may be attributed to site-effects by reasons of topography or surface geology, or else due to the presence of their specificities connected to the source or to the path.

3. Characterization of the Studied Data

A total of 1315 events taken from the SIVISA database (BULLETINS, 1998–2000) were used in the computations, Figure 2, producing 3412 records for the Z component, 3643 for the N-S component and 3614 for the E-W. In many cases there are no simultaneous records of all three components.

The records correspond to events with epicentral distances from 0 to 800 km and magnitudes $2.0 < M_L < 5.6$ (Fig. 3). This figure shows two main groups, one with events with epicentral distances up to 30 km, corresponding to the Faial/Pico seismic

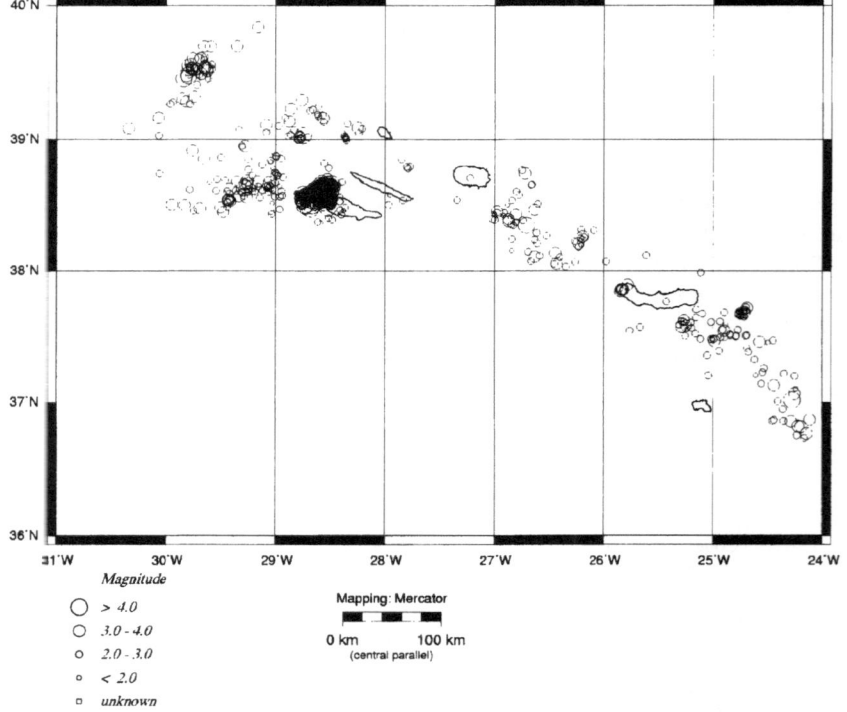

Figure 2
Map of epicenters of studied events (1 June 1998 to 30 June 2000, $M_L \geq 2.0$).

sequence in 1998, and the other for the events with epicentral distances reaching roundly 300 km.

The analysis of data in Figures 2 to 3 shows that we are in the presence of a significant set of events. They represent the seismic activity presently under way in the Azores Archipelago, with epicenters displayed along most of the seismic active arch and with magnitudes M_L up to 5.6. With the current seismic network, within a few years it will be possible to increase the size of the sample very significantly.

In the following, we emphasize a few other characteristics of the events under study, namely the ratio of peak values among components (NS)/(EW) and (NS)/(Z) (in velocity), presenting in Table 2 the values of the correlation coefficients for those ratios. We verify that the amplitudes of the horizontal components are approximately twice larger than the vertical, and that for the entire Archipelago, the EW component is in average 90% of NS, denoting differences in amplitudes between radial and transverse components. However, caution should be exercised because ratios present some dispersion.

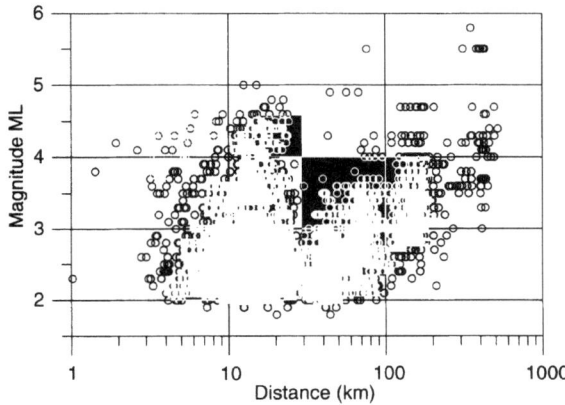

Figure 3
Magnitude (M_L) — Epicentral distances (km) under analysis.

Table 2

Amplitude ratios: correlation values ($y = Ax$)

	A	R^2
NS/Z	2.164	0.864
EW/Z	2.041	0.837
EW/NS	0.910	0.920

4. Theoretical Formulation for the Computation of the Local Magnitude (M_L)

The computation of local magnitude M_L is made according to equation (1), LAY and WALLACE (1995):

$$M_L = \log(A) - \log(A_0) \tag{1}$$

where A is half of the peak-to-peak amplitude in (mm) of the Wood-Anderson instrument response (using an amplification factor of 2800) and A_0 is a correcting factor taking into consideration the epicentral distance Δ in km ($\log \equiv \log_{10}$). By definition, the amplitude of 1 mm recorded in a station at an epicentral distance of $\Delta = 100$ km, corresponds to a magnitude $M_L = 3$. Another alternative way of defining local magnitude is considering the reference distance equal to 17 km, instead of 100 km, for which a magnitude 2.0 takes place (HUTTON and BOORE., 1987). However, the present configuration of the network, strongly limited by the morphology of the Archipelago and the location of epicenters, recommends the use of the standard 100 km as the reference distance for magnitude definition.

The correcting factor, $-\log(A_0)$, for the situation described initially, is given by an expression of the type

$$-\log(A_0) = a \times \log(\Delta/100) + b \times (\Delta - 100) + 3 + c_i \tag{2}$$

where a and b are empirical constants and c_i is a station correction applicable to each component of each station i function of its behavior relative to the average of all the other stations. The physical meaning of the a coefficient is related to the geometrical spreading, whereas b is related to the anelastic attenuation coefficient (BAKUN and JOYNER, 1984).

For California, HUTTON and BOORE (1987) obtained $a = 1.110$ and $b = 0.00189$. In other seismic locations, and neglecting the term b, constant a varies from 1.00 in southern California (BAKUN and JOYNER, 1984) to 1.41 in Portugal Mainland (CARRILHO, 1997), and 1.71 in Korea (KANG *et al.*, 2000).

Having used the values for California as initial estimates for a and b (here named CALIF), we proceed to the minimization of residuals between the corresponding value of magnitudes estimated using the California coefficients and those for other values of a and b. A 2-step approach was followed in the minimization procedure, solving first for a and b (and taking $c_i = 0$), and then solving for c_i. The minimization procedure, based on the algorithm SOLVER (Excel®), lead to the estimates of a and b presented in Table 3 (these determinations were made for all 1315 events).

In a second step, taking the values of Table 3 for a and b and repeating the procedure of eq. (2) for the data of each station, it was possible to estimate station corrections c_i, for the 8 studied stations, Table 4. Station corrections, which can be strongly connected to site effects, can reach absolute values quite significant, up to 0.4, not neglectable.

Table 3

a and b values from the minimization of residuals

	Z	N-S	E-W
a	0.300	1.000	1.000
b	0.00393	0.00322	0.00286

An alternative procedure (here named ALT) adapted from GONZÁLEZ *et al.* (2000) was also essayed. It consists in developing eq. (2) into a system of as many equations as existing data for all components, having as unknowns M_L, a and b. In this procedure, c_1, c_2, ..., c_8 are a set of imposed station corrections:

$$M_{L_i} = \sum_{j=1}^{m} \left(\log(A_{i,j}) + a \times \log\left(\frac{\Delta_{i,j}}{100}\right) + b \times (\Delta_{i,j} - 100) + 3.0 + c_j \right) \quad (3)$$

with
- M_{L_i} local magnitude of event i;
- A_{ij} amplitude measured in station j for event i;
- Δ_{ij} hypocentral distance for station j and event i;
- c_j station correction fixed as –0.3 for HOR and 0.0 for all other stations;
- m stations and n events;
- in case of stations with more than one reading, the contribution to the value of magnitude is weighted as a function of the number of components with amplitude values.

Objective:
- To determine a and b that minimizes

$$\sigma^2 = \sum_{i=1}^{n} \sum_{j=1}^{m} \left(\log(A_{i,j}) + a \times \log\left(\frac{\Delta_{i,j}}{100}\right) + b \times (\Delta_{i,j} - 100) + c_j + 3.0 - M_{L_i} \right)^2 \quad (4)$$

Table 4

c_i estimates for station corrections

	Z	N-S	E-W
HOR Ci	−0.274	−0.090	−0.120
PCED Ci	0.060	0.172	0.183
PCAN Ci	0.035	0.016	0.034
PGRA Ci	0.410	0.271	0.130
PMAN Ci	0.125	−0.273	−0.213
PSBA Ci	0.191	−0.147	−0.240
PSCM Ci	0.121	−0.034	−0.126
PSMA Ci	0.248	−0.005	0.013

A gric search technique was implemented for the minimization procedure.

Criteria for event selection:

- Minimum of 4 stations (among 8) with amplitude readings;
- Amplitude of vertical component A_Z converted into horizontal component A_H through $A_H = 2.1 \times A_Z$, where the value 2.1 was obtained from the average values in table 2;
- Minimum ratio for epicentral distances: $d\text{max}/(d\text{max}-d\text{min}) < 3$ ($d\text{max}$ – maximum epicentral distance and $d\text{min}$ - minimum epicentral distance, for each event);
- Maximum separation between M_L (California) and M_L (ALT) less than 0.50 for the total number of events used in the fitting.

In the ALT version only 397 events were considered.

Results:

$$a = 0.89$$
$$b = 0.00256$$

In Figure 4 we present the results correspondent to the differences between the corrections obtained in this study and the ones proposed for southern California by HUTTON and BOORE, (1987) (through the algorithm SOLVER and through the alternative procedure, ALT). These results are similar not only to the values obtained with SOLVER but also with the California proposals. While for short distances, the corrections for the NS and EW components are similar to those in California, for longer distances the attenuation correction does not match with California, reaching a maximum difference of 0.45 for 500 km (NS component), while 0.2 for 300 km (see Fig. 4). The correction for the Z component was obtained without consideration of the empirical relation between vertical and horizontal amplitudes; this can explain the larger values for closer distances, where experimentally we observe a larger ratio A_H/A_Z (Table 2). The alternative method considers all components and contains all filtering procedures above described, and the results are those which better agree with the values determined by HUTTON and BOORE (1987).

One should note that all values obtained give attenuations larger than in California, at least for distances exceeding 300 km.

5. Testing New Values

In order to check the effect of the newly determined coefficients and station corrections on the local magnitude estimations, all magnitudes and corresponding standard deviations (σ) were computed. It was decided to use only horizontal components due to the fact that, presently, the routine for magnitude determinations in SIVISA do not account for the vertical component.

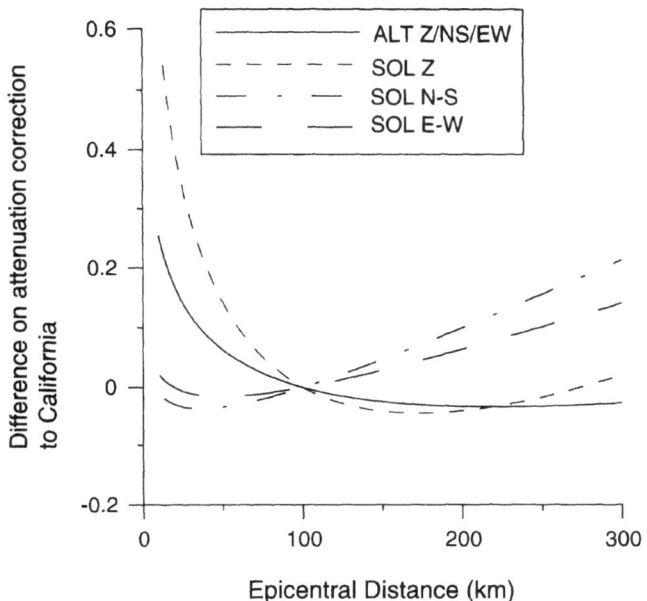

Figure 4
Comparison of attenuation corrections computed herein with the ones given for Southern California (HUTTON and BOORE, 1987).

For each event, the magnitude and the standard deviation were computed and the global standard deviation was considered as the mean value taken from the single event determinations. All comparisons were based on the global standard deviations (σ_{global}). The standard deviations were computed considering two approaches: in the first one, separate components were used in the computations; in the second one, only the mean magnitude, computed from the horizontal components, was used.

Several case studies were analyzed: CALIF — designates the old coefficients (HUTTON and BOORE, 1987) without station corrections; SOL — represents the case where coefficients listed in Table 3 are used without station corrections; SOL + C — stands for the case in which coefficients listed in Table 3 are used together with station corrections presented in Table 4; ALT + C — represents the case in which the set of coefficients and station corrections were obtained by the process described in the second part of section 4.

In Table 4 we present the global standard deviations obtained for the cases described above.

From Table 5, it can be seen that, as expected, all new sets of coefficients and station corrections lead to smaller global standard deviations and from these, the SOL + C case presents the best results.

Table 5

Mean values for dispersions for different hypothesis

Case	Separate 2-horizontal components	Mean 2-horizontal components
CALIF	0.251	0.241
SOL	0.249	0.239
ALT + C	0.226	0.215
SOL + C	0.208	0.195

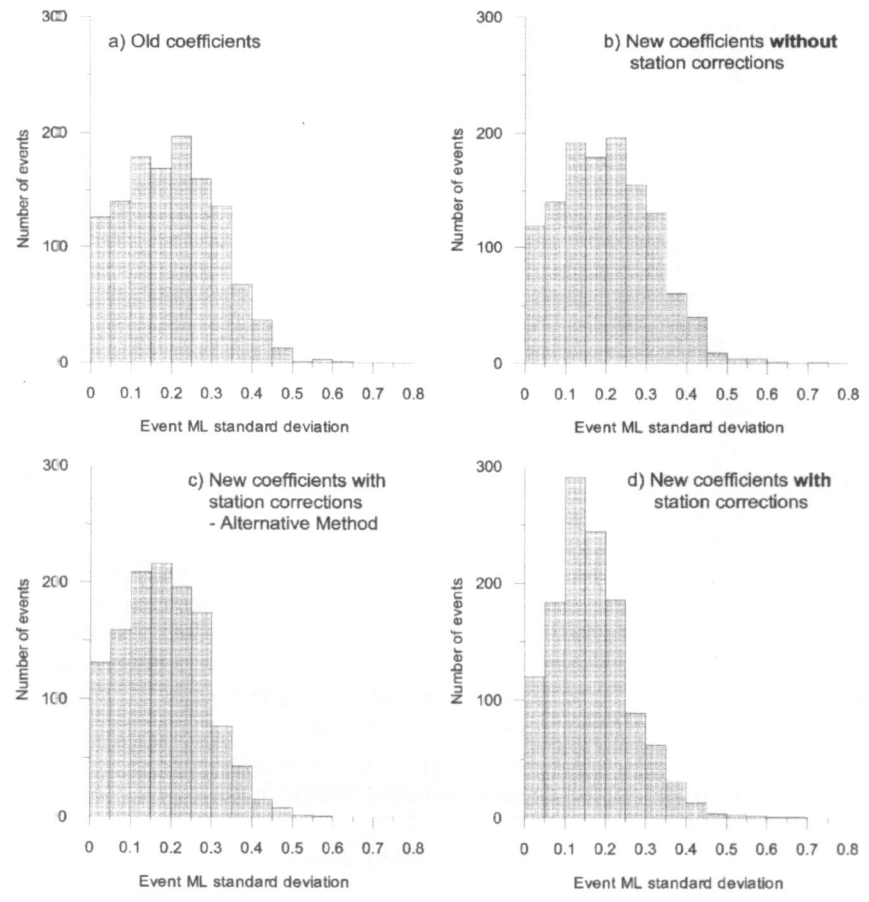

Figure 5
Distribution of standard deviations σ for three different sets of parameters: a) CALIF parameters; b) SOL; c) ALT + C parameters; d) SOL + C parameters.

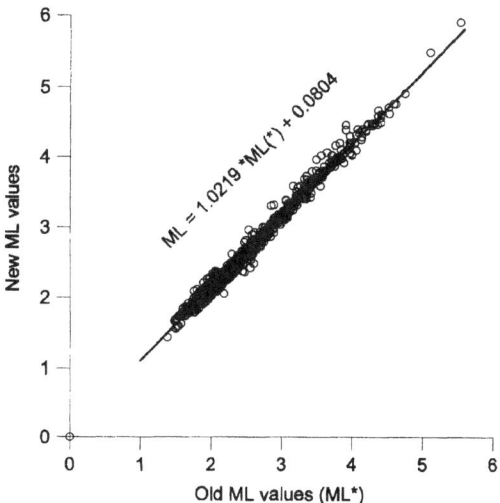

Figure 6
Correspondence between new magnitude values (M_L), computed with SOL + C parameters, and the previous ones (M_L^*), computed with CALIF.

For better understanding the impact of the new coefficients on magnitude estimations, histograms in Figures 5a-d present the distribution of the event standard deviations for three cases (CALIF, ALT + C and SOL + C). Clearly, the SOL + C case produces the best statistical distribution. In the near future, with the increase of recordings that the present stations will produce, it will be possible to more precisely tune station corrections.

The difference between solution SOL + C and solution CALIF can also be seen on Figure 6, where the relation between new M_L magnitudes and the older ones (M_L^*) are fitted by a simple linear equation:

$$M_L = 1.0219 \times M_L^* + 0.0804. \tag{6}$$

This means that, in general, new M_L values are slightly higher than the previous ones. One consequence of the new calibrated values is the new magnitude estimated for the largest earthquake recorded in the network, the July 9, 1998 Faial/Pico event. It was corrected from 5.6 to 5.9 which is in better agreement with the 6.1 M_W and 6.0 M_S reported by the Harvard CMT Catalog.

6. Final Considerations

This study shows that the coefficients used to estimate the magnitude M_L, based on the computed Wood-Anderson amplitudes observed in various digital stations in the Azores Archipelago within a 150-km radius, are similar to the ones recommended for California, even though large systematic variations do exist from station to station. These variations were quantified in statistical terms and they have been introduced in the seismological evaluation routine at the SIVISA.

For epicentral distances exceeding 150 km, the differences in the correction values for the attenuation, in relation to California values, became larger. For the analyzed data, and depending on the method used, for the Azores the values of M_L should be increased $(+0.1)$ to $(+0.5)$ when the distance of the station is of the order of 500 km. This fact seems to indicate that for longer distances, the attenuation in the Azores is slightly larger than in southern California.

In the near future with the same digital network it will be possible to increase the size of the sample very significantly, allowing a more precise tuning of station corrections. Also, a magnitude calibration anchored to the reference distance equal to 17 km should be analyzed for better definition of magnitude of events recorded at short distances. A third improvement would consist in allowing constrains such as $M_L = n_b$, obtained from teleseismic measurements (in other outside networks) when available for the largest events.

We also intend to use other correlations for the characterization of the seismic source such as moment magnitude M_w, seismic moment M_O and to study the influence of the azimuth.

Acknowledgements

This work was partially supported by PPERCAS/Praxis "*Projecto Praxis para o Estudo do Risco/Casualidade Sísmica do Grupo Central do Arquipélago dos Açores*", Ciência e Tecnologia 3/3.1/CEG/2531/95", and by DISPLAZOR/Praxis "*Controlo de deslocamentos tectónicos, vulcânicos e escorregamentos no Faial, Pico e S. Jorge (Açores) usando GPS*", CTA/32444/99-00.

REFERENCES

BAKUN, W. H. and JOYNER, W. B. (1984), *The M_L Scale in Central California*, Bull. Seismol. Soc. Am. 74, 5, 1827–1843.
BAUMBACH, M., BINDI, D., GROSSER, H., MILKEREIT, C., PAROLAI, S., KARAKISA, S., and ZÜNBÜL, S. (2002), *M_L Scale in Northwestern Turkey from 1999 Izmit Aftershocks*, Abstract, XXVIII General Assembly of the European Seismological Commission (ESC), Genova, pp. 99.
Boletim Sismológico Preliminar dos Açores (1998–2000), SIVISA, Ponta Delgada.
CARRILHO, F. (1997), *personal communication*.
FERRETTI, G., SPALLAROSSA, D., SOLARINO, S., and EVA, C. (2002), *A Local Magnitude Scale for the RegionalSeismic Network of Lunigiana — Garfagnana (Tuscani, Italy)*, Abstract, XXVIII General Assembly of the European Seismological Commission (ESC), Genova, pp. 98.

GONZÁLEZ, M. (1999), *Definición de una Ley de Magnitud Adaptada a Catalunya: Cálculo de las Magnitudes Wood-Anderson (periodo 92–97) y Comparación com magnitudes LDG*, Institut Cartogràfic de Catalunya, Informe núm: GS-118/99. (in Spanish).

GONZÁLEZ, M., SECANELL, R., SUSAGNA, T., and GOULA, X. (2000), *Inversión de Amplitudes de Registros Sísmicos para la Definición de M_L"*, 2ª Asamblea Hispano-Portuguesa de Geodesia y Geofísica, Lagos, Febrero 2000 (in Spanish).

HARVARD CMT CATALOG, http://www.seismology.harvard.edu/projects/CMT.

HAVSKOV, J., and OTTEMÖLLER, L. (1999), *SEISAN: The Earthquake Analysis Software*,Version 7.0, Norway.

HUTTON, L. K., and BOORE, D. M. (1987), *The M_L Scale in Southern California*, Bull. Seismol. Soc. Am. *77*, 6, 2074–2094.

KANG, I. B., JUN, M-S., and SHIN, J. S. (2000), *Research on Local Magnitude (M_L) Scale in and near the Korean Peninsula*, Annali di Geofisica *43*, 5, 1011–1020.

KIM, W. Y. (1998), *The M_L Scale in Eastern North America*, Bull. Seismol. Soc. Am. *88*, 935–951.

LAY, T., and WALLACE, T. C. (1995), *Modern Global Seismology* (Academic Press, London).

RICHTER, C. F. (1935), *An Instrumental Earthquake Magnitude Scale*, Bull. Seismol. Soc. Am. *25*, 1–32.

SECANELL, R., SUSAGNA, T., GOULA, X., and ROCA, A. (1996), *Contribution to a Definition of the M_L Scale in Catalonia*, XXIV General Assembly of the European Seismological Commission (ESC), Reykjavik, 479–484.

SPALLAROSSA, D., BINDI, D., CATTANEO, M., AUGLIERA, P., and EVA, C. (2002), *M_L Scale in Northwestern Italy from 1982–2001, RSNI Seismic Network Database*. Abstract, XXVIII General Assembly of the European Seismological Commission (ESC), Genova, pp. 98.

(Received April 2, 2002, revised October 9, 2002, accepted October 17, 2002)

To access this journal online:
http://www.birkhauser.ch

Pure appl. geophys. 161 (2004) 661–681
0033–4553/04/030661–21
DOI 10.1007/s00024-003-2468-z

❙ Pure and Applied Geophysics

Geodetic Measurements of Crustal Deformation in the Western Mediterranean and Europe

J.-M. Nocquet[1,*] and E. Calais[2]

Abstract — Geodetic measurements of crustal deformation over large areas deforming at slow rates (< 5 mm/yr over more than 1000 km), such as the Western Mediterranean and Western Europe, are still a challenge because (1) these rates are close to the current resolution of the geodetic techniques, (2) inaccuracies in the reference frame implementation may be on the same order as the tectonic velocities. We present a new velocity field for Western Europe and the Western Mediterranean derived from a rigorous combination of (1) a selection of sites from the ITRF2000 solution, (2) a subset of sites from the European Permanent GPS Network solution, (3) a solution of the French national geodetic permanent GPS network (RGP), and (4) a solution of a permanent GPS network in the western Alps (REGAL). The resulting velocity field describes horizontal crustal motion at 64 sites in Western Europe with an accuracy on the order of 1 mm/yr or better. Its analysis shows that Central Europe behaves rigidly at a 0.4 mm/yr level and can therefore be used to define a stable Europe reference frame. In that reference frame, we find that most of Europe, including areas west of the Rhine graben, the Iberian peninsula, the Ligurian basin and the Corsica-Sardinian block behaves rigidly at a 0.5 mm/yr level. In a second step, we map recently published geodetic results in the reference frame previously defined. Geodetic data confirm a counterclockwise rotation of the Adriatic microplate with respect to stable Europe, that appears to control the strain pattern along its boundaries. Active deformation in the Alps, Apennines, and Dinarides is probably driven by the independent motion of the Adriatic plate rather than by the Africa-Eurasia convergence. The analysis of a global GPS solution and recently published new estimates for the African plate kinematics indicate that the Africa-Eurasia plate motion may be significantly different from the NUVEL1A values. In particular, geodetic solutions show that the convergence rate between Africa and stable Europe may be 30–60% slower than the NUVEL1A prediction and rotated 10–30° counterclockwise in the Mediterranean.

Key words: Plate motions, plate boundary zone, crustal deformation, Mediterranean, geodesy, GPS.

Introduction

The boundary between the African and Eurasian plates in the Mediterranean area consists of a broad zone of deformation, extending from north Africa to central Europe (Fig. 1). Active deformation in the Africa-Eurasia plate boundary zone is generally interpreted as the result of the convergence between the African and

---

[1] Institut Géographique National, Laboratoire de Recherche en Géodésie, Marne-la-Vallée, France.
[2] Purdue University, Department of Earth and Atmospheric Sciences, West Lafayette, IN 47907-1397, USA.
* Now at: Department of Earth Sciences, University of Oxford, UK.

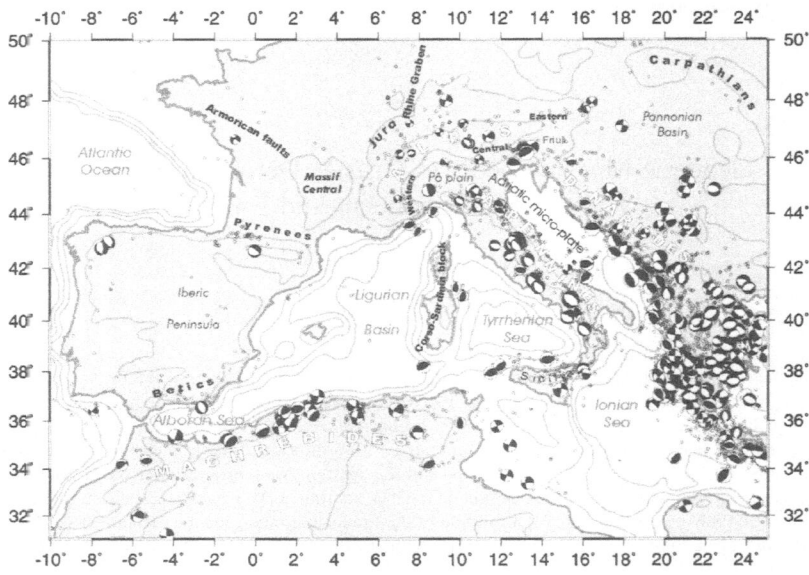

Figure 1

Seismicity map of Europe and the Western and Central Mediterranean. Epicenter locations for the 1973–2002 period (M$_s$ > 3.0, NEIC catalog). Earthquake focal mechanism from Harvard (http://www.seismology.harvard.edu/CMTsearch.html) and Swiss tensor Moment Project (http://seismo.ethz.ch/info/mt.html) for the 1997–2002 period, M > 5.

Eurasian plates, ranging from ∼10 mm/yr at the longitude of Turkey to ∼4 mm/yr in the Gibraltar Strait according to the NUVEL1A global plate kinematic model (DeMets *et al.*, 1990, 1994). In addition, geodetic measurements in Greece and Turkey reveal the existence of a separate Anatolian plate, moving west to southwestward at 20–30 mm/yr with respect to Eurasia (e.g., McClusky *et al.*, 2000). In contrast with the eastern part of the Africa-Eurasia plate boundary zone in Greece and Turkey, with large velocities and strain rates, little is known yet regarding the kinematics of crustal deformation in the Western Mediterranean.

The Western Mediterranean area is part of the plate boundary between Africa and Eurasia (Fig. 1). It is surrounded by Alpine mountain ranges (Betics, Atlas and Maghrebides, Apennines, Dinarides, Alps), usually thought to accommodate the Africa-Eurasia plate convergence. However, the Western Mediterranean-Alps domain also includes significant strike-slip (e.g., Western Swiss Alps, Eva and Solarino, 1998) and extensional (e.g., Apennines, D'Agostino *et al.*, 2001) active tectonic features. In addition, several aseismic domains are embedded in the plate boundary zone, interpreted either as rigid blocks or microplates (Corsica-Sardinian block, Adriatic and Iberian microplates), or as undeformed sedimentary basins (Ligurian and Pannonian basins). The relatively low plate motion and strain rates in the Western Mediterranean and the long recurrence time for large earthquakes make seismological and geomorphological indicators of active deformation scarce and difficult to interpret. As a consequence, strain distribution across the Africa-Eurasia plate boundary in the Western Mediterranean and strain accumulation on the major seismogenic structures are still largely unknown.

In addition to active deformation in the Africa-Eurasia plate boundary zone, instrumentally recorded seismicity shows a moderate but non-negligible activity in some intraplate areas in Europe such as the Rhine graben and the Armorican faults (Fig. 1). Both structures are located hundreds to thousands of kilometers away from the major plate boundary and penetrate deeply inside the Eurasian plate interior, suggesting that some of the Africa-Eurasia convergence could be transferred as far north as Central Europe. Measuring possible intraplate deformation of the Eurasian plate in Western Europe is therefore important to understand how stress is transferred across plate boundaries. Assessing the level of rigidity of the Eurasian plate in Europe is also important to define an unbiased plate-fixed reference frame for mapping geodetic velocities.

In this study, we present a newly combined velocity field for Western Europe and the Western Mediterranean and a critical analysis of recent geodetic results published for this area.

A Combined Velocity Field for Western Europe and the Western Mediterranean

In the past decade, geodetic measurements have been widely used to monitor crustal motions, from tectonic plates to local surveys of active faults, with precision levels on the order of 2–3 mm/yr (horizontally) routinely achieved. In the last few years, the increasing accuracy and density of space geodetic measurements have also permitted the testing of plate rigidity at a 2 mm/yr level (ARGUS and GORDON, 1996; DIXON et al., 1996; DEMETS and DIXON, 1999; KOGAN et al., 2000). However, 1–2 mm/yr of motion within several hundreds of kilometers can still lead to significant deformation over recent geological times. In particular, it may result in sufficient elastic stress accumulation to cause moderate to large earthquakes on faults with long recurrence intervals (1000 years and more, e.g., NEWMAN et al., 2001). In addition, 1–2 mm/yr of internal deformation within a block or plate chosen as a "stable" reference frame for the purpose of a geophysical interpretation may introduce a significant bias in the velocity field, in particular in areas deforming at very slow rates (DIXON et al., 1996). However, the determination of a dense and consistent velocity field at a continental scale, accurate at a sub-millimeter per year level, still remains a challenge.

This study is based on a newly combined velocity field for Western Europe and the Western Mediterranean. Indeed, combining the results from several networks and/or analysis centers provides a number of advantages over the analysis of each solution independently. First, it minimizes possible systematic errors associated with each processing strategy taken individually. Second, sites shared by several solutions provide a way to tie these solutions into a single and consistent velocity field and permit comparison of individual solutions for the detection of outliers. Third, reference frame constraints applied in individual geodetic solutions can significantly affect velocities (and positions), making direct comparisons between such solutions inadequate (SILLARD and BOUCHER, 2001).

We use a combination methodology that handles reference frame constraints simultaneously for all individual solutions in a rigorous way (e.g., BROCKMANN, 1997; DAVIES and BLEWITT, 2000; ALTAMIMI *et al.*, 2002). Because we use 14-parameter transformations and minimally constrained solutions in the combination and no additional constraints, relative positions and velocities of individual solutions are not affected by the reference frame definition. Finally, we apply a weighting scheme that rescales the variance-covariance matrices of each individual solution and provides realistic formal errors. The combination methodology is similar to the one used for the determination of the ITRF. It is fully described in ALTAMIMI *et al.* (2002) and NOCQUET and CALAIS (2003). We include in the combination (1) a selection of 36 sites from the ITRF2000 solution, (2) a solution from a subset of sites of the European Permanent GPS Network (EUREF-EPN), (3) a solution of the French national geodetic permanent GPS network (RGP), and (4) a solution of a permanent GPS network in the western Alps (REGAL; CALAIS *et al.*, 2000). The input data to the combination consist of individual solutions with minimal constraints applied. The combination model consists in estimating simultaneously, a position at a reference epoch and a velocity for each site, and a 14-parameter transformation between the individual and the combined solution (see ALTAMIMI *et al.*, 2002 for details). The reference frame definition in the combination is implemented by imposing the 14–parameter transformation between ITRF2000 and the combined solution to be zero (no translation, scale factor, or rotation and no rate of change of these parameters). The resulting velocity field is therefore expressed in the ITRF2000 reference frame. From this preliminary combination, an *a posteriori* variance factor σ_s^2 for each individual solution s is estimated in the inversion, which is then applied to the variance-covariance matrix of the corresponding individual solution iteratively until both individual σ_s^2 and the global *a posteriori* variance factor equals 1. Normal residuals in the combination are used for outlier detection. For most of the sites common to several individual solutions we find an agreement in horizontal velocities on the order of 0.5 mm/yr, with formal errors in horizontal velocities less than 1 mm/yr. The best determined sites have a formal error of about 0.2 mm/yr on horizontal velocities. Once the combined velocity field is obtained, we analyze it for plate rigidity following a procedure fully described in NOCQUET *et al.* (2001). We first use an algorithm that searches, over all possible site combinations, the subset of four sites whose velocities best fit a rigid rotation. We use χ^2 tests and minimal variance criteria to rank the site subsets according to their fit to a rigid rotation. We then progressively augment this initial subset of sites by adding one site at a time and testing the consistency of the new site subset with a rigid rotation using χ^2 and F-ratio tests. Following this procedure, we find a 29-site subset that defines a rigid domain extending from Central Europe to the westernmost part of Europe, including Spain and Sardinia (Fig. 2). This 29-site subset defines a stable Europe-ITRF2000 rotation pole located at 56.0°N/–101.5°W with an angular rate of $0.25 \pm 0.003°/\text{Ma}$. The weighted rms of the residuals at the sites used to define stable Europe is 0.4 mm/yr. It makes use of all the available geodetic techniques and of the full covariance matrix of the ITRF2000 SINEX file. This approach is independent from the NNR-NUVEL-1A plate motion model and benefits from the consistency of our velocity field combination over Europe. Although the 29-sites rigid subset extends to the east as far as Moscow, we cannot assert that the rigid motion defined by this subset is representative of the whole Eurasian plate motion. We therefore use the expression "stable Europe" rather than "Eurasia" to refer to the region encompassing our 29 sites hereafter in the text. We find no significant residual motion with respect to central Europe at the 0.5 mm/yr level at the continuous GPS

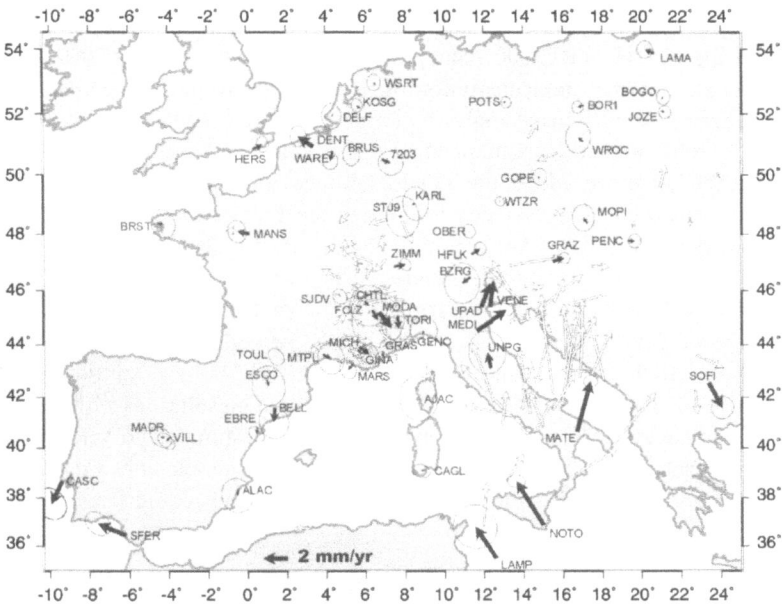

Figure 2

Combined velocity field for Western Europe and the Western Mediterranean, with respect to stable Eurasia (definition in the text). Black arrows show velocities at continuous sites. Error ellipses are 95% confidence and account for the variance of the data and the variance of the Eurasia Euler parameters. White arrows show velocities at campaign sites.

sites located in France, outside of the seismically active areas of the Alps and Jura (BRST, SJDV, TOUL, MTPL). This indicates that the stable part of France (outside the Alps and Jura) is rigidly attached to Central Europe and places an upper bound of 0.5 mm/yr (1 sigma) on possible horizontal motion across the Rhine graben.

In addition to solutions from continuous GPS networks, we also use results from GPS campaign measurements because they provide denser spatial sampling of the actively deforming areas. We used a selection of sites from (1) ANZIDEI et al. (2001; 1991–1998 surveys in Italy and North Africa), (2) ALTINER (2001; 1994–1996 surveys in the Dinarides), (3) VIGNY et al. (2002; 1993–1998 surveys in the Western Alps), and (4) GRENERCZY et al. (2000; 1994–1997 surveys in Central Europe and the Friuli area). We had access to VIGNY et al.'s (2002) full solutions (estimated parameters and full covariance) and were therefore able to combine them rigorously with the continuous GPS solutions, following the methodology described above. Since we did not have access to the full solution (including reference frame constraints information and full covariance matrix) for the other campaign results, we chose to account for the fact that they use different processing strategies and/or definitions of the stable Europe-fixed frame, by estimating a rotation between each solution and the ITRF2000, using sites common with the ITRF and verifying the obtained residuals. We find no significant rotation and residuals between GRENERCZY et al.'s (2000) results and ITRF2000. We find a significant rotation between ANZIDEI et al.'s (2001) results and the ITRF2000 and a poor agreement with the ITRF2000 velocities, suggesting an incorrect implementation of the reference frame in the GPS analysis. This is confirmed by a new analysis of this data set (SERPELLONI et al., 2001). ALTINER's (2001) velocities are mapped with respect to the IGS station GRAZ (Graz, Austria). We therefore added the ITRF2000 velocity for GRAZ and imposed a continuity constraint of the velocity field along the Trimiti line where Altiner's data merge Anzidei et al.'s data. The resulting velocity field is shown in Figure 1.

We emphasize the fact that, besides Vigny et al.'s solution, the campaign velocities used here are not derived from a homogeneous processing or a rigorous geodetic combination. In addition, the simplistic procedure used here to join the campaign solutions does not allow us to produce reliable uncertainties. However, when rigorously combining VIGNY et al.'s (2002) campaign solutions with continuous data, we found that the campaign-only results underestimate the velocity formal errors by a factor of 2 to 5. Consequently, the campaign velocities presented here should be interpreted with caution and only general trends could be considered as reliable.

Africa-Eurasia Convergence in the Western Mediterranean

The relative motion between the African and Eurasia plates is the kinematic boundary condition of the Alpine-Mediterranean active deformation system. It is therefore a key parameter for understanding the dynamics of active deformation in the Africa-Eurasia plate boundary zone. Because of the lack of geodetic sites on the African plate, its current motion is usually taken from plate models derived from oceanic magnetic anomalies and transform azimuths directions along the mid-Atlantic ridge, such as ARGUS *et al.*'s model (1989) or, more often, the NUVEL1A global plate model (DeMETS *et al.*, 1990, 1994). Only recently, new permanent GPS stations installed on the African plate have allowed for the first direct estimates of its rotation parameters (Fig. 3).

We computed rotation parameters of Africa with respect of stable Europe using the IGS02P09 combined IGS solution, updated for GPS-week 1155 (March 2nd, 2002). This solution is a combination of weekly global solutions provided by the 7 IGS data analysis centers. It contains positions and velocities in ITRF2000 for 11

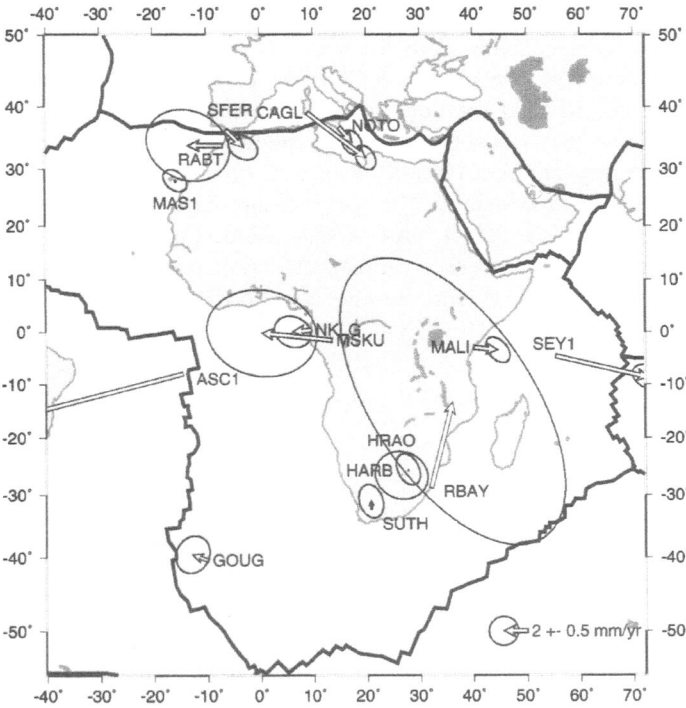

Figure 3

Residual velocities in Africa and some surrounding sites, after removing the rigid rotation defined by GOUG, HARB, NKLG, MAS1, HRAO, SUTH (M2 model, Table 1). This solution based on the IGS combined solution IGS02P09. The 95% confidence ellipses account for the data variance and the variance of the African Euler parameters. A rigid rotation estimated for (GOUG, HARB, NKLG, MAS1) (M1 model, Table 1) shows no significant difference with the one presented here.

continuous GPS sites in Africa and the surroundings (Fig. 3), with the full associated covariance matrix. We first select the sites that best represent stable Africa, following the procedure described above for Europe (NOCQUET et al., 2001). We find the "most rigid" site subset to be GOUG, HARB, NKLG, and MASP, with residual velocities less than 0.7 mm/yr (Model IGS02P09-M1 in Table 1). Residual velocities are 2 to 8 mm/yr at the sites located east of the East African rift on the Somalian plate and can therefore not be used to estimate a rigid rotation for the African plate. SUTH, located southwest of HARB, has a residual velocity of 1.8 mm/yr. Adding this site to the estimation of the African plate rotation parameters has a negligible impact on predicted velocities in the Mediterranean area (< 0.5 mm/yr, Model IGS02P09-M2 in Table 1, Fig. 3). Fisher and χ^2 tests indicate that both NOTO (Sicily, 1.4 ± 0.5 mm/yr residual velocity with respect to Africa) and SFER (Southern Spain, 2.0 ± 0.7 mm/yr residual velocity with respect to Africa) can be distinguished from Africa at the 95% confidence level but not at the 99% level. CAGL (Cagliari, Sardinia) shows a large residual (6.0 ± 1.5 mm/yr) clearly indicating that Sardinia does not belong to the African plate.

We then subtract the stable Europe-ITRF2000 rotation parameters defined above from the Africa-ITRF2000 rotation parameters from our model (IGS02P09-M2, Table 1) to obtain the Africa-stable Europe rotation parameters. Figure 4 shows predicted velocities along the Africa-Eurasia plate boundary in the Mediterranean, according to these Africa-stable Europe rotation parameters and to recently published Africa-Eurasia models. All models besides ALBARELLO et al. (1995) indicate N0 to N45W convergence between the African and Eurasian plate in the Mediterranean, from convergence in the eastern part of the plate boundary,

Table 1

Recently published Africa-Eurasia rotation parameters

Source	Lat. (degrees)	Long.	Ang. vel. (deg./Myr)	Data used
ARGUS et al. (1989)	18.8	−20.3	0.10 ± 0.02	Magnetic anomalies and transform azimuths directions along the mid-Atlantic ridge
DEMETS et al. (1994)	21.0	−20.6	0.13 ± 0.02	Same as above plus plate circuit closure
CRETAUX et al. (1998)	26.1	20.2	0.139 ± 0.03	4 sites, DORIS solution
ALBARELLO et al. (1995)	41.6	−11.8	0.117	Same as ARGUS et al., (1989)
SELLA et al. (2002)	18.2	−20.0	0.060 ± 0.005	5 sites, GPS solution
KREEMER and HOLT (2001)	2.6	−21.0	0.036 ± 0.005	7 sites, GPS solution
This study (IGS02P09 M1)	2.1	−20.0	0.07 ± 0.02	4 sites (GOUG, MAS1, HARB, NKLG), GPS IGS global solution
This study (IGS02P09 M2)	7.7	−18.3	0.07 ± 0.02	6 sites (same as above + HRAO, SUTH), GPS IGS global solution

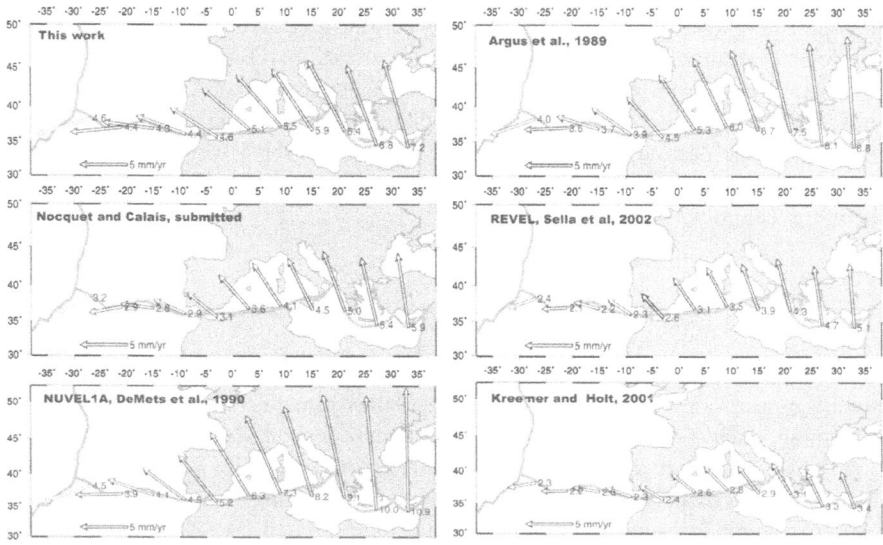

Figure 4

Predicted velocities for the African plate with respect to stable Europe along the Africa-Eurasia plate boundary in the Mediterranean (Table 1). Panel labelled "this work" corresponds to the IGS02P09 M1 solution of Table 1. Numbers by the arrows are velocities in mm/yr.

progressively transitioning westward to transpression (Tunisia-Algeria-Morocco), strike-slip (Gloria transform fault), and transtension (Azores). The convergence velocity varies between models in the 3–10 mm/yr range for the eastern Mediterranean, and in the 3–8 mm/yr range for the Western Mediterranean. The comparison between geological and geodetic models shows an Africa-stable Europe convergence rate 30 to 60% slower for the geodetic models. Also, the geodetic estimates are rotated 10° to 30° counterclockwise compared to the geological model values. The differences among the geodetic estimates themselves are significant, ranging from 3 to 5.7 mm/yr at the longitude of Sicily to 2.5 to 4.5 mm/yr in the Gibraltar Strait. In particular, we find a significant difference between our IGS02P09-derived Euler parmeters and those of SELLA et al. (2002) (Table 1).

The differences among the geodetic estimates may be due to the different data span and data processing strategy used, or to the list and geographic distribution of the sites used to invert for the Eurasian and African plate rotation parameters. In order to separate these two effects, we have used the IGS02P09 solution and inverted for the Eurasia and Africa plate motion using the same site selection as SELLA et al. (2002). We find Euler parameters that are statistically indistinguishable from SELLA et al. (2002) and velocity predictions in the Mediterranean that differ by less than 0.5 mm/yr. However, we find that the Sella et al.'s rotation parameters for Eurasia lead to significant residual velocities in the IGS02P09 solution at the sites located on the Eurasian plate (up to 3 mm/yr). We therefore suspect that the difference between the Sella *et al.* and IGS02P09 parameters for the Africa–Eurasia relative motion results primarily from the choice of sites used to define Eurasia. Our tests indeed show that the definition of the Eurasian plate (i.e., the subset of sites chosen to represent stable Eurasia) influences geodetic velocities in the Mediterranean by up to 2.5 mm/yr. This is 50% of the expected signal in the Western Mediterranean and, therefore, cannot be neglected.

The Adriatic Microplate and its Boundaries

The Adriatic indenter (or Apulian promontory) is a prominent feature in the Mediterranean area. It is a relatively aseismic domain, bounded by actively deforming areas (Apennines, Alps, Friuli, Dinarides). ANDERSON and JACKSON (1987) first proposed that the Adriatic indenter may actually be an independent microplate, detached from the African plate and rotating counterclockwise with respect to stable Europe around a pole located at 45.8°N/10.2°E. Using VLBI results at MATE and MEDI, WARD (1994) reached a similar conclusion but proposed a rotation pole located at 46.8°N/6.3°E and an angular rate of 0.30 ± 0.06°/Ma. WESTAWAY (1990) used tectonic information and earthquake focal mechanisms to infer a rotation of the Adriatic microplate at 0.3°/Ma around a pole at 44.5°N/9.5°E. More recently, CALAIS *et al.* (2002) inverted simultaneously geodetic velocities from a combination of permanent GPS arrays in Western Europe and ANDERSON and JACKSON's (1987) slip vectors of major earthquakes in Italy. They find a rotation pole at 45.36°N/9.10°E and an angular rate of 0.52°/Ma (Fig. 5). Figure 6 compares, in a stable Europe-fixed frame, the observed velocities from the combined solution described above with velocities predicted by each of these models. We find that the geodetic data generally agree with a counterclockwise rotation of the Adriatic plate with respect to stable Europe. However, the fit of the GPS-derived velocities to a rigid plate model is fair, at best. This is due, in part, to the lower quality of the campaign solutions in the combination presented here. It may also reflect internal deformation of the Adriatic microplate. ANDERSON and JACKSON (1987), WESTAWAY (1990), and CALAIS *et al.* (2002) models fit the GPS data equally well.

These models imply NE-SW extension in the Apennines at a rate that increases from North (1–2 mm/yr) to South (4–6 mm/yr), in good agreement with the extension rates derived from independent geodetic studies (ANZIDEI *et al.*, 2001; D'AGOSTINO *et al.*, 2001; HUNSTAD and ENGLAND, 1999). This rotation of the Adriatic microplate implies NE-SW shortening in the Dinarides and the N-S compression in the Friuli area, consistent with recently published focal mechanisms (MONTONE *et al.*, 1999; PONDRELLI *et al.*, 2002) and neotectonic studies (BENEDETTI *et al.*, 2000). The velocity of UPAD and VENE with respect to stable Europe (Fig. 5)

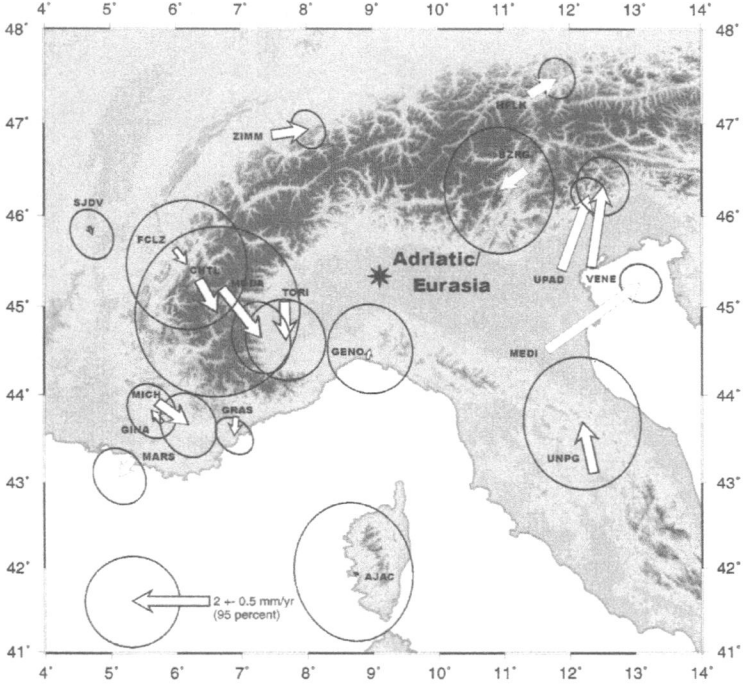

Figure 5

Close-up on the Alps and the northern part of the Apennines and the Adriatic plate, velocities derived from permanent GPS networks with respect to stable Eurasia (definition in the text). Error ellipses are 95% confidence and account for the variance of the data and the variance of the Eurasia Euler parameters. The star indicates the location of the Adriatic/Eurasia Euler pole computed from UPAD and TORI velocities and earthquake slip vector data used by ANDERSON and JACKSON (1987).

indicates NS shortening at ∼2 mm/yr (2.2 ± 0.6 mm/yr at UPAD). The corresponding strain rate is on the order of 10^{-8} yr^{-1}, consistent with GRENERCZY et al.'s (2000) result of 8.6 ± 2.5 × 10^{-9} yr^{-1} in the same area. Further west, the rotation of the Adriatic microplate implies NW-SE compression, transitioning to dextral shear between 8°E and 10°E, consistent with seismotectonic data (MAURER, 1997; EVA and SOLARINO, 1998). In the Western Alps (west of 8°E), the rotation of the Adriatic microplate, together with the arcuate shape of the contact between the Po plain and the Alps, implies dextral shear kinematic boundary conditions, with an additional divergent component in their central part and in Switzerland, and a convergent component in their southern part. Earthquake focal mechanisms in the western Alps show right-lateral motion on NE-SW and NS trending faults, combined with extension in the northern and central parts (EVA et al., 1997; EVA and SOLARINO, 1998; SUE et al., 1999), and compression in the southern part (MADDEDU et al., 1997;

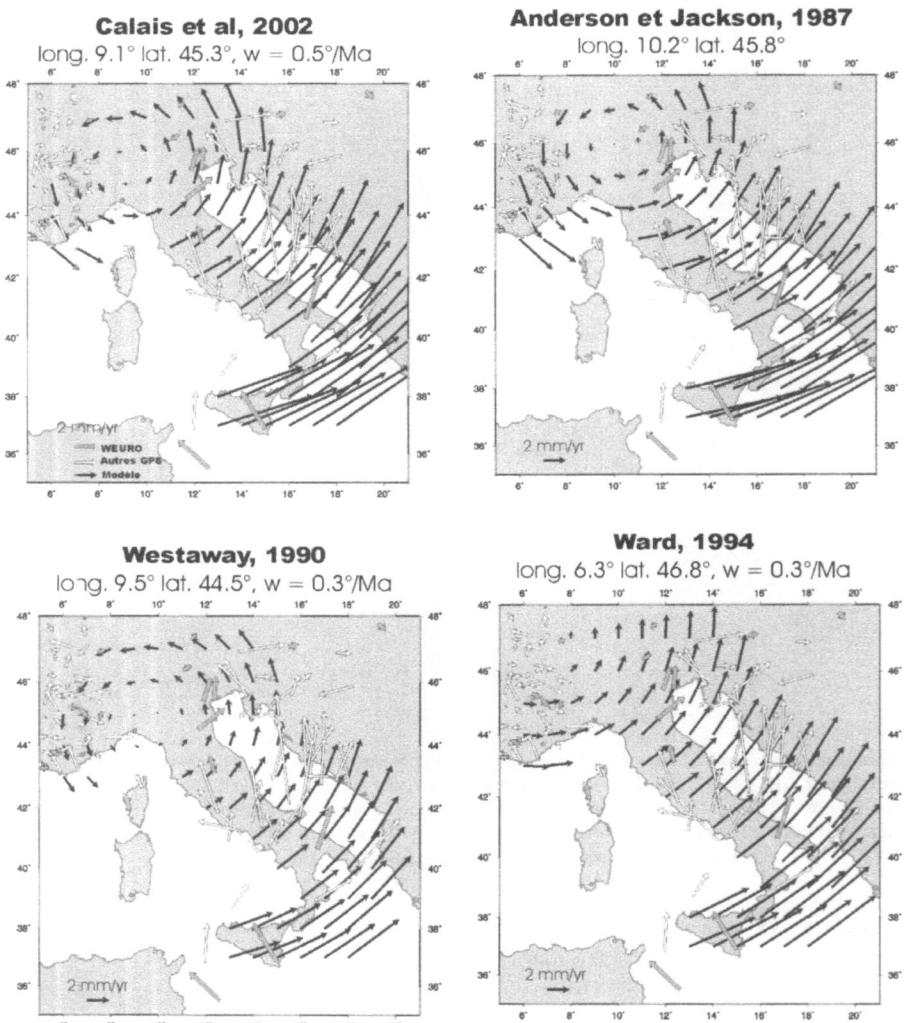

Figure 6

Kinematics of the Adriatic microplate. Black arrows show the velocities predicted by four different models. Grey arrows show observed velocities from continuous GPS stations. White arrows show GPS campaign results from ANZIDEI *et al.* (2001), ALTINER *et al.* (2001), GRENERCZY *et al.* (2000) and VIGNY *et al.* (2002).

BAROUX *et al.*, 2001). In addition, continuous GPS data in the western Alps show east to southeastward velocities with respect to stable Europe, indicating a strain regime that combines right-lateral shear with E-W extension in the northern and central part of the range, and N-S to NW-SE compression in the southern part

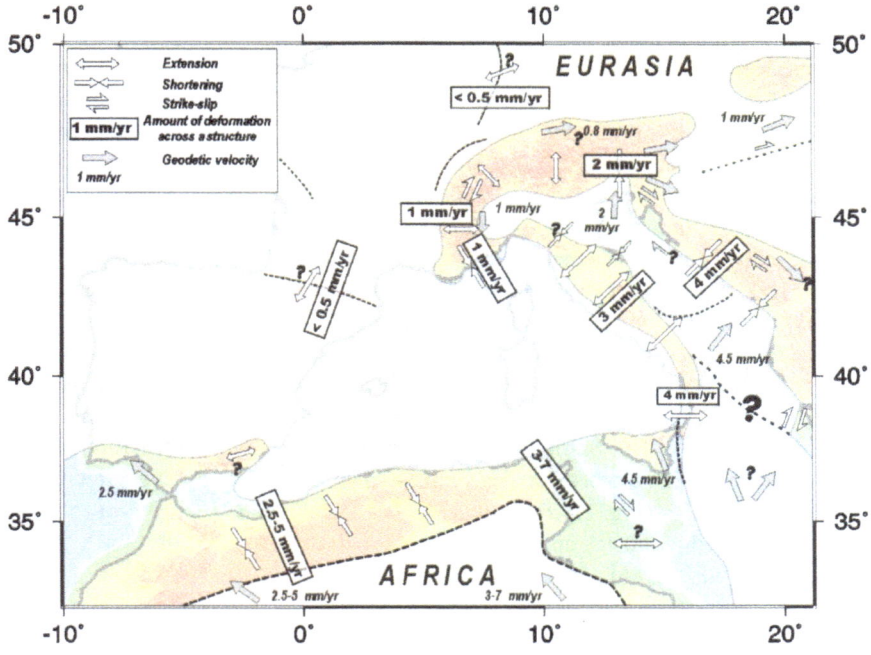

Figure 7

Schematic map of the Africa-Eurasia plate boundary zone in the Western Mediterranean. Areas of active deformation are highlighted, they correspond to the main seismically active areas. Numbers indicate amounts of deformation across highlighted areas. Open arrows indicate strain regime, grey arrows indicate direction of motion with respect to the Eurasian plate.

(Fig. 5; CALAIS *et al.*, 2002). The velocity gradient across the Western Alps does not exceed 2 mm/yr, corresponding to strain rates on the order of 10^{-8} yr^{-1} or less.

The Corsica-Sardinia Block

Corsica and Sardinia constitute a narrow continental domain, relatively aseismic, bounded by two oceanic basins, the Ligurian basin to the west and the Tyrrhenian Sea to the east. Corsica and Sardinia have sometimes been interpreted as an independent tectonic block, moving northward and colliding with continental Europe. This hypothesis is derived from earthquake focal mechanisms and microtectonic observations of neogene faulting, that show N-S shortening along the northern margin of the Ligurian basin in the southern part of the French Alps, and in Provence (RITZ, 1992; MADDEDU *et al.*, 1997; BAROUX *et al.*, 2001). The current geodetic results do not support that conclusion (Figs. 2 and 5). The best determined geodetic site in this domain is CAGL (Cagliari, Sardinia, Fig. 2). Its velocity is determined from SLR and continuous GPS measurements since 1995. All the geodetic solutions used here exhibit a residual velocity for CAGL less than 0.5 mm/yr in a stable Europe-fixed frame as found by DEVOTI et al. (2002). Similarly, continuous GPS site AJAC, although less well determined than CAGL, shows a negligible residual velocity (< 0.3 mm/yr) with respect to stable Europe (Figs. 2 and 5).

It therefore appears that Corsica and Sardinia, and probably the entire Ligurian basin, are rigidly attached to stable Europe. We propose that the N-S shortening observed along the northern margin of the Ligurian basin, in the southern part of the French Alps and in Provence, results from the counterclockwise rotation of the Adriatic microplate, as explained above, rather than from the motion of an independent Corsica-Sardinia block. Furthermore, the fact that Corsica-Sardinia and the Ligurian basin are part of stable Europe implies that the convergence between Africa and stable Europe in the Western Mediterranean must be accommodated south of Sardinia, most probably by active deformation in the Maghrebides ranges in North Africa. This is also suggested by the velocities at CAGL and LAMP (Fig. 2), that indicate between 3 and 5 mm/yr of NW-SE shortening, possibly taken up by transpressional deformation in North Africa (MEGHRAOUI et al., 1986).

The Iberian Peninsula

The Iberian peninsula is a relatively aseismic area separated from the rest of Europe by the Pyrenées mountain range. The Pyrenées have a moderate seismicity with some instrumentally recorded earthquakes reaching magnitude 5, mostly located in the northwestern part of the range (DELOUIS et al., 1993; SOURIAU and PAUCHET, 1998). Geodetic VLBI observations in Europe, including the VLBI station of Madrid, have been performed on a regular basis since the early 1990s. They indicate a negligible velocity of Madrid with respect to Wettzel, located on stable Europe (0.1 ± 0.3 mm/yr; HAAS et al., 2000). The analysis of the ITRF2000 velocity field, that includes these VLBI measurements, confirms this result (Fig. 2). In addition, the continuous GPS site VILL (Villafranca, near Madrid) shows a residual velocity less than 1.1 mm/yr relative to stable Europe (Fig. 2). This result is confirmed by an independent GPS-only solution recently published by SELLA et al. (2002). Residual velocities at other continuous GPS sites in Spain show no significant motion with respect to stable Europe at their uncertainty level (Fig. 2). Together with the velocity at site TOUL, located in France about 100 km north of the Pyrenées, these results imply an upper bound of 0.5 ± 1.5 mm/yr (95% confidence) for possible motion across the Pyrenées. We therefore consider the Iberian peninsula as rigidly attached to stable Europe. Consequently, stable Europe (at the ~0.5 mm/yr level) spans from central Europe to France, and includes the Iberian Peninsula, the Ligurian basin, Corsica, and Sardinia.

The southern part of the Iberian peninsula is characterized by a diffuse seismicity, well expressed in the Betics Cordillera and in southern Portugal. The velocity at SFER, the best determined site in the southern part of the Iberian peninsula, is 2.4 ± 1.1 mm/yr in a westward direction with respect to stable Europe (95% confidence, Fig. 2). We therefore suggest that the Africa-Eurasia plate boundary at the longitude of Iberia involves a relatively broad domain, encompassing the Betic Cordillera and possibly southern Portugal.

Summary and Discussion

African Plate

The motion of the African plate with respect to stable Europe is not fully established yet. Geodetic estimates agree on a N45W ± 20 ° convergence at the longitude of Sicily, transitioning progressively to a more E-W convergence direction toward the Gibraltar Strait. Estimates of convergence rates range between 3 to 7 mm/yr at the longitude of Sicily, decreasing westward to 2 to 5 mm/yr at the longitude of the Gibraltar Strait (Figs. 4 and 7). In spite of the fairly large scatter between the geodetic solutions, they consistently indicate an Africa-stable Europe convergence rate slower than the prediction of geological models by a factor ranging from 1.2 to 2. These differences between the geodetic solution and geological model may be due to inaccuracies in the model and/or in the geodetic data. CHU and GORDON (1999), found that splitting Africa into two plates (Nubia and Somalia) and removing the plate circuit closure condition in the Pacific significantly modified the angular velocity of India relative to Eurasia, compared to the original NUVEL-1A value. The updated version of NUVEL1 (MORVEL) is currently being produced and will include a Nubian and a Somalian plate. However, a preliminary version indicates a velocity of 6.9 mm/yr in a N341 direction at NOTO with respect to Eurasia (C. DEMETS, personal communication, 2002). If this was confirmed, it therefore seems unlikely that splitting Africa into two plates will reduce the mismatch between geodetic observations and global kinematic models in the Mediterranean.

North Africa

The lack of deformation in the Western Mediterranean area north of the African coastline indicated by the seismicity distribution and the geodetic results, together with the NW-ward motion of the African plate with respect to stable Europe, imply that the Africa-stable Europe motion in the Western Mediterranean is essentially accommodated by deformation in North Africa and southern Iberia. Very little strain, if any, seems to be transferred to the north, into the Eurasian plate.

East of about 2°W, the spatial distribution of the seismicity shows that most of the large earthquakes are concentrated along a relatively narrow crustal strip following the northern coast of Africa (Fig. 1). The NW-SE oblique convergence between Africa and Eurasia in the Western Mediterranean is consistent with the earthquake focal mechanisms, that show mostly reverse motion on NE-SW trending, and usually NW-dipping, faults, often combined with a strike-slip component. Geodetic and seismological data therefore suggest that the strain induced by the Africa-Eurasia oblique convergence concentrates at the northern edge of Africa. The sharp rheological transition between the oceanic crust of the Ligurian basin to the north, and the north of the African continental crust to the south may contribute significantly in focusing stress and strain, similarly as proposed by LOWRY and SMITH (1995) to explain the concentration of seismicity at the boundary between the Colorado Plateau and the Basin and Range province in the western United States. Interestingly, west of about 2°W, where this sharp rheological transition vanishes, active deformation seems to affect a broader area encompassing the Moroccan Riff as well as the southernmost part of the Iberian peninsula, including the Betics cordillera and possibly southern Portugal (MEGHRAOUI et al., 1996). However, other processes such as mantle delamination could also play a significant role in the Alboran Sea area (SEBER et al., 1996). Finally, independent estimates from geological observations (MEGHRAOUI et al., 1996) or seismic moment summation (KIRATZI and PAPAZACHOS, 1995) indicate a shortening rate of 1.0–2.5 mm/an in the Algerian Tell Atlas and the Moroccan Riff. Since the total Africa-Eurasia plate convergence rate is likely to be (slightly) higher, it could indicate that some of this convergence is accommodated further in the Saharian Atlas in Morocco and Algeria. However, there is no evidence for active deformation in these areas, where the level of seismicity is low.

Further east, the role played by the deforming area located between Sicily and Tunisia and extending southward to the Lybian coast is unclear. It could accommodate a differential motion between Africa and the Ionian Sea, with a different convergence direction (with respect to Europe) for the Ionian Sea.

Adriatic Microplate

Geodetic results clearly indicate that the Adriatic domain is not part of the African plate. They support, to the first order, the hypothesis of a counterclockwise rotation of an Adriatic microplate. This rotation implies 2.5 to 4–5 mm/yr of NE-SW extension along the Apennines from north to south, ~4 mm/yr of NE-SW shortening in the Dinarides, transitioning to ~2 mm/yr of NS shortening in the Friuli area, and ~1 mm/yr of combined extension and right-lateral shear in the Swiss and Western Alps. Geodetic and seismotectonic data in the Alps, Apennines, and Dinarides are consistent with this model. This suggests that active deformation in the Alps, Apennines, and Dinarides is controlled, and possibly dynamically driven, by the motion of the Adriatic microplate rather than by the convergence between Africa and Eurasia, as usually assumed. These results raise the issue of the driving mechanism for the motion of the Adriatic plate, which does not appear to be related to the African plate motion in a simple manner. A definite answer to this question is out of the scope of this study and would require dynamic geophysical models incorporating realistic boundary conditions (Africa–Eurasia plate motion and Aegean subduction, located just east of our study area), stresses induced by gravitational potential energy variations (e.g., MOLNAR and LYON-CAEN, 1988) and/ or by the peri-Adriatic subducted slabs (WORTEL and SPAKMAN, 2000) and by the westward push of the Anatolian-Aegean-Balkan system (TAPPONNIER, 1977; MANTOVANI et al., 2002).

The definition of the southern boundary of the Adriatic microplate also remains unclear. There has to be a tectonic discontinuity between southern Italy, with NE-ward velocities w.r.t. Europe (e.g., site MATE), and southern Sicily and the rest of the African plate, with NW-ward velocities w.r.t. Europe (e.g., sites NOTO and LAMP). WESTAWAY (1990) inferred that the southern boundary of the Adriatic microplate corresponds to an alignment of earthquakes along the "Trimiti line" between the Gargano and Dubrovnik. This is consistent with the velocity of NOTO, which does not fit that of the Adriatic microplate, therefore suggesting a major tectonic discontinuity between Sicily and continental Italy. However, the well determined velocity of MATE (Matera), located south of the Trimiti line, shows a velocity that fits reasonably well that of the rigid Adriatic microplate defined above. The southern boundary of the Adriatic plate should therefore be located south of Matera and north of Noto. Although there are active faults in that area, no active fault is known offshore across the Ionian Sea that could constitute the southern boundary of the Adriatic microplate. Data from additional continuous GPS sites in Italy, combined in a rigorously implemented reference frame, are necessary in order to understand the transition from the southern part of the Adriatic plate to the Ionian Sea and Sicily.

Eastern Alps and Pannonian Basin

In the eastern Alps, we find a significant eastward residual velocity for GRAZ with respect to stable Europe (Fig. 2, 1.0 ± 0.4 mm/yr, 95% confidence level), in agreement with a previous study by GRENERCZY et al. (2000; 1.7 ± 0.8 mm/yr). We also find a similar residual velocity at station PENC, in the Pannonian basin (Fig. 2). These velocities, together with the campaign results of GRENERCZY et al. (2000) support the idea of an eastward motion of the easternmost part of the Alps and the northern Pannonian basin. This eastward motion has been interpreted as an extrusion process, in response to the collision of the Adriatic indenter in the Friuli region (BADA, 1999; GRENERCZY et al., 2000). We suggest that the deforming areas in continental Greece and at the Aegean subduction may constitute a low friction boundary facilitating east and southeastward motions in southern central Europe. This far-field effect of the Aegean subduction may be indicated by the 2.4 mm/yr SE-ward residual velocity at SOFI (Bulgaria) with respect to stable Europe (Fig. 2).

Conclusion

The combination procedure used here is an efficient and rigorous way to integrate several geodetic solutions into a single and consistent reference frame. The rigorous combination of campaign with continuous solutions is possible, but requires the availability of the associated statistical information (full variance-covariance matrix and explicit description of reference frame constraints) which was missing for most of the data used in the present work. We therefore stress that the most reliable results in this study are obtained using continuous GPS data. We also propose a rigorous definition of stable Europe, based on the combination of a redundant set of solutions from continuous geodetic networks in Europe. We then map velocities in this stable Europe reference frame.

Among the results presented here, we find that geodetic data consistently indicate an Africa-stable Europe convergence rate slower than the prediction of geological models. This, if confirmed by more geodetic data on the African plate and longer observation time series, has significant implications for the rigidity of the African plate and/or may imply recent variations of the African plate motion. Also, we find that the Africa-stable Europe oblique convergence in the Western Mediterranean is mostly accommodated by transpressional deformation in north Africa and southern Iberia. Very little strain, if any, appears to be transferred into the Eurasian plate. Active deformation in the Alps, Apennines, and Dinarides is probably driven by the independent motion of the Adriatic plate rather than by the Africa-Eurasia convergence. However, the forces responsible for the motion of the Adriatic plate and the relationship between extension in the Apennines and convergence between Africa and Eurasia remain open questions.

The kinematic description of crustal deformation in the Western Mediterranean and Western Europe presented here provide boundary conditions and validation data for studies aimed at modelling lithospheric-scale deformation processes in the Mediterranean (e.g., JIMENEZ-MUNT et al., 2001a, b, in press). It remains however critical to increase the amount of geodetic observations in Africa in general, and North Africa in particular, in order to better estimate the kinematics of the African plate and complete the determination of strain distribution across the Africa-Eurasia plate boundary zone in the Mediterranean area.

Acknowledgments

We thank all the individuals and institutions participating in the EUREF-EPN network. We thank P. Nicolon (IGN) for providing us the weekly RGP solutions and the agencies operating and maintaining the permanent GPS stations of the REGAL network (Univ. Savoie, IPSN, Univ. Montpellier, LDG, Géosciences Azur). This work was funded by the French national program "Géofrance 3D" (BRGM, MENR, INSU) and by the ACI "Catastrophes Naturelles" (MRT).

REFERENCES

ALBARELLO, D., MANTOVANI, E., BABBUCCI, D., and TAMBURELLI, C. (1995), *Africa-Eurasia Kinematics: Main Constraints and Uncertainties*, Tectonophysics *243*, 25–36.

ALTAMIMI, Z., SILLARD, P., and BOUCHER, C. (2002), *ITRF2000, A New Release of the International Terrestrial Reference Frame for Earth Science Applications*, J. Geophys. Res., in press.

ALTINER, Y. (2001), *The Contribution of GPS Data to the Detection of the Earth's Crust Deformations Illustrated by GPS Campaigns in the Adria Region*, Geophys. J. Int. *145*, 550–555.

ANDERSON, H. A. and JACKSON, J. A. (1987), *Active Tectonics of the Adriatic Region*, Geophys. J. R. Astron. Soc. *91*, 937–983.

ANZIDEI, M., BALDI, P., CASULA, G., GALVANI, A., MANTOVANI, E., PESCI, A., RIGUZZI, F., and SERPELLONI, E. (2001), *Insights into Present-day Crustal Motion in the Central Mediterranean Area from GPS Surveys*, Geophys. J. Int. *146*, 98–110.

ARGUS, D. F. and GORDON, R. G. (1996), *Tests of the Rigid-plate Hypothesis and Bounds on Intraplate Deformation Using Geodetic Data from the Very Long Baseline Interferometry*, J. Geophys. Res. *101*, 13,555–13,572.

ARGUS, D. F., GORDON, R. G., DEMETS, C., and STEIN, S. (1989), *Closure of the Africa-Eurasia-North America Plate Motion Circuit and Tectonics of the Gloria fault*, J. Geophys. Res. *94*, 5585–5602.

BADA, G., HORVATH, F., GERNER, P., and FEJES, I. (1999), *Review of the Present-day Geodynamics of the Pannonian Basin: Progress and Problems*, Geodynamics *27*, 501–527.

BAROUX, E., BÉTHOUX, N., and BELLIER, O. (2001), *Analyses of the Stress Field in Southeastern France from Earthquake Focal Mechanisms*, Geophys. J. Int. *145*, 336–348.

BENEDETTI, L., TAPPONNIER, P., KING-GEOFFREY, C. P., MEYER-BERTRAND, B., and MANIGHETTI, I. (2000), *Growth Folding and Active Thrusting in the Montello Region, Veneto, Northern Italy*, J. Geophys. Res. *105*, 739–766.

BROCKMANN, E. (1997), *Combination of Solutions for Geodetic and Geodynamics Applications of the Global Positioning System*, Ph.D. Thesis, University of Bern.

CALAIS, E. et al. (2000), *REGAL: A permanent GPS Network in the French Western Alps, Configuration and First Results*, C.R. Acad. Sci. Paris *331*, 435–442.

CALAIS, E., NOCQUET, J.-M., JOUANNE, F., and TARDY, M. (2002), *Current Extension in the Central Part of the Western Alps from Continuous GPS Measurements, 1996–2001*, Geology *30-7*, 651–654.

CHU, D. and GORDON, R. G. (1999), *Evidence for Motion between Nubia and Somalia along the Southwest Indian Ridge*, Nature *398*, 64–67.

CRÉTAUX, J.-F., SOUDARIN, L., CAZENAVE, A., and BOUILLÉ, F. (1998), *Present-day Tectonic Plate Motions and Crustal Deformations from the DORIS Space System*, *103*, 30,167–30,181.

D'AGOSTINO, N., GIULIANI, R., MATTONE, M., and BONCI, L. (2001), *Active Crustal Extension in the Central Apennines (Italy) Inferred from GPS Measurements in the Interval 1994–1999*, Geophys. Res. Lett. *28*, 2121–2124.

DAVIES, P. and BLEWITT, G. (2000), *Methodology for Global Geodetic Time Series Estimation, A New Tool for Geodynamics*, J. Geophys. Res. *105*, 11,083–11,100.

DELOUIS, B., HAESSLER, H., CISTERNAS, A., and RIVERA, L. (1993), *Stress Tensor Determination in France and Neighbouring Regions*, Tectonophysics *221*, 413–438.

DEMETS, C. and DIXON, T. H. (1999), *New Kinematic Models for Pacific-north American Motion from 3 Ma to Present*, 1: *Evidence for Steady Motion and Biases in the NUVEL-1A Model*, Geophys. Res. Lett. *26*, 1921–1924.

DEMETS, C., GORDON, R. G., ARGUS, D. F., and STEIN, S. (1990), *Current Plate Motions*, Geophys. J. Int. *28*, 2121–2124.

DEMETS, C., GORDON, R. G., ARGUS, D. F., and STEIN, S. (1994), *Effect of Recent Revisions to the Geomagnetic Reversal Time Scale on Estimates of Current Plate Motions*, Geophys. Res. Lett. *21*, 2191–2194.

DEVOTI, R., FERRARO, C., GUEGUEN, E., LANOTTE, R., LUCERI, V., NARDI, A., PACIONE, R., RUTIGLIANO, P., SCIARRETTA, C., and VESPE, F. (2002), *Geodetic Control on Recent Tectonic Movements in the Central Mediterranean Area*, Tectonophysics *346*, 151–167.

DIXON, T. H., MAO, A., and STEIN, S. (1996), *How Rigid is the Stable Interior of the North American Plate?* Geophys. Res. Lett. *23*, 3035–3038.

EVA, E. and SOLARINO, S. (1998), *Variations of Stress Directions in the Western Alpine Arc*, Geophys. J. Int. *135*, 438–448.

EVA, E., PASTORE, S., and DEICHMAN, N. (1998), *Evidence for Ongoing Extensional Deformation in the Western Swiss Alps and Thrust Faulting in the Southwestern Alpine Foreland*, J. Geodynamics *26*, 27–43.

GRENERCZY, G., KENYERES, A., and FEJES, I. (2000), *Present Crustal Movement and Strain Distribution in Central Europe Inferred from GPS Measurements*, J. Geophys. Res. *105*, 21,835–21,846.

HAAS, R. E., GUEGUEN, H.-G., SCHERNECK, NOTHNAGEL, A., and CAMPBELL, J. (2000), *Crustal Motion Results Derived from Observations in the European Geodetic VLBI Network*, Earth Planets Space *52*, 759–764.

HUNSTAD, I. and ENGLAND, P. (1999), *An Upper Bound on the Rate of Strain in the Central Apennines from Triangulation Measurements between 1869 and 1963*, Earth Planet. Sci. Lett. *169*, 261–267.

JIMENEZ-MUNT, I., FERNANDEZ, M., TORNE, M., and BIRD, P. (2001a), *The Transition from Linear to Diffuse Plate Boundary in the Azores–Gibraltar Region: Results from a Thin Sheet Model*, Earth Planet. Sci. Lett. *192*, 175–189.

JIMENEZ-MUNT, I., BIRD, P., and FERNANDEZ, M. (2001b), *Thin–shell Modeling of Neotectonics in the Azores– Gibraltar Region*, Geophys. Res. Lett. *28*, 1083–1086.

JIMENEZ-MUNT, I., SABADINI, R., GARDI, A., and BIANCO, G. *Active Deformation in the Mediterranean from Gibraltar to Anatolia Inferred from Numerical Modeling, Geodetic, and Seismological Data*, J. Geophys. Res., in press.

KIRATZ, A. and PAPAZACHOS, C. (1995), *Active Crustal Deformation from the Azores Triple Junction to the Middle East*, Tectonophysics *243*, 1–24.

KOGAN, M. G., STEBLOV, G. M., KING, R. W., HERRING, T. A., FROLOV, D. I., EGOROV, S. G., LEVIN, V. Y., LERNER-LAM, A., and JONES, A. (2000), *Geodetic Constraints on the Rigidity and Relative Motion of Eurasia and North America*, Geophys. Res. Lett. *27*, 2041–2044.

KREEMER, C. and HOLT, W. E. (2001), *A No-net-rotation Model of Present Day Surface Motion*, Geophys. Res. Lett. *28*, 4407–4410.

LOWRY, A. and SMITH, R. B. (1995), *Strength and Rheology of the Western US Cordillera*, J. Geophys. Res. *100*, 17,947–17,963.

MADDEDU, B., BÉTHOUX, N., and STÉPHAN, J. F. (1997), *Champ de contrainte post-pliocène and déformations récentes dans les Alpes sud-occidentales*, Bull. Soc. Géol. Fr. *167*, 797–810.

MANTOVANI, E., ALBARELLO, D., BABBUCCI, D., TAMBURELLI, C., and VITI, M. (2002), *Trench arc-back arc systems in the Mediterranean area: examples of extrusion tectonics*, In *Reconstruction of the Evolution of the Alpine-Himalayan Orogeny*, J. Virtual Explorer.

MAURER, H., BURKHARD, M., DEICHMANN, N., and GREEN, G. (1997), *Active Tectonism in the Central Alps: Contrasting Stress Regimes North and South of the Rhone Valley*, Terra Nova *9*, 91–94.

MCCLUSKY et al. (2000), *Global Positioning System Constraints on Plate Kinematics and Dynamics in the Eastern Mediterranean and Caucasus*, J. Geophys. Res. *115*, 5695–5719.

MEGHRAOUI, M., CISTERNAS, A., and PHILIP, H. (1986), *Seismotectonics of the Lower Chéliff Basin: Structural Background of the El Asnam (Algeria) Earthquake*, Tectonics *5(6)*, 809–836.

MEGHRAOUI, M., MOREL, J-L., ANDRIEUX, J., and DAHMANI, M. (1996), *Tectonique plio-quaternaire de la chaîne tello-rifaine and de la mer d'Alboran - une zone complexe de convergence continent-continent*, Bull. Soc. Géol. France *167*, 141–157.

MOLNAR and LYON-CAEN (1988)

MONTONE, P., AMATO, A., and PONDRELLI, S. (1999), *Active Stress Map of Italy*, J. Geophys. Res. *104*, 25,595–25,610.

NEWMAN, A., STEIN, S., WEBER, J., ENGELN, J., MAO, A., and DIXON, T. (2001), *Slow Deformation and Lower Seismic Hazard at the New Madrid Seismic Zone*, Science *284*, 619–621.

NOCQUET, J. M. and CALAIS, E. (2003), *The Crustal Velocity Field in Western Europe from Permanent GPS Array Solutions, 1996–2001*, Geophys. J. Int. *154*, 72–99.

NOCQUET, J.-M., CALAIS, E., ALTAMIMI, Z., SILLARD, P., and BOUCHER, C. (2001), *Intraplate Deformation in Western Europe Deduced from an Analysis of the ITRF97 Velocity Field*, J. Geophys. Res. *106*, 11,239–11,258.

PONDRELLI S., MORELLI, A., EKSTRÖM, G., MAZZA, S., BOSCHI, E., and DZIEWONSKI, A. M. (2002), *European-Mediterranean Regional Centroid-Moment Tensors: 1997–2000*, Phys. Earth and Planet. Int. *130*, 71–101.

RITZ, J.-F. (1992), *Tectonique récente and sismotectonique des Alpes du sud: Analyse en termes de contraintes*, Quaternaire *3*, 111–124.

SEBER, D., BARAZANGI, M., IBENBRAHIM, A., and DEMNATI, A. (1996), *Geophysical Evidence for Lithospheric Delamination beneath the Alboran Sea and Rif-Betic Mountains*, Nature *379*, 785–790.

SELLA, G. F., DIXON, T. H., and MAO, A. (2002), *REVEL, A Model for Recent Plate Velocities from Space Geodesy*, J. Geophys. Res.

SERPELLONI, E., ANZIDEI, M., BALDI, P., SHEN, Z., CASULA, G., GALVANI, A., and PESCI, A. (2001), *Combination of Permanent and Non-Permanent GPS Networks for the Analysis of the Strain-Rate Field in the Mediterranean Area*, EOS Trans. AGU *82(47)*, Fall Meet. Suppl., Abstract G41A-0207.

SILLARD, P. and BOUCHER, C. (2001), *A Review on Algebraic Constraints in Terrestrial Reference Frame Datum Definition*, J. Geod. *75*, 63–73.

SOURIAU, A. and PAUCHET, H. (1998), *A New Synthesis of the Pyrenean Seismicity and its Tectonic Implications*, Tectonophysics *290*, 221–244.

SUE, C., THOUVENOT, F., FRÉCHET, J., and TRICART, P. (1999), *Widespread Extension in the Core of the Western Alps Revealed by Earthquake Analysis*, J. Geophys. Res. *104*, 25,611–25,622.

TAPPONNIER P. (1977), *Evolution tectonique du système alpin en Méditerranée: poinçonnement et écrasement rigide-plastique*, Bull. Soc. Géol. Fr. *19*, 437–460.

VIGNY, C. et al (2002), *GPS Network Monitors the Western Alps, Deformation over a Five-year Period: 1993–1998*. J. of Geod. *76*, 63–76.

WARD, S. N. (1994), *Constraints in the Seismotectonics of the Central Mediterranean from very Long Baseline Interferometry*, Geophys. J. Int. *117*, 441–452.

WESTAWAY, R. (1990), *The Tripoli, Libya, Earthquake of September 4, 1974: Implications for the Active Tectonics of the Central Mediterranean*, Tectonics *9*, 231–248.

(Received June 24, 2002, revised December 6, 2002, accepted December 23, 2002)

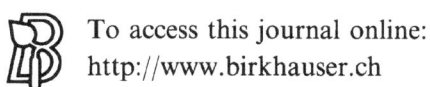

To access this journal online:
http://www.birkhauser.ch

Pure appl. geophys. 161 (2004) 683–699
0033–4553/04/030683–17
DOI 10.1007/s00024-003-2469-y

❘ Pure and Applied Geophysics

Recent Geodetic Results in the Azores Triple Junction Region

R. M. S. Fernandes[1,3], L. Bastos[2], B. A. C. Ambrosius[1],
R. Noomen[1], S. Matheussen[1], and P. Baptista[2]

Abstract — GPS (Global Positioning System) observations started to be carried out in the Azores region under the scope of the TANGO (TransAtlantic Network for Geodesy and Oceanography) project in 1988. The measurements carried out between 1993 and 2000 (five campaigns) on nine GPS sites (one per island) were reprocessed using two state–of–the–art software packages. Different methodologies were applied to compute each campaign solution and the derived velocity field. The velocity fields, including the motions of two permanent stations, recently installed in the Azores, were computed within the most recent geodetic reference frame, ITRF2000 (International Terrestrial Reference Frame, solution 2000). They are compared with the motions of the stable rigid tectonic plates using as reference DEOS2k, a global tectonic model developed using geodetic data. The relative motions between the Western and Central groups of islands yield to evaluate the opening rate of the Mid-Atlantic Ridge (boundary between the North American plate and the Eurasian and African plates). Concerning the boundary between the Eurasian and African plates, the motion of the TANGO sites in the Central and Eastern groups clearly identifies the transition pattern between those two plates. Two of the sites are considered to be located in the stable part of these plates, whereas the remaining five are within the deformation region of the Eurasia-Africa boundary. The conclusions are analyzed in view of the different deformation models, derived from geodynamic or geophysical data that have been proposed for the region.

Key words: Azores Triple Junction, GPS processing, plate tectonics.

1. Introduction

The Azores Archipelago is located in the junction area of three major tectonic plates: Eurasia, Africa and North America. The complex tectonic processes caused by the interaction between these plates put permanently at risk the life and assets of the thousands of inhabitants of the Azores Archipelago. The most recent example is the 9 July 1998 earthquake that struck the island of Faial causing nine casualties.

---

[1] Delft Institute for Earth-Oriented Space Research (DEOS), Kluyverweg 1, 2629 HS Delft, The Netherlands. E-mail: rui@deos.tudelft.nl

[2] Astronomical Observatory, Faculty of Sciences, University of Porto (AOUP), Monte da Virgem, 4430-146 V. N. Gaia, Portugal.

[3] also at DI – UBI, 6201-001 Covilhã, Portugal.

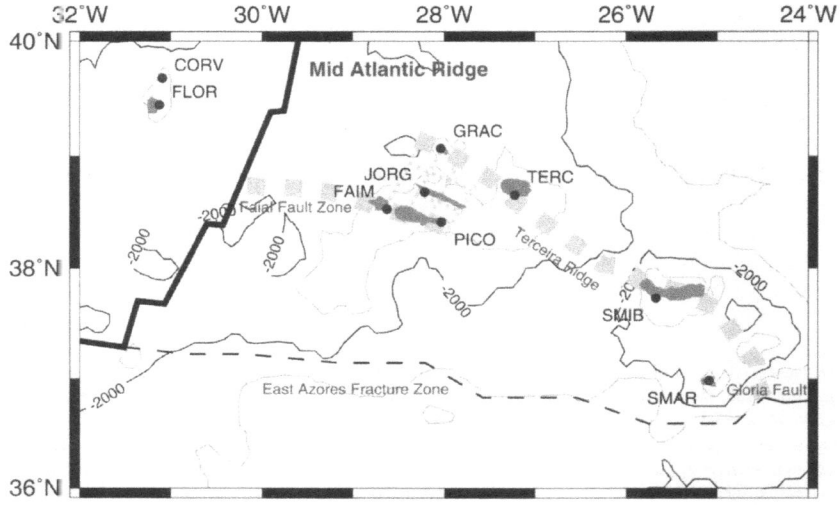

Figure 1
Major geodynamic features in the Azores Archipelago. Gray lines show the major directions of stress in the region according to recent published models. Black circles show the location of the GPS sites being monitored since 1988.

Figure 1 shows the major tectonic settings of the Azores Triple Junction region. The boundary between North America and the other two plates is well defined by the Mid-Atlantic Ridge (MAR). As for the western segment of the Eurasia-Africa plate boundary, the exact location and the features in the region of the Azores Plateau (submarine platform roughly limited by the 2000 m bathymetric contour (LOURENÇO *et al.*, 1998)) are not yet well determined. During the last two decades, several models have been proposed based upon various geological and geophysical analyses: the Terceira Rift model (BUFORN *et al.*, 1988); the Azores Microplate model (FREIRE LUÍS *et al.*, 1994); the Leaky Transform model (MADEIRA and RIBEIRO, 1990). Recently, different works, based on bathymetric (LOURENÇO *et al.*, 1998), gravimetric (LUIS *et al.*, 1998), and seismic (MIRANDA *et al.*, 1998) data have proposed a new model for the Azores Plateau region, called the Azores Blocks model. These authors state that, presently, the Azores domain is a narrow diffuse plate boundary consisting of several tectonic blocks limited by two sets of faults, oriented in the directions N120E and N150E. This area acts simultaneously as an oblique ultra slow spreading center and as a transfer zone that accommodates the differential shear movement between the Eurasian and African plates from the MAR until the beginning of the Gloria fault (a fairly straight dextral transform fault well mapped by bathymetry (ARGUS *et al.*, 1989)).

2. GPS Observations

In 1988, one of the first GPS networks for regional geodynamic studies was established in the Azores-Gibraltar region, within the scope of the TANGO project (BASTOS et al., 1991). The initial network comprised 17 sites, which were located in the Azores Archipelago (9, cf. Figure 1), the Madeira Island (1), the Portuguese mainland (2), the Gibraltar Strait (2) and the Canary Archipelago (3). Reobservations of this network have been carried out in 1991, 1993, 1994, 1997, 1999 and 2000. During the 1999 campaign, 27 new sites were installed in the Central Group of the Azores Archipelago in order to allow more detailed geodynamic studies (FERNANDES et al., 2000b). Table 1 shows the availability and the average time span of the observations for each campaign taken in the Azores Archipelago.

The 1988 campaign was not considered for this analysis due to the low quality of the data. At the time, the GPS satellite constellation was still very incomplete and the short observation period (cf., Table 1) limited the quality of the results. As for the 1991 campaign, it was not possible to obtain results with the desired accuracy level due to different reasons: (1) the peak of solar activity in 1991 had a direct negative impact on the quality of the observations (KLOBUCHAR, 1996); (2) the receivers (Trimble 4100 SLD and Ashtech XII) only provided the measurement of half of the L2 phase; and (3) the small number of reliable global continuous stations prevented the computation of orbits with good accuracy.

The first available permanent GPS station in the Azores was installed on the Island of São Miguel (PDEL) by the Portuguese mapping agency (IGP) in December 1999. This station is located in the vicinity ($\approx$1.6 km) of the SMIB site (cf., Fig. 1) and the tectonic motions can be considered the same, which allows linkage of the two sites in the future (continuation of time series). In October 2000, the DEOS and AOUP groups, in collaboration with the Instituto de Meterologia, installed a

Table 1

Available GPS observations in the Azores Archipelago since 1988

Campaign	Time			Sites								
	Month	Days	Session Length	SMAR	SMIB	TERC	GRAC	JORG	PICO	FAIM	FLOR	CORV
1988	Oct	2 to 4	3h	X	X	X	X	X	X	X	X	X
1991	Sep	5	8h	X	X	X	X	X	X	X	X	X
1993	Dec	3 to 4	14h		X	X				X	X	
1994	Oct	4 to 6	14h	X	X	X	X	X	X	X	X	X
1997	May	6	24h	X	X	X	X	X	X	X	X	X
1999	Sep	3 to 11	24h	X	X	X	X	X	X	X	N	X
2000	Oct	5 to 11	24h	X	X	X	X	X	X	X	N	X

N – New site (previous site destroyed in 1998).

permanent GPS site at the TANGO marker on Faial Island (FAIM). This location was chosen because of the higher seismic activity recorded since 1998. Data from the two permanent sites were processed until May 2001.

3. Data Processing and Analysis

3.1 Daily Processing

The daily observations were processed with two state-of-the-art software packages: GIPSY (WEBB and ZUMBERGE, 1995) and GAMIT (KING and BOCK, 1999), which use different algorithms to process the GPS observations. GAMIT, as most of the GPS processing software packages, is based on the double difference formulation, in which satellite and receiver clock errors are eliminated. On the contrary, GIPSY handles undifferenced observations by estimating corrections to the satellite and receiver clocks using a very stable Kalman filter algorithm.

Two strategies can be adopted using GIPSY: PPP (Precise Point Positioning) and FN (Free Network solution). The PPP strategy implies that consistent precise corrections to satellite clocks and orbits are known and kept fixed, which are daily provided by JPL (Jet Propulsion Laboratory). This approach allows solving for positions on a station-by-station basis (ZUMBERGE *et al.*, 1997).

In the FN strategy, the entire network is processed simultaneously. No tight constraints are imposed to any of the station coordinates, allowing slight variations of the entire network (including satellite orbits) from day to day (HEFLIN *et al.*, 1992).

Tests carried out by FERNANDES *et al.* (2000a) using data of the 1997 campaign have shown slightly better repeatabilities with the PPP strategy. In addition, the PPP strategy has the major advantage of saving an enormous amount of computer resources when compared with the FN strategy. Consequently, with GIPSY, campaign and permanent station data were processed using PPP.

The double difference formulation in the GAMIT software automatically implies the use of the FN approach.

3.2 Velocity Field Estimation

The final geodetic product derived from the Azores GPS network is the velocity field estimated in a known reference frame. Observations are combined to accurately estimate the position and velocity of stations in a unique global reference system. Several realizations of the ITRS (International Terrestrial Reference System) (McCARTHY, 1996), named ITRF, have been produced during the last decade, with a continuous improvement in the number of sites and in the quality of the solutions. The latest ITRF solution, ITRF2000, was released in the first quarter of 2001.

The estimation of the velocities in a no-net-rotation reference frame, as ITRF2000 (ALTAMIMI *et al.*, 2002), is fundamental in order to analyze the observed motions in the frame of absolute global plate tectonics.

The procedures used to derive the velocity field in ITRF2000 using GIPSY and GAMIT are significantly different. Therefore, for each application, a detailed explanation of the steps taken is given.

3.2.1 GIPSY

The analysis is basically carried out in two steps: first, for each campaign, the station position is estimated in ITRF2000 from the daily solutions. Next, the linear velocities are derived from the epoch campaign positions.

Figure 2 shows the scheme of the procedures carried out to estimate the velocity field using GIPSY. In order to map the TANGO stations into ITRF2000, the network has been expanded with a subset of IGS sites, located around the Azores-Gibraltar region. Site selection has been made considering location and quality. As shown in Figure 3, the network is better constrained in the Eastern (Europe) and

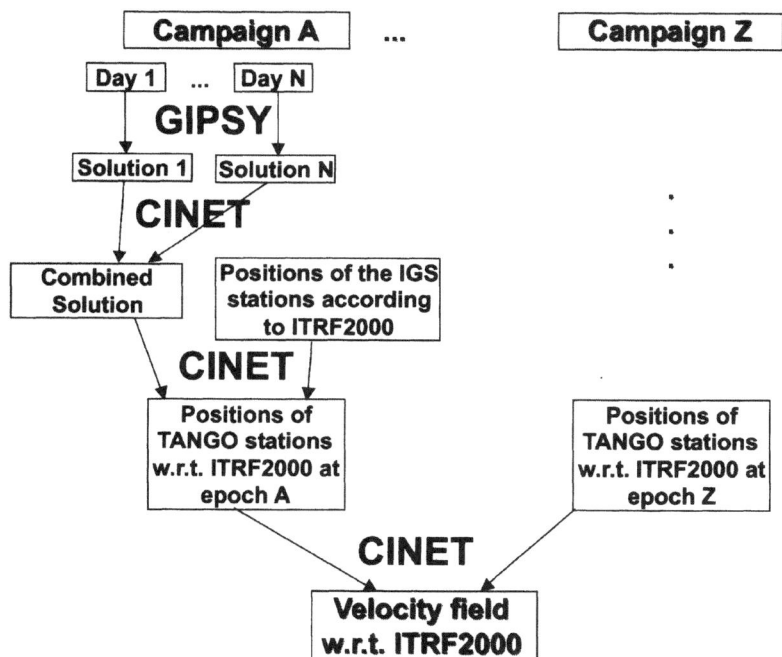

Figure 2
Scheme of the data processing in order to obtain the velocity field using the GIPSY and CINET applications.

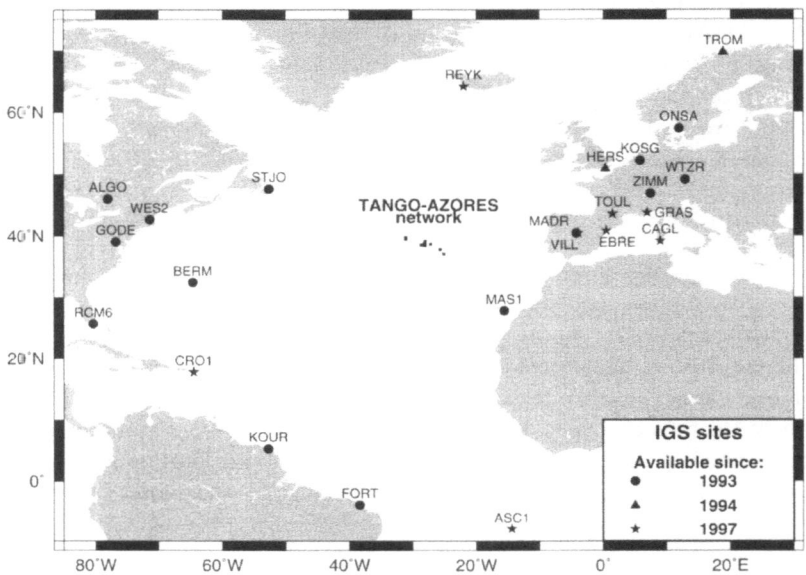

Figure 3

IGS stations used to map the TANGO network into ITRF2000 with CINET. Some of the stations do not have available data for all campaigns.

Western (North America) than in the Southern part, where the stations are, in general, located further away and have a shorter observation time-span. In addition, stations located in the equatorial region, such as FORT (Fortaleza) and KOUR (Kourou), show a degraded data quality, mainly caused by stronger ionospheric activity at these low latitude observations (KLOBUCHAR, 1996). This was visible in the daily solution's repeatabilities. However, tests carried out have shown that the advantages of a better overall network geometry outweigh the disadvantages of the reduced quality of these stations.

For each campaign, the PPP daily solutions of the entire network (TANGO and IGS stations) were combined in a unique multi-day averaged solution using CINET, an enhanced version of the 3DMOTION software (NOOMEN *et al.*, 1993). This program computes the statistically averaged position of all stations based on the daily network solutions. Systematic differences between the daily solutions are eliminated by applying a 7-parameter Helmert transformation between each solution and a fixed one, arbitrarily chosen. Although GIPSY also includes tools to transform the daily solutions into ITRF2000, CINET provides more options to handle the data.

The combined solution is not yet in a known reference frame. The orbits were computed using a non-fiducial approach that introduces slight daily variations in the reference frame of the orbits (which, in turn defines the frame where the stations are referred to). To project the combined solution into ITRF2000, the ITRF2000 positions of the IGS stations are first propagated to each campaign epoch. The common stations (IGS) are used to estimate the systematic differences between the combined and the ITRF2000 solutions. The estimated Helmert parameters are then applied to the combined solution in order to obtain the ITRF2000 coordinates of the TANGO network for each epoch.

Finally, CINET was used again to estimate the velocity field of the TANGO network. The three velocity components are fitted through the time series of campaign solutions using a standard least-squares approach. Due to the unrealistic formal errors obtained from PPP and the modeling of just white noise (MAO et al., 1999), the velocity uncertainties were scaled with a factor of 20. This value has been estimated by averaging the fractions between the sum of the r.m.s. of the daily repeatabilities plus the r.m.s. of the mapping residuals with the formal uncertainties for each site/campaign.

For the permanent sites FAIM and PDEL, an extensive check of the quality of each daily solution and its mapping into ITRF2000 is not feasible due to the enormous amount of data. Therefore, scripts were created to automatically process and map the solutions into ITRF2000 with the applications included in the GIPSY software package. For each week, a combined solution was estimated using the daily solutions. Outliers were automatically removed by evaluating the three residuals of the position. Finally, all weekly solutions were combined in order to derive the linear velocity estimates.

3.2.2 GAMIT

With GAMIT, a totally different approach has been used in order to obtain the velocity field. Instead of computing campaign combined solutions and using them to derive the velocity field, the GAMIT applications estimate the station motion by fitting all the available daily solutions.

Figure 4 shows the scheme of the processing using the applications part of the GAMIT and GLOBK software packages.

For each campaign, two different networks were computed: a regional network containing the TANGO sites plus a set of IGS stations surrounding the region of interest. The latter, called fiducial sites, were also processed in a second run together with a set of globally distributed IGS stations (cf. Figure 5). No external global solutions publicly available are used so as to enable a thorough check and to ensure the consistency in the processing setup of the regional and global networks. The fiducial sites allow the connection between both networks. The division in two networks is necessary due to limitations on computer resources. The global sites have two functions: first, most of them are used to map the entire network into ITRF2000 in a global approach, constraining stations on the major plates; second, they allow the initially non-fiducial orbits to be adjusted.

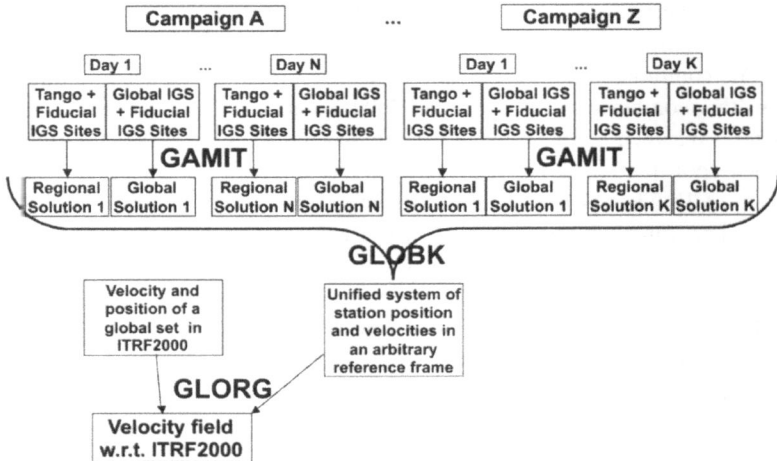

Figure 4

Scheme of the data processing in order to obtain the velocity field using the GAMIT, GLOBK and GLORG applications.

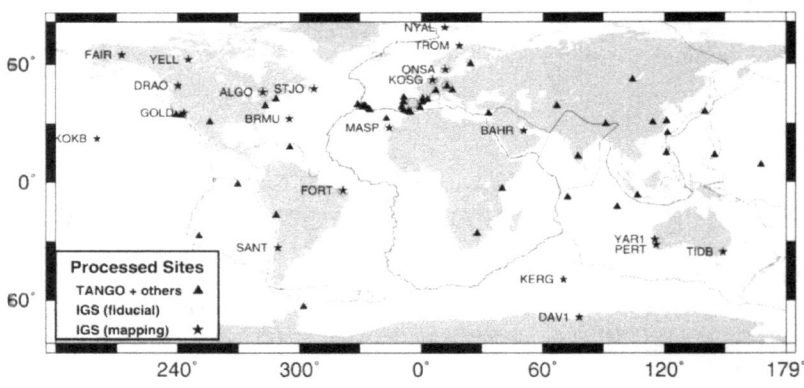

Figure 5

IGS stations used to map the TANGO network into ITRF2000 with GLOBK and GLORG. Some of the stations do not have available data for all campaigns.

All daily solutions are merged using GLOBK (DONG *et al.*, 1998). The output is a unique set of positions and velocities for the entire network, in an arbitrary reference frame. Next, the loosely constrained GLOBK system is scaled, rotated and translated to ITRF2000 in the *glorg* stabilization process by constraining positions and velocities of a set of globally distributed IGS stations (MCCLUSKY *et al.*, 2000).

3.3 Results

Figure 6 shows the estimated velocity fields for the horizontal component. The vertical component is not considered: due to the geometric constraints of the GPS system, this component has less accuracy. In addition, the "campaign" type of measuring is more sensitive to errors in the vertical component. However, errors in the antenna setup do not influence significantly the horizontal estimates due to the small antenna heights relative to the TANGO markers. The difference in the size of the ellipses results from the different weighting schemes adopted in each processing, whereas their west-east shape is due to not fixing the phase carrier ambiguities.

For the entire network, the GIPSY and GAMIT derived velocities are identical within the 95% confidence region (error ellipses in Fig. 6). This is clearly verified in Table 2, where most magnitude differences fall within the 1σ uncertainty level. The formal uncertainties at FAIM and FLOR are larger due to the amount of data: only the first three campaigns (from 1993 to 1997), whereas for the remaining stations 4–5 campaign observations (from 1993–1994 to 2000) have been used. The FLOR marker was destroyed in 1998. The motion of FAIM was influenced by the 9 July 1998 earthquake, where a significant coseismic horizontal motion of about 6 cm was reported by FERNANDES et al. (2002a). In consequence, the steady-state velocity of Faial is based on the campaign data until 1997. This estimate is compared with the velocity derived from the data of the permanent site together with the 1999 observations (cf. Fig. 6 and Table 2). A significant difference between the pre- and the post-seismic motions is observed: 4.4 ± 6.2 mm/y. The short data time-span of the permanent data (≈1.6 years), which causes the large associated uncertainty, does

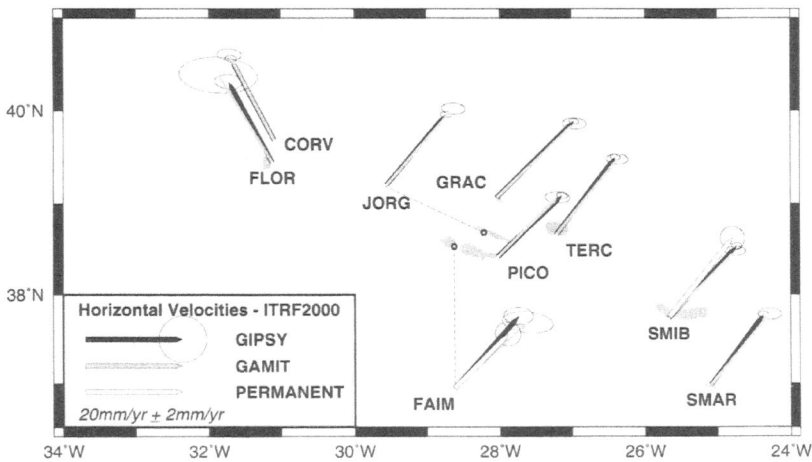

Figure 6
Horizontal velocities with respect to ITRF2000 (error ellipses represent 95% confidence interval).

Table 2

Magnitude of the differences between the velocity solutions presented in Figure 6: GAMIT and GIPSY solutions for TANGO network and the solution from the processing of the two continuous sites

Site	GAMIT-GIPSY		PERMANENT-GIPSY	
	Magnitude (mm/yr)	1σ (mm/yr)	Magnitude (mm/yr)	1σ (mm/yr)
FLOR	2.5	3.9		
CORV	0.9	1.3		
SMIB	1.1	1.1	1.7	5.8
SMAR	1.3	1.2		
TERC	0.6	1.1		
GRAC	0.5	1.2		
JORG	2.0	1.2		
PICO	1.1	1.3		
FAIM	2.3	2.8	4.4	6.2

not allow a conclusion as to whether the change in the observed motion is caused by strong tectonic stress in this region, revealed by the occurrence of a significant number of seismic events of small magnitude since 1988, or whether it is just a data feature. Since the velocity of the permanent station near SMIB shows a good agreement with that of the TANGO site, post-earthquake relaxation at the FAIM site is a probable hypothesis.

The derived velocity fields can be considered as very reliable due to the fact that identical estimations have been obtained with two different methodologies (software and strategies). In the subsequent analysis just the GIPSY velocity field is considered.

4. Discussion

4.1. Plate Models

The absolute velocity field gives the directions and magnitudes of the motions in a global reference frame. However, to fully comprehend the results, it is necessary to consider the relative motions with respect to the three major plates in the Triple Junction. Therefore, plate models specifying Euler vectors must be taken as reference. Currently, the most commonly known model is NUVEL-1A-NNR (DEMETS *et al.*, 1994), based on geophysical and geological data (magnetic anomalies, transform fault azimuths and earthquake slip vectors) covering approximately 3 million years.

However, Figure 7, where the residuals between the ITRF97 and ITRF2000 solutions and NUVEL-1A predictions are presented, exhibits significant discrepancies between space-geodetic motions and NUVEL-1A predictions. Not all used IGS stations (cf., Fig. 3) are plotted due to: (1) clarity of representation (namely in

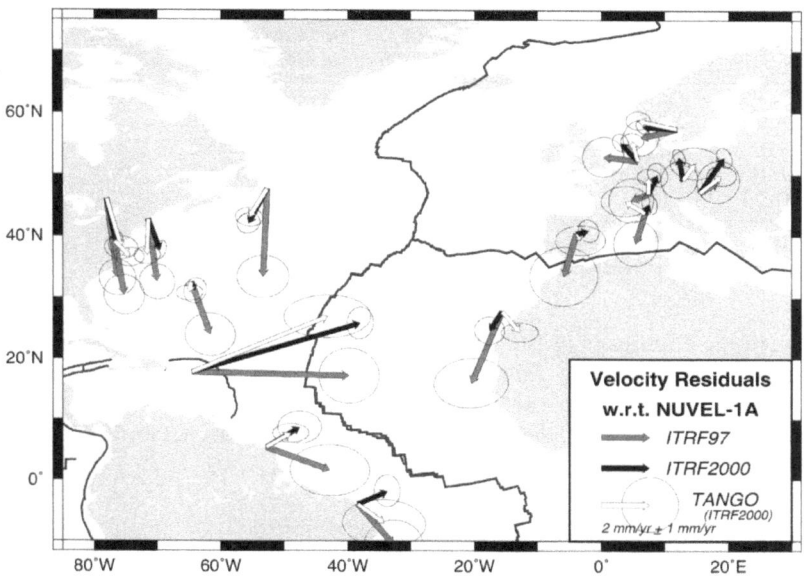

Figure 7
Differences between geodetic motions and NUVEL-1A-NNR predictions for IGS sites used for mapping the TANGO network onto ITRF2000. It is presented the ITRF97 (gray) and ITRF2000 (black) solutions and the estimated motions in ITRF2000 derived from the TANGO campaign solutions (white).

Europe, *e.g.*, HERS); (2) being located in deformation regions (*e.g.*, REYK); (3) were only used in 1–2 campaigns (*e.g.*, RCM6). The discrepancies are more evident for the ITRF97 residuals, especially for the North American plate. These differences are less significant when the ITRF2000 solution is considered, but still can be observed: the patterns of the residuals are discussed in FERNANDES *et al.* (2002b) for a larger set of stations.

In particular for the Eurasian plate, NOCQUET *et al.* (2001) has recently shown significant discrepancies between the motions predicted by NUVEL-1A-NNR and the geodetic motions derived from ITRF97. A significantly different relative pole between North America and Eurasia was also derived by KOGAN *et al.* (2000).

In order to obtain a model that better describes the present-day motions, several models have been developed based on the ITRF solutions (*e.g.*, SILLARD *et al.*, 1998) or own computed solutions (*e.g.*, LARSON *et al.*, 1997). The DEOS2k model (FERNANDES *et al.*, 2002b), based on ITRF2000, is used here as reference to analyze the relative motions. Table 3 shows the angular velocities for Eurasia, Africa and North America as predicted by DEOS2k.

Figure 7 also shows the motions as estimated using the GIPSY campaign solutions in ITRF2000 with respect to NUVEL-1A. Basically, the differences

Table 3

*Angular velocities according to the DEOS2k model for the three plates: EURA, AFRC and NOAM that form
the Azores Triple Junction (extracted from FERNANDES et al., 2002b)*

Plate	Lat.°N	Lon.°E	$\omega°$/Myr	σ^*_{max}	σ^*_{min}	ξ^{**}_{max}	$\sigma_\omega°$/Myr
EURA	53.59	254.72	0.245	1.67	0.44	38.1	0.003
AFRC	49.40	278.50	0.270	2.04	1.23	94.0	0.005
NOAM	-6.02	276.89	0.192	1.82	0.53	-2.4	0.004

* σ_{max}, σ_{min}: semi-major and semi-minor axes of the 1 σ error ellipse.
**ξ_{max}: azimuth of the semi-major axis reckoned clockwise from north.

between these residuals (white arrows) and the ITRF2000 residuals (black arrows) provide an estimate of the quality of the mapping. It is observed that, for most of the stations, the differences are at a sub-mm/year level. As expected, due to the reasons described in section 3.2.1, the largest discrepancies are observed in the southern part of the network.

4.2. Opening of the Mid-Atlantic Ridge

With respect to North America, CORV and FLOR have residual velocities, which are smaller than 1.5 mm/yr in magnitude and thus clearly equal to zero at the 95% confidence level. Therefore, the Western group of islands is considered to be located on the stable North American plate. The relative motions of the Central islands with respect to CORV and FLOR give an immediate estimate of the local spreading rate of the MAR (cf., Table 4). These estimates (23 mm/yr) agree extremely well with motions derived from magnetic anomalies observed north of the Azores (between 23 and 24 mm/yr) (ARGUS et al., 1989).

Table 4

*Differences in the magnitude and azimuth between the estimated (GIPSY) and predicted (DEOS2k)
velocities for the TANGO sites (azimuths counted from east, positive counter-clockwise)*

Site	Eurasia		Africa		North America	
	δMag (mm)	δAzimuth (degrees)	δMag (mm)	δAzimuth (degrees)	δMag (mm)	δAzimuth (degrees)
FLOR	23.2	177	18.8	174	1.0	8
CORV	23.1	176	18.7	174	1.2	23
SMIB	1.8	-153	2.9	-21	22.9	-7
SMAR	5.2	-167	1.7	-123	19.6	-9
TERC	2.9	172	1.6	14	21.5	-3
GRAC	1.2	5	5.6	2	25.4	-3
JORG	2.9	-157	2.0	-29	21.8	-7
PICO	3.3	-120	3.9	-46	23.1	-11
FAIM	1.7	-165	2.8	-5	22.8	-5

4.3. Eurasia-Africa Relative Motion

Figures 8 and 9 show the residual velocities of the TANGO sites in the Central and Eastern groups with respect to stable Eurasia and stable Africa, respectively. They are obtained simply by subtracting the predicted motion given by DEOS2k (considered as absolute reference) from the observed velocity. To compare, the predicted DEOS2k and NUVEL-1A-NNR motions for an arbitrary point on the stable part of the plates are also plotted.

No significant differences exist between GPS residual velocities as predicted by DEOS2k and NUVEL-1A-NNR. However, there is a significant difference between the relative motions predicted by these models for the stable plates: the magnitude is similar (4 mm/yr), but the predicted azimuth differs considerably. With respect to the principal orientation of the islands (roughly N120E), NUVEL-1A-NNR suggests that the mechanism in this boundary area is mainly a ridge opening (supported by the Terceira Rift model), whereas DEOS2k shows a more oblique motion (proposed by the "Leaky Transform" and the "Azores Block" models).

Considering the behavior of the stations individually (cf., Table 4), two of them clearly agree with the stable plate motions. GRAC belongs to the stable Eurasian plate, whereas SMAR follows the African plate motion with a larger residual with respect to Eurasia.

The remaining stations are clearly in the deformation zone, which seems to be diffuse. No considerations are made for FAIM due to the magnitude of the

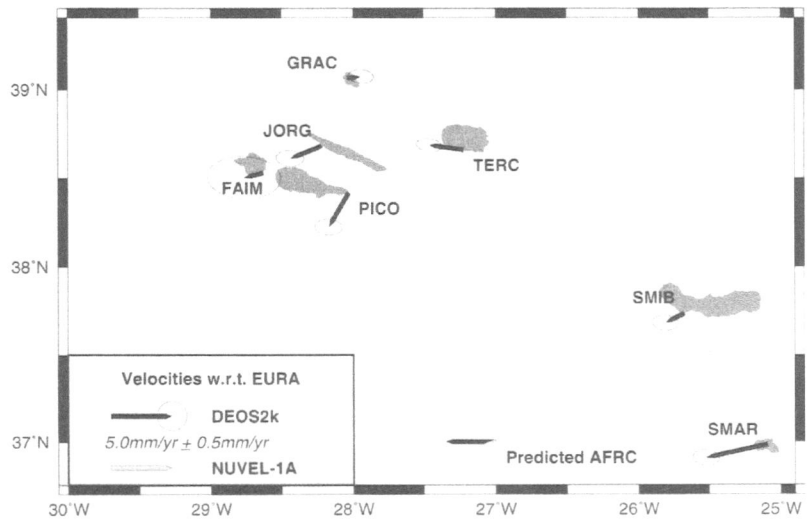

Figure 8
Residuals with respect to stable Eurasia (EURA) as predicted by DEOS2k. Also plotted is the relative predicted motion of Eurasia (EURA) with respect to AFRC by DEOS2k and NUVEL-1A-NNR.

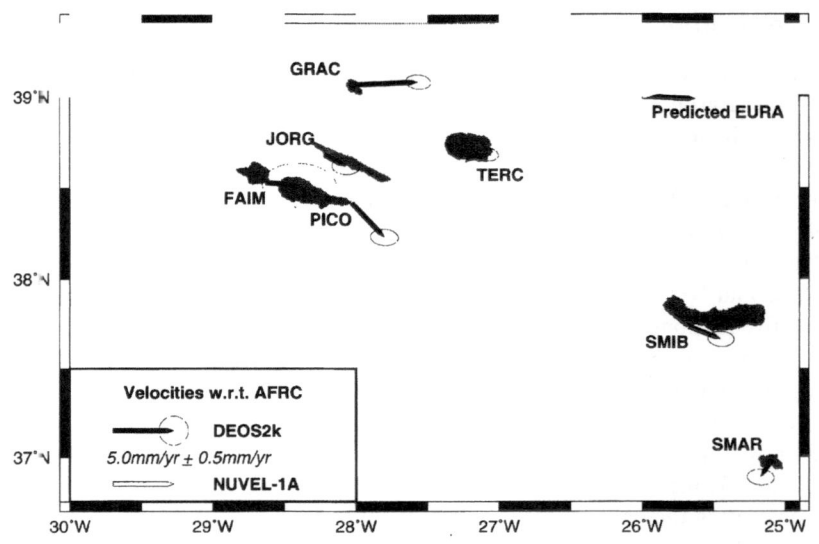

Figure 9

Residuals with respect to stable Africa (AFRC) as predicted by DEOS2k. Also plotted the relative predicted motion of Eurasia (AFRIC) with respect to EURA by DEOS2k and NUVEL-1A-NNR.

uncertainty. All sites, except PICO, show residuals that are within the range of the differential motion between both stable plates in magnitude and azimuth.

SMIB is slightly closer to Eurasia, which suggests that a large part of the deformation occurs between SMIB and SMAR. Oppositely, the TERC motion is slightly closer to Africa, suggesting that part of the boundary zone is located north of the TERC station. The densification of the GPS network on the island of Terceira can contribute to the determination of whether the deformation is concentrated within this island or not. The behavior of JORG is similar to that of TERC, which may also mean that part of the boundary is between JORG and GRAC. The large southern component of the PICO residual might be explained by the possible existence of an active fault, such as the ones suggested by the Azores Block model. Also here, the densification of the TANGO network will lead to a better understanding of the observed behavior.

5. Conclusions

The results presented here are based on GPS observations covering a time-span of seven years. This allowed derivation of a reliable velocity field, as proven by comparing the GIPSY and GAMIT solutions.

The GPS observations confirm that the Western group of the Azores Archipelago belongs to the North American plate. In the vicinity of the Central and Eastern groups, the relative motion between Africa and Eurasia is very small (4 mm/yr), making it difficult, at this stage, to draw conclusions concerning the exact location of the boundaries. Nevertheless, the analysis of the velocity field confirms the existence of a diffuse boundary zone, limited in the Northwest by Graciosa and in the southeast by Santa Maria.

The new sites in the Central group will considerably expand the available information for this region. Additionally, the recent increase of the number of available permanent stations in the archipelago will fundamentally contribute to the comprehension of the mechanisms that drive the geodynamic behavior of the Azores region.

Acknowledgments

The authors wish to express thanks to the Institute of Geodesy and Navigation, University FAF Munich, for their decisive contribution to the set-up of the TANGO project and collaboration in the first campaigns; to the Secretaria de Equipamento e Obras Públicas dos Açores for the excellent logistical support provided to all campaigns; and also to the observers that collaborated with extreme dedication in the realization of the different campaigns.

The figures have been produced using the GMT software (WESSEL and SMITH, 1991).

The TANGO project, from 1988 to 2000, was funded within the scope of different national Portuguese projects funded by FCT (Fundाçāp para a Ciência e Tecnologia). The most recent campaigns were further funded by the Netherlands research center ISES (Integrated Solid Earth Science).

REFERENCES

ALTAMIMI, Z., SILLARD, P., and BOUCHER C. (2002), ITRF2000: *A New Release of the International Terrestrial Reference Frame for Earth Science Applications*, J. Geophys. Res. doi:10.1029/2001JB000561.

ARGUS, D., GORDON, R., DeMETS, C., and STEIN, S. (1989), *Closure of the Africa-Eurasia-North America Plate Motion Circuit and Tectonics of the Gloria Fault*, J. Geophys. Res. *94*, 5585–5602.

BASTOS, L., OSÓRIO, J., LANDAU, H., and HEIN, G. (1991), *The Azores GPS Network*, Arqchipélago, Life and Earth Sciences *9*, 1–9.

BUFORN, E., UDÍAS, A., and COLOMBÁS, M.A. (1988), *Seismicity Source Mechanisms and Tectonics of the Azores–Gibraltar Plate Boundary*, Tectonophysics *152*, 89–118.

DeMETS, C., GORDON, R.G., ARGUS, D.F., and STEIN, S. (1994), *Effect of Recent Revisions to the Geomagnetic Reversal Time Scale on Estimate of Current Plate Motions*, Geophys. Res. Lett. *21*, 2191–2194.

R. M. S. Fernandes *et al.*   Pure appl. geophys.,

DONG, D., HERRING, T.A., and KING, R.W. (1998), *Estimating Regional Deformation from a Combination of Space and Terrestrial Geodetic Data*, J.Geodesy *72*, 200–214.

FERNANDES, R., SIMONS, W., AMBROSIUS, B., and BASTOS, L. (2000a), *Análise de diferentes metodologias no processamento de campanhas GPS*, Proc. 2ª Assembleia de Geodesia e Cartografia Luso, Portugal, IPCC, Lisboa, 92–99.

FERNANDES, R.M.S., BASTOS, L., AMBROSIUS, B., OSÓRIO, J., and BAPTISTA, P. (2000b), *The TANGO99 GPS campaign: Description and results*, Proc. 2nd Assembly Portuguese-Spanish of Geodesy and Geophysics, Lagos, Portugal, 35–36, 2000b.

FERNANDES, R.M.S., MIRANDA, J.M., CATALÃO, J., LUIS, J.F., BASTOS, L., and AMBROSIUS, B.A.C. (2002a), *Coseismic Displacements of the $M_W = 6.1$, July 9, 1998, Faial Earthquake (Azores, North Atlantic)*, Geophys. Res. Lett. 29, 16, doi:10.1029/2001GL014415, 2002a.

FERNANDES, R.M.S., AMBROSIUS, B.A.C., NOOMEN, R., BASTOS, L., WORTEL, M.J.R., SPAKMAN, W., and GOVERS, R. (2002b), *The Relative Motion between Africa and Eurasia as Derived from ITRF2000*, Geophys. Res. Lett., submitted for publication.

FREIRE LUÍS, J., MIRANDA, J.M., GALDEANO, A., PATRIAT, P., ROSSIGNOL, J.C., and MENDES VICTOR, L.A. (1994), *The Azores Triple Junction Evolution since 10 Ma from an Aeromagnetic Survey of the Mid-Atlantic Ridge*, Earth and Planet. Sci. Lett. *125*, 439–459.

HEFLIN H., BERTIGER, W., BLEWITT, G., FREEDMAN, A., HURST, K., LICHTEN, S., LINDQWISTER, U., VIGUE, Y., WEBB, F., YUNCK, T., and ZUMBERGE, J. (1992), *Global Geodesy Using {GPS} Without Fiducial Sites*, Geophys. Res. Lett. *19*, 131–134.

KING, R. W. and BOCK, Y. (1999), Documentation for the GAMIT GPS Software Analysis Version 9.9, Mass. Inst. of Technol., Cambridge.

KLOBUCHAR, J., *Ionospheric effects on GPS*. In *Global Positioning System: Theory and Applications I* (eds. B. Parkinson and J. Spliker), (AIAA, Washington, D.C., 1996) pp. 485–516.

KOGAN M., STEBLOV, G., KING, R., HERRING, T., FROLOV, D., EGOROV, S., LEVIN, V., LERNER-LAM, A., and JONES, A. (2000), *Geodetic Constraints on the Rigidity and Relative Motion of Eurasia and North America*, Geophys. Res. Lett. *27*(14), 2041–2044.

LARSON K., FREYMUELLER, J., and PHILIPSEN, S. (1997), *Global Plate Velocities from the Global Positioning System*, J. Geophys. Res. *102*, 9961–9982.

LOURENÇO, N., MIRANDA, J., LUIS, J., RIBEIRO, A., MENDES VICTOR, L., MADEIRA, J., and NEEDHAM, H. (1998, *Morpho-tectonic Analysis of the Azores Volcanic Plateau from a new Bathymetric Compilation of the Area*, Marine Geophys. Res. *20*, 141–156.

LUIS, J., MIRANDA, J., GALDEANO, A., and PATRIAT, P. (1998), *Contraints on the Structure of the Azores Spreading Center from Gravity Data*, Marine Geophys. Res. *20*, 157–170.

MADEIRA, J. and RIBEIRO, A. (1990), *Geodynamic models for the Azores Triple Junction: A Contribution from Tectonics*, Tectonophysics *184*, 405–415.

MAO A., HARRISON, C.G.A., and DIXON, T.H. (1999), *Noise in GPS Coordinate Time Series*, J. Geophys. Res. *104*, 2797–2816.

McCARTHY, D. (1996), IERS Conventions (1996), IERS Technical Note 21, Observatoire de Paris, Paris, 1996.

McCLUSKY, S., BALASSANIAN, S., BARKA, A., DEMIR, C., ERGINTAV, S., GEORGIEV, I., GURKAN, O., HAMEURGER, M., HURST, K., KAHLE, H., KASTENS, K., KEKELIDZE, G., KING, R., KOTZEV, V., LENK, O., MAHMOUD, S., MISHIN, A., NADARIYA, M., OUZOUNIS, A., PARADISSIS, D., PETER, Y., PRILEPIN, M., REILINGER, R., SANLI, I., SEEGER, H., TEALEB, A., TOKSOZ, M., and VEIS, G. (2000), *Global Positioning System Constraints on Plane Kinematics and Dynamics in the Eastern Mediterranean and Caucasus*, J. Geophys. Res. *105*, 5695–5719.

MIRANDA, J., MENDES VICTOR, L., SIMÕES, J., LUIS, J., MATIAS, L., SHIMAMURA, H., SHIOBARA, H., NEMOTO, H., MOCHIZUKI, H., HIRN, A., and LÉPINE, J. (1998), *Tectonic Setting of the Azores Plateau Deduced from a OBS Survey*, Marine Geophys. Res. *20*, 171–182.

NOCQUET, J.M., CALAIS, E., ALTAMIMI, Z., SILLARD, P., and BOUCHER, C. (2001), *Intraplate Deformation in Western Europe Deduced from an Analysis of the ITRF97 Velocity Field*, J. Geophys. Res. *106*, 11239–11258.

NOOMEN, R., AMBROSIUS, B., and WAKKER, K. *Crustal Motions in the Mediterranean Region determined from Laser Ranging to Lageos.* In *Contributions of Space Geodesy to Geodynamics: Crustal Dynamics,* (Am. Geophys. Union, Washington D.C., 1993) pp. 331–346.

SILLARD, P., ALTAMIMI, Z., and BOUCHER, C. (1998), *The ITRF96 Realization and its Associated Velocity Field,* Geophys. Res. Lett. *25,* 3223–3226.

WEBB, F. and ZUMBERGE, J., *An Introduction to GIPSY/OASIS-II, JPL D-11088* (California Institute of Technology, Pasadena, California, July 17, 1995).

WESSEL, P. and Smith, W. H. F. (1991), *Free Software Helps Map and Display Data,* EOS Trans. AGU **72,** 441.

ZUMBERGE, J., HEFLIN, M., JEFFERSON, D., WATKINS, M., and WEBB, F. (1997), *Precise Point Positioning for the Efficient and Robust Analysis of GPS Data from Large Networks,* J. Geophys. Res. *102,* 5005–5017.

(Received February 21, 2002, revised November 26, 2002, accepted January 28, 2003)

To access this journal online:
http://www.birkhauser.ch

Pure appl. geophys. 161 (2004) 701–722
0033–4553/04/030701–22
DOI 10.1007/s00024-003-2470-5

❙ Pure and Applied Geophysics

New Palaeomagnetic Data from the Betic Cordillera: Constraints on the Timing and the Geographical Distribution of Tectonic Rotations in Southern Spain

M. L. Osete[1], J. J. Villalaín[2], A. Palencia[1], C. Osete[1],
J. Sandoval[3], and V. García Dueñas[4]

Abstract — A palaeomagnetic investigation has been carried out at 14 sites on Jurassic red nodular limestones from the central and eastern part of the External Zones of the Betic Cordillera (Subbetic and Prebetic Zones). Progressive thermal demagnetisation of samples from the Subbetic Zone reveals the presence of two stable magnetic components of the natural remanent magnetisation: 1) a secondary Neogene syn-folding component and 2) the original Jurassic magnetisation. As similar characteristics have been reported in Jurassic limestones from the western Subbetic Zone, a widespread remagnetisation event took place within $< 10^6$ years in the entire Subbetic region during Neogene times. In contrast, in the Prebetic region, no evidence for a secondary overprint has been detected. Palaeomagnetic Jurassic declinations indicate variable and locally very large clockwise rotations (35°–140°), but the two sites in the north-westernmost part of the investigated region are not rotated. The use of both components of magnetisation and the incremental fold-test results allowed the timing of block rotations in the Subbetic Zone to be constrained. Rotations in the western Subbetic occurred after the acquisition of the secondary overprint, whereas in the central part of the Subbetic Zone they were completed by the time of the remagnetisation event.

Key words: Palaeomagnetism, Betic Cordillera, remagnetisation, rotation, Jurassic, Neogene.

Introduction

Palaeomagnetism is a very useful tool for studying the rotational component of the kinematics of a deformed region. Most commonly, palaeomagnetic declination is used to determine the component of vertical axis rotation, which is generally undetectable using conventional structural analysis. The Betic Cordillera is the

---
[1] Dep. Física de la Tierra I. F. CC. Físicas, Universidad Complutense, 28040 Madrid, Spain.
E-mail: mlosete@fis.ucm.es
[2] Departamento de Física, Escuela Politécnica Superior, Universidad de Burgos, Avda, Cantabria s/n, 09006 Burgos, Spain.
[3] Departamento de Estratigrafía y Paleontología, Universidad de Granada, Avda, Fuentenueva s/n, 18071 Granada, Spain.
[4] Departamento de Geodinámica, Universidad de Granada, Avda, Fuentenueva s/n, 18071 Granada, Spain.

northern branch of the Betic-Rifean orogen, an arc-shaped mountain belt bordering the Alboran Sea, that constitutes the westernmost segment of the Mediterranean Alpine orogenic system. The entire chain developed in response to the collision between Africa and Eurasia since the late Mesozoic. The Betic-Rifean orogen can be divided into four tectonic domains (BALANYÁ and GARCÍA-DUEÑAS, 1987): the Alboran domain (Internal Zones), the Southiberian and Maghebrian domains (External Zones) and the allochthonous Flysch trough. The Alboran domain is made up of several thrusts that have been grouped into three main tectonic complexes (the Nevado-Filabride, Alpujarride and Malaguide complexes) and mainly consist of metamorphic rocks of Paleozoic and Triassic age. The Southiberian and Maghebrian domains represent the paleomargins of the Iberian and African plates respectively, and comprise mostly unmetamorphosed Mesozoic and Tertiary sediments. The External Betic Zones are divided into the Prebetic Zone (the most external) and the Subbetic Zone (itself being differentiated into External, Middle and Internal Subbetic).

Over the last 15 years, the Betic Cordillera has been the subject of several palaeomagnetic studies. Early tectonic studies were performed in Jurassic volcanic and sedimentary rocks from the Subbetic Zone (OSETE *et al.*, 1988, 1989). They showed that systematic dextral block rotations took place in the central part of the Subbetic (to the north of Granada) and a very large rotation was found at one site in the eastern Betics. PLATZMAN and LOWRIE (1992), PLATZMAN (1992) and PLATT *et al.* (1994) carried out extensive palaeomagnetic studies in sedimentary rocks of Jurassic and Cretaceous age around the Gibraltar Arc. They observed clockwise block rotations in the western Subbetic and counterclockwise rotations in the Rif Mountains of Morocco. A systematic palaeomagnetic study of mostly Jurassic sedimentary rocks from the eastern External Betic Zone was conducted by ALLERTON *et al.* (1993, 1994). They observed a more heterogeneous behavior with mainly clockwise rotations, sometimes very large, along with some regions that had experienced no rotation at all. VILLALAÍN *et al.* (1994) carried out a palaeomagnetic study in grey oolitic limestones and grey and red nodular limestones of upper Jurassic age from the western Subbetic. This revealed that the natural remanent magnetisation (NRM) of these rocks is dominated by a widespread and pervasive remagnetisation of Neogene age. This work also demonstrated that the remagnetisation is coeval with the deformation by folding in the Betics. Later, VILLALAÍN *et al.* (1996) presented an evaluation of the consequences of an incorrect interpretation of the nature (primary or secondary) of the NRM. They found that heterogeneous rotational patterns could be observed if the secondary magnetisation is erroneously interpreted as a primary magnetisation. The palaeomagnetic study carried out by KIRKER and McCLELLAND (1996) on upper Jurassic grey micrites from the western Subbetic also revealed a multicomponent remanence, including a syn-deformational component of magnetisation.

The existence of a strong secondary overprint in Jurassic limestones gives rise to doubts about the reliability of previous palaeomagnetic studies carried out in the Betics in which clockwise but heterogeneous rotations have been found (OSETE et al., 1989; PLATZMAN and LOWRIE, 1992; ALLERTON et al., 1993). This may also be true in northern Africa, where a complex pattern can be seen (PLATZMAN, 1992). At present the Neogene remagnetisation has been well documented only in the western Subbetic. Although there is evidence indicating that rocks from the central and eastern Subbetic could also be remagnetised (OSETE et al., 1988; ALLERTON et al., 1993), it is not possible, with the presently available data, to extrapolate the existence of an intensive Neogene remagnetisation event to the rest of the Betics. This study has three goals: 1) to determine if Jurassic nodular limestones from the central and eastern Subbetic Zone are remagnetised; 2) to quantify the block rotations at selected sites where it is clearly demonstrated that the primary component is present and correctly isolated and 3) to investigate the distribution of rotational deformation across the central Subbetic.

Sampling Strategy and Palaeomagnetic Methods

Sixteen sedimentary sites of Jurassic age were investigated in the central and eastern part of the External Betic Zone that were grouped into nine localities to enable field-tests (Fig. 1). Most palaeomagnetic samples were drilled in the field using a portable, two-stroke, hand-held drill. Orientation was achieved using a magnetic compass (the low magnetisation of the sediments had no effect on the compass orientation). From two sites (BRJ sites) oriented hand samples were obtained that were drilled and cut into standard palaeomagnetic specimens in the laboratory. Typically between 10 and 36 cores were taken at each site. Most of the sites are within the External and Middle Subbetic, with three sites (CAZ sites) in the Prebetic.

Previous palaeomagnetic studies in the External Betics have shown that the most favorable lithology for palaeomagnetic purposes are the red nodular limestones (ammonitico rosso facies) of Jurassic age. Therefore sampling was concentrated on this facies (14 sites). In addition grey micritic limestones of Toarcian and Tithonian age (CAZ2 and CYT sites) were sampled at two sites, however these exhibited very weak intensities and unstable direction behaviour and were excluded in the following discussion.

The sampling strategy was planned taking account of the characteristics of the Neogene remagnetisation affecting the western Subbetic, in order to detect if the secondary magnetisation is present in Jurassic sediments in the central and eastern Subbetic and so to be able to interpret properly the NRM components. These characteristics were (VILLALAÍN et al., 1994): 1) the remagnetised component was exclusively of normal polarity, and 2) the remagnetisation occurred during differing stages of Neogene deformation by folding of the Subbetic (pre-, syn- or post- folding

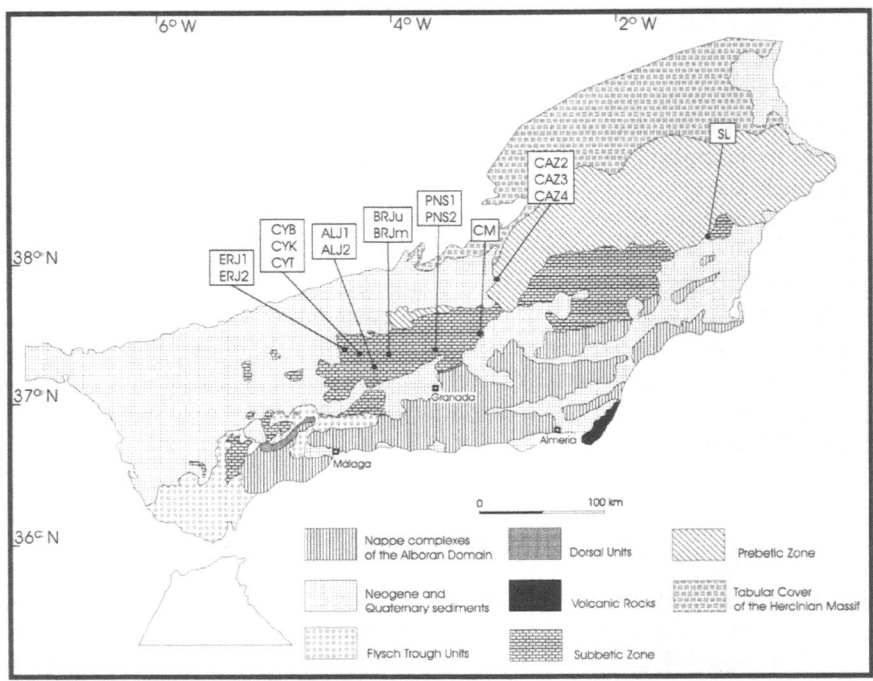

Figure 1
Simplified geological map of southern Spain showing the location of sampling sites.

in different folds in the western Subbetic). As the Jurassic is a period of mixed polarity of the Earth's magnetic field (GRADSTEIN *et al.*, 1994), cores were taken from continuous and well expanded sections to ensure sampling of more that one polarity interval. Therefore, the original Jurassic component, if present, should exhibit both normal and reversed polarities, in contrast to the secondary component of normal polarity. Three cylindrical folds of kilometric scale were sampled at the localities ALJ, PNS and BRJ in order to perform fold-tests. In the BRJ locality, due to outcrop constrains, upper Jurassic (Kimmeridgian-Tithonian) red nodular limestones were sampled in the northern limb of a SW-NE oriented fold. In the southern limb middle Jurassic red nodular limestones (ammonitico rosso facies) were sampled. At ALJ and PNS the same Jurassic stage was sampled on both limbs of the fold. Two SW-NE oriented anticlines were sampled at sites ALJ and PNS.

A further consideration was that an angular difference of about 20° is observed between the upper and the middle-lower Jurassic poles for Iberia (SCHOTT *et al.*, 1981; JUÁREZ *et al.*, 1996; GIALANELLA, 1999). Therefore some rotational scatter could be introduced by larger scale apparent polar wander if the age of the investigated sites is not well controlled. Only outcrops with very good biostrati-graphic control were sampled. The age of sites was based on ammonites assemblages and ranges from the Toarcian up to the Tithonian (SEQUEIROS, 1974; OLÓRIZ, 1978; JIMÉNEZ and RIVAS, 1979; SANDOVAL, 1983; LINARES and SANDOVAL, 1993). In order to investigate if there is a gradient in the rotational deformation in the central Subbetic, sampling sites are also located along an E-W oriented transect (see Fig. 1).

Magnetic measurements were carried out at the palaeomagnetic laboratories of Madrid and Zurich universities. The NRM was measured using JR5 spinner and 2G cryogenic magnetometers. Progressive thermal demagnetisation was performed using Schonstedt TSD-1 and ASC furnaces. In addition, some pilot samples were demagnetised with a GSD-5 Schonstedt alternating field demagnetiser. Bulk magnetic susceptibility was measured after each thermal demagnetisation step to monitor magnetic mineral alteration during heating. The component structure of NRM was analyzed on orthogonal plots using standard least-squares routines.

Palaeomagnetic Results

The intensity of the NRM varied between $5 \ 10^{-3}$ and $4 \ 10^{-4}$ A/m. After a pilot study, thermal treatment was found to be more effective than alternating field demagnetisation in isolating different magnetic components, and was then system-atically applied to all remaining samples. These samples were heated from room temperature up to 600°C or 700°C in temperature intervals varying between 20°C and 125°C. Special care (temperature steps of 20°–50°C) was necessary at high temperatures (over 400°C) to better constrain the magnetic components. Thermal demagnetisation treatment reveals two distinct types of behaviour, and sites have been grouped accordingly.

Group 1. This comprised most sites (ERJ, ALJ, CM, CYB, CYK, PNS, SL, BRJu and BRJm). After removing a viscous magnetisation, two stable components could be identified (Figs. 2a–n). A low-temperature component with a maximum unblocking temperature of 450–475°C was isolated after heating to 200–350°C. This component was always of normal polarity. A subsequent high-temperature component was removed by 550–575°C. The high-temperature component showed normal and reversed polarities. This magnetic behaviour is similar to that described

Figure 2
Orthogonal vector plots showing thermal demagnetisation of representative samples from each sampling locality. Directions are plotted in geographic coordinates. Solid symbols are for the horizontal projection and open symbols for the vertical projection. Component A always shows normal polarity. Component B shows both normal and reversed polarities.

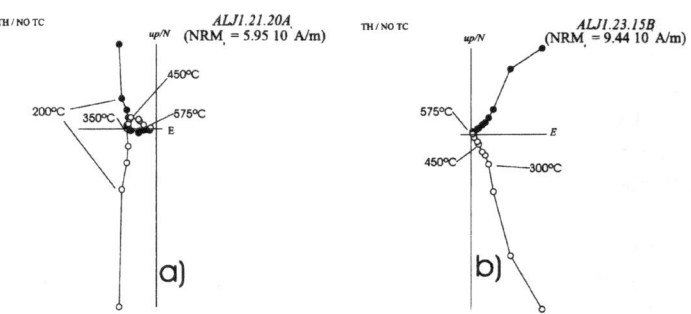

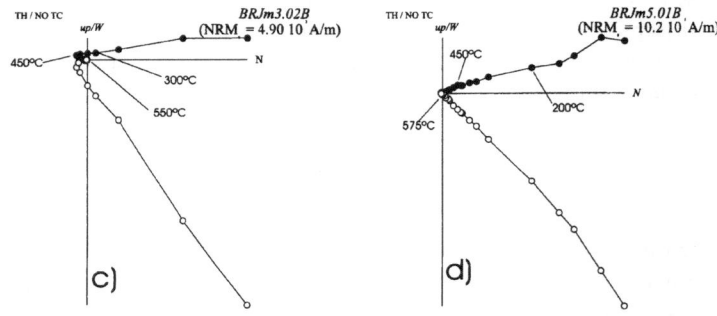

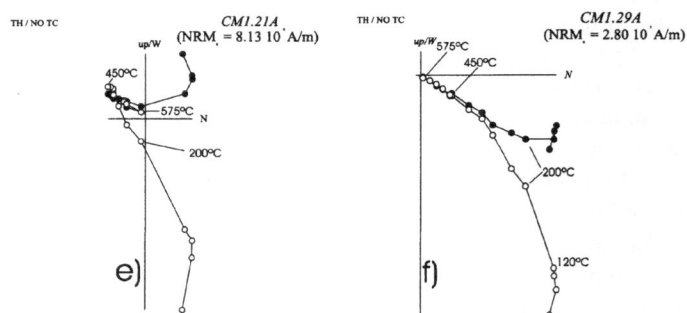

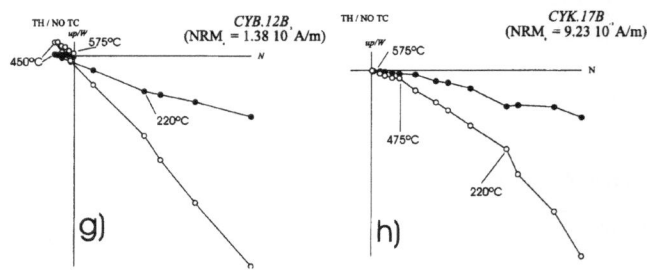

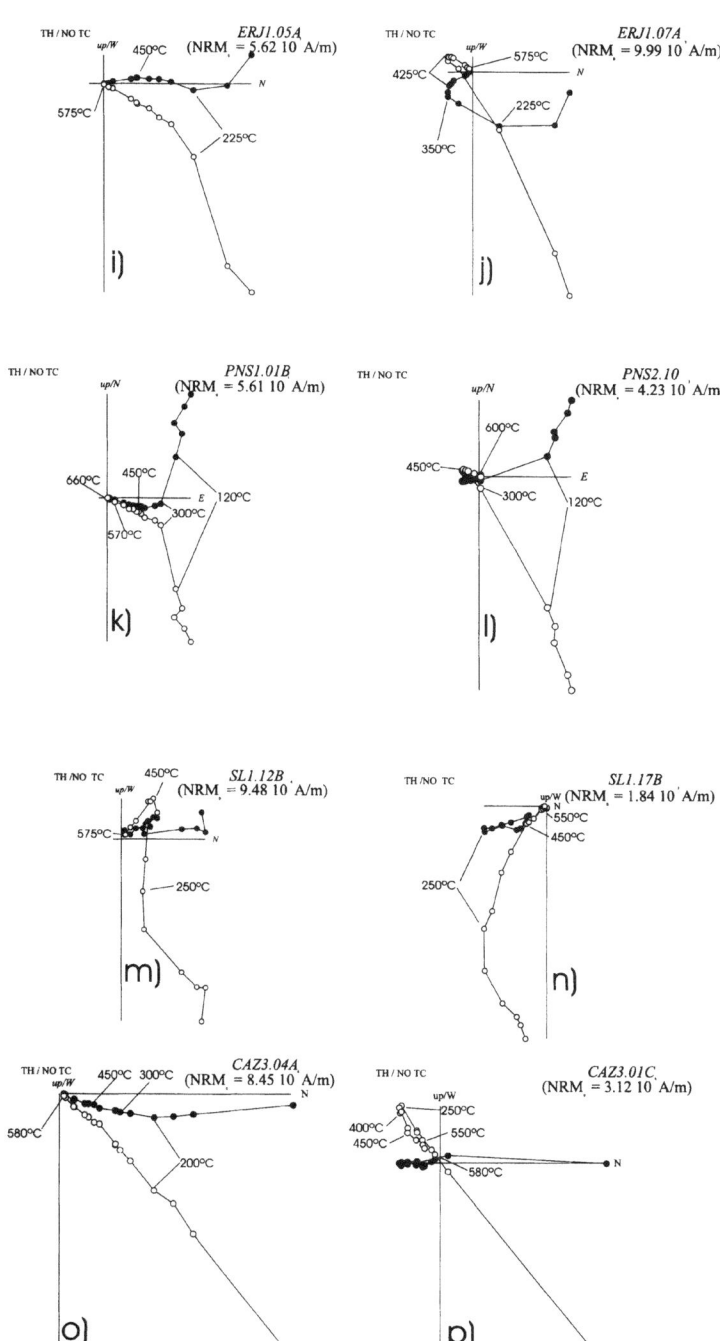

by VILLALAÍN *et al.* (1994) in the western Subbetic. Following the nomenclature used in the previous studies, the low- and high-temperature component are referred to as components A and B, respectively. Both components could be isolated in most of the sites. At one site, PNS, component A could not be properly isolated due to technical problems leading to the loss of data between 120°C and 300°C. In some samples from other sites the overlap between components A and B made the proper isolation of directions difficult. These samples were rejected in the subsequent calculations. At BRJ sites component B had much lower intensities and the NRM was strongly dominated by component A (see Figs. 2c and 2d). Nonetheless it was possible to determine directions of component B with acceptable accuracy.

Group 2. This group comprised two sites: CAZ3 and CAZ4, located in the Prebetic Zone, close to the locality of Cazorla. The NRM was composed of a viscous component, removed after heating to 200–250°C, and a high-temperature component (maximum unblocking temperature of 550–575°C) exhibiting normal and reversed polarities (Figs. 2o and 2p). The characteristics of this high temperature component were similar to those of component B of Group 1, whereas component A was either not present or very weak.

Component B exhibited normal and reversed directions at all localities (Fig. 3). A consistency between the polarity of the magnetisation and the stratigraphic position of the samples was also observed (detailed magnetostratigraphic studies are in progress in some of these sections). The mean directions and statistical parameters of component B are summarized in Table 1, which includes results of a reversal test. This test was statistically positive at different degrees of probability (95% or 99%) for all sites, indicating that component B has been well isolated statistically. The locality mean directions have $\alpha_{95} \leq 13°$ and can be considered for tectonic purposes.

Component A is present in all sites from the Subbetic Zone, always with a normal polarity magnetisation. Its relative intensity varies depending on the sites, but usually carries more than 50% of the non-viscous fraction of the NRM intensity. Mean site directions of component A are shown in Table 2.

A fold test was performed at three localities (ALJ, BRJ and PNS) with results for each component shown in Table 3 and Fig. 4. The statistical parameter used to estimate the significance of fold-test was determined using the McFADDEN and JONES (1981) method. Fold-test results are significantly different for components A and B: the best grouping of directions of component B is observed after tectonic correction, whereas component A clearly fails the fold-test. The fold-test is positive for component B at the 95% level of confidence at sites ALJ and PNS. At BRJ, the best grouping of this component is achieved after tectonic correction, although it is not statistically significant at the 95% level of confidence. This is probably due to the higher scatter of directions observed in these sites. This scatter is produced by the Upper to Middle Jurassic apparent polar wander (Upper Jurassic rocks were

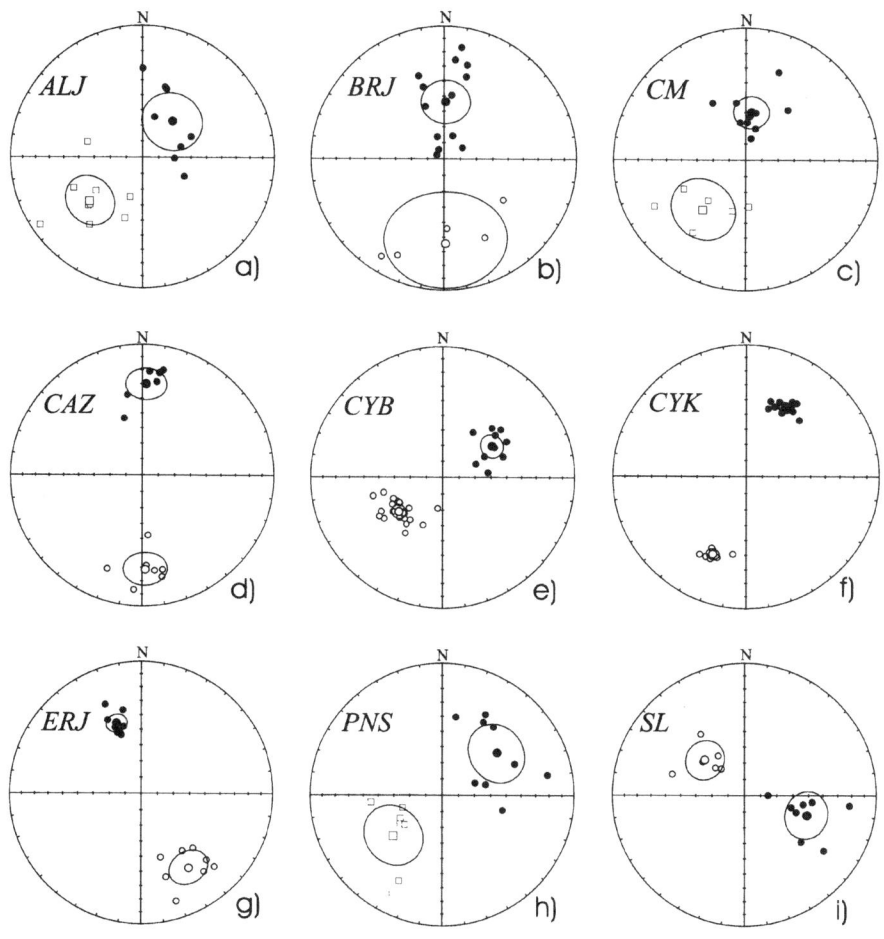

Figure 3

Equal area projections showing directions of component B for each locality after bedding correction. Black (white) dots mean upper (lower) hemisphere. Mean directions and 95% confidence circle are also represented for both normal and reversed populations.

sampled at BRJu and Middle Jurassic at BRJm). In addition, the lower intensity of component B with respect to component A could also introduce additional scatter. In any case, the best grouping is achieved after tectonic correction and the statistical parameter f and its $F_{99\%}$ significance level value are very close ($F_{99\%} = 0.334$ and $f = 0.345$). Therefore this fold-test is also considered positive.

The directions of component A in ALJ and BRJ have a distribution which is statistically different at the 95% and 99% level of confidence for both pre-folding

M. L. Osete *et al.*

Table 1

Palaeomagnetic directions and statistical parameters of component R

Site	Site Latitude	Site Longitude	polarity	n	Before T.C. Dec	Inc	k	α95	After T.C. Dec	Inc	k	α95	Reversal Test Results γ0	γc (95%)	γc (99%)
ALJ1				8	86.4	43.0	26.6	10.9	32.3	49.2	38.7	9.0			
ALJ2	37.32	−4.17		9	23.8	−4.3	8.9	18.3	56.7	55.8	7.1	20.8			
ALJ			N	8	25.8	10.0	6.4	29.0	39.2	60.6	10.0	18.5			
			R	9	251.2	−28.6	5.0	25.6	230.2	−46.6	12.3	15.3			
			N + R	17	49.0	21.6	3.7	21.6	43.6	53.1	10.9	11.3	16.4652	18.1348 pos c	23.1348 pos ind.
BRJu				11	21.3	83.0	13.1	13.1	353.8	38.8	16.3	11.6			
BRJm				7	349.6	32.0	5.2	29.4	22.8	64.4	6.3	26.0			
BRJu + BRJm	37.41	−4.02	N	13	345.8	67.6	10.0	13.8	1.7	53.7	9.4	14.3			
			R	5	178.3	−47.1	2.2	67.8	178.8	−35.5	6.0	34.0			
			N + R	18	355.9	65.4	5.0	17.3	0.8	48.8	8.1	13.0	18.309	29.1468 pos. ind	37.3544 pos. ind
CAZ3				8	13.7	40.4	26.7	10.9	1.7	29.5	26.7	10.9			
CAZ4				6	6.9	34.5	31.9	12.0	358.4	28.7	31.9	12.0			
CAZ	37.96	−2.93	N	7	13.9	38.8	25.0	12.3	2.5	29.9	31.0	11.0			
			R	7	187.5	−36.9	30.6	11.1	178.1	−28.4	27.1	11.8			
			N + R	14	10.7	37.9	28.8	7.5	0.3	29.2	30.7	7.3	4.1714	15.1271 pos. c	19.4460 pos. c
CM			N	10	32.5	28.7	21.9	10.6	6.7	59.6	21.9	10.6			
	37.60	−3.25	R	6	226.7	−10.8	12.7	19.5	220.2	−48.4	12.7	19.5			
			N + R	16	38.1	22.3	14.0	10.2	21.2	56.6	14.0	10.2	22.3434	18.9537 neg.	24.2877 pos indet.
CYB			N	10	11.4	47.5	46.1	7.2	57.3	53.6	46.1	7.2			
	37.44	−4.29	R	21	186.8	−46.0	53.6	4.4	232.2	−55.4	53.6	4.4			
			N + R	31	8.3	46.5	51.7	3.6	53.9	54.8	51.7	3.6	3.5840	7.8205 pos. b	9.8392 pos. b

Site	Lat	Long	Pol	n	Dec	Inc	k	α95	Dec (T.C.)	Inc (T.C.)	k (T.C.)	α95 (T.C.)	γo	γc(95%)	γc(99%)
CYK	37.44	−4.29	N	12	8.1	23.4	143.0	3.6	28.8	39.1	143.0	3.6			
			R	9	185.8	−18.3	206.3	3.6	202.7	−36.2	206.3	3.6			
			N + R	21	7.1	21.2	143.1	2.7	26.1	37.9	143.1	2.7	5.5504	5.0438 neg.	6.3932 pos. b.
ERJ1				5	334.9	25.3	47.1	11.3	322.8	37.7	47.1	11.3			
ERJ2				11	348.4	27.6	21.5	10.1	338.6	38.1	32.4	8.1			
ERJ	37.47	−4.37	N	8	351.9	31.7	68.3	6.8	340.3	42.9	86.6	6.0			
			R	8	156.6	−19.5	14.7	15.0	147.6	−33.0	25.6	11.2			
			N + R	16	344.1	27.1	23.9	7.7	333.6	38.2	31.7	6.7	14.0840	11.9467 neg	15.2815 pos. c
PNS1				11	90.7	26.4	13.6	12.9	52.0	46.9	11.9	13.8			
PNS2				5	358.3	17.5	8.9	27.1	50.1	50.4	7.7	29.5			
PNS	37.43	−3.55	N	9	41.7	30.7	3.4	33.0	52.4	46.4	9.5	17.7			
			R	7	268.9	−25.6	5.4	28.7	230.2	−49.9	11.6	18.5			
			N + R	16	65.0	30.7	3.4	23.5	51.4	48.8	11.0	11.7	3.8137	24.491 pos ind	31.4468 pos. ind.
SL	38.25	−1.20	N	8	131.7	50.7	16.5	14.0	107.9	48.9	16.5	14.0			
			R	6	337.8	−50.1	32.2	12.0	311.8	−57.1	32.2	12.0			
			N + R	14	143.1	51.1	18.2	9.6	117.2	53.0	18.2	9.6	16.4652	18.1348 pos c	23.3567 pos ind

n, number of samples; N, normal polarity; R, reversed polarity; N + R, normal and reversed polarity; Dec. and Inc., declination and inclination; k and α95, statistical parameters (Fisher, 1953); γo; γc(95%); γc(99%), Statistical parameters of reversal test (McFadden and McElhinny, 1990); T.C., tectonic correction.

Table 2

Palaeomagnetic directions and statistical parameters of component A

Site	n	Before T.C.				After T.C.			
		Dec.	Inc.	k	α_{95}	Dec.	Inc.	k	α_{95}
ALJ1	17	26.2	65.1	11.1	11.2	349.0	28.9	12.1	10.7
ALJ2	8	10.2	21.3	13.6	15.6	119.2	70.7	11.1	17.4
BRJu	14	19.9	64.9	23.0	8.5	5.2	24.6	50.3	5.7
BRJm	8	344.1	48.9	34.0	9.6	43.9	80.3	135.0	4.8
CM	18	28.2	48.4	25.7	6.9	324.8	69.4	25.7	6.9
CYB	36	24.7	48.1	360.1	1.3	66.0	46.5	360.1	1.3
CYK	22	19.4	46.8	26.3	6.2	61.2	48.7	26.3	6.2
ERJ1	7	25.2	53.0	78.1	6.9	25.0	73.9	81.7	6.7
ERJ2	12	2.9	35.5	35.1	7.4	359.0	47.3	14.8	11.7
SL	11	159.4	64.5	108.6	4.4	112.2	68.6	108.9	4.4

Same symbols as in Table 1.

and post-folding configurations (Table 3). Incremental fold-tests were performed for these two localities (Figs. 4d and 4e). Table 3 gives the unfolding values for which maximum clustering were obtained (20% of unfolding for BRJ and 30% for ALJ). Mean locality directions for component A, calculated after these percentages of full bedding correction, are also given in Table 3.

Component B is considered to be the primary Jurassic component, based on: 1) the presence of normal and reversed polarities of magnetisation which is consistent with the pattern of reversals of the Earth's magnetic field for the Jurassic period (GRADSTEIN *et al.*, 1994); 2) the stratigraphic consistency of polarity of the magnetisation; 3) fold test results that indicate a pre-folding acquisition of the magnetisation and 4) the inclination values obtained after bedding correction that are in agreement with the expected Jurassic inclinations (SCHOTT *et al.*, 1981; STEINER *et al.*, 1985; GALBRUN *et al.*, 1990; JUÁREZ *et al.*, 1996, 1998 and GIALANELLA, 1999). The inclination variability of component B is related with the Jurassic path of the Iberian APWp, which is in agreement with the Jurassic segment of the North-American APWp proposed by VAN FOSSEN and KENT (1990).

The secondary origin of component A is deduced from the negative results of the fold-test and from the incremental-fold-test data. Considering the folding deformation history of the Subbetic (e.g., GARCÍA-HERNÁNDEZ *et al.*, 1980; BALANYÁ and GARCÍA-DUEÑAS, 1987; PLATT and VISSERS, 1989; BANKS and WARBURTON, 1991; GARCÍA-DUEÑAS *et al.*, 1992; ALLERTON, 1994) and the syn-folding character of the remagnetisation, it is possible to conclude that the remagnetisation process took place in Neogene times. In addition, inclination values of the secondary component in BRJ and ALJ sites, after partial bedding correction (Table 3), are in agreement

Table 3

Fold test results for components B and A

Locality		N	Before Tectonic Correction					After Tectonic Correction					F95%	F99%	fold-test result % of confidence
			D	I	k	a95	f	D	I	k	a95	f			
Component B	ALJ	17	49.0	21.6	3.7	21.6	2.433	43.6	53.1	10.9	11.3	0.1174	0,221	0.359	Prefolding (95%)
	BRJ	18	355.9	65.4	5.0	17.3	0.7460	0.8	48.8	8.1	13.0	0.3448	0.206	0.334	Prefolding (close to 99%)
	PNS	16	65.0	30.7	3.4	23.5	2.400	51.4	48.0	11.0	11.7	0.1542	0,239	0.389	Prefolding (95%)
Component A	ALJ	25	18.0	51.5	6.9	11.9	0.7597	357.9	51.5	4.1	16.5	1.660	0,139	0.222	Synfolding acquisition after 30 %unf
			After 30% Unfolding:					9.9	52.8	11.2	9.0	0.02346	0,139	0.222	
	BRJu+BRJm	22	3.0	60.3	17.1	7.7	0.5851	8.8	45.0	7.4	12.2	7.637	0.1616	0.2589	Synfolding acquisition after 20 %unf
			After 20% Unfolding:					3.8	57.5	26.1	6.2	0.2544	0.1616	0.2589	

Same symbols as in Table 1: f, McFADDEN and JONES (1981) fold test statistical parameters; $F_{95\%}$ and $F_{99\%}$, significance level value of f.

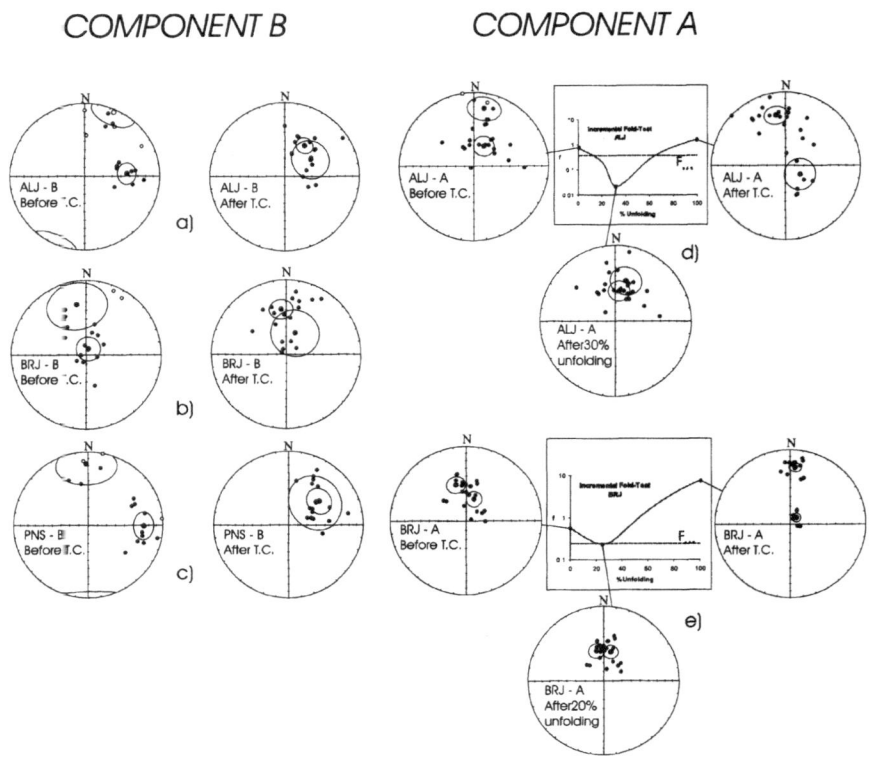

Figure 4

Fold-test results of ALJ, BRJ and PNS localities. a), b), c): Equal area projections showing directions for component B before and after bedding correction. Mean direction and 95% confidence circle are also represented for each limb (site) of the fold. d), e): Equal area projections showing directions for component A for ALJ and BRJ sites. Results of the incremental fold test are also included: the MCFADDEN and JONES (1981) parameter f of Component A as a function of the percentage unfolding of bedding tilt about the line of strike. Dashed line indicates the critical value of f at the 95% confidence level ($F_{95\%}$). An additional equal area projection with the directions corresponding to the best-clustered configuration is also presented for both sites.

with the expected Tertiary inclination values for Iberia (DIJKSMAN, 1977; BÓGALO *et al.*, 1994; BARBERÁ *et al.*, 1996).

In summary, all sites investigated in the Subbetic show a similar magnetic behaviour: The NRM is composed of two stable components: 1) Component A which is interpreted as a secondary syn-folding magnetisation of Neogene age, and 2) Component B which is considered to represent the original Jurassic magnetisation. In contrast, in the two sites studied in the Prebetic region, the only stable component identified is component B.

Discussion

Remagnetisation

The magnetic properties of the red nodular limestones in the central and eastern part of the Subbetic Zone are similar to those found in Jurassic limestones from the western Subbetic (VILLALAÍN *et al.*, 1994). All sites from the Subbetic show a very stable secondary component. Consequently, the principal characteristic of the remagnetisation phenomenon in the Subbetic is its widespread character. It is most probably related to an important event in the geological history of the rocks which produced massive fluid migrations and mineralisations, significant heating or all of these.

The occurrence of remagnetisations in rocks has been known for many years, nonetheless during the last ten years, remagnetisation has been recognised as considerably more common than previously supposed. These studies have demonstrated the existence of remagnetisations ranging from local episodes to very widespread events (e.g., McCABE and ELMORE, 1989). To explain remagnetisations affecting extensive areas, which were tectonically active during the times of remagnetisation, the formation of authigenic magnetite associated with tectonically driven fluid migration has been proposed (McCABE *et al.*, 1983; OLIVER, 1986; SUK *et al.*, 1990, 1993). A thermoviscous mechanism related with heating has also been suggested (e.g., DOBSON and HELLER, 1992; VILLALAÍN, 1995; JUÁREZ *et al.*, 1998).

The maximum unblocking temperature of 450°C of the secondary component is anomalous (it does not correspond to a Curie temperature of the magnetic minerals commonly found in sediments) though very constant throughout the whole of the studied region. These very curious properties of the remagnetised component observed in the Subbetic are not well documented in other investigated widespread remagnetisations, which should be related with similar remagnetisation process. The study of the overprint in the Subbetic can offer clues to the understanding of the remagnetisation mechanism associated with active tectonics. The secondary component always exhibits normal polarity. On the basis of fold-test results, the age of the remagnetisation has been proposed as Neogene, a period of mixed polarity. It can thus be concluded that this secondary component was acquired in a relatively short time-span. 10^6 years is the maximum length of any of the normal polarity chrons during the Neogene (CANDE and KENT, 1995). In contrast to the Subbetic Zone, the secondary component could not be identified in the two sites investigated in the Prebetic. This means that this region is not strongly remagnetised or not remagnetised at all, therefore the mechanism operating with the Subbetic was apparently confined to that zone.

Rotations

Table 4 gives the palaeomagnetic declinations of the considered original Jurassic magnetisation (component B), the expected declinations values from stable Iberia and the rotation calculated for each locality. The expected declinations values have

Table 4

Bulk rotation calculated from component B

Locality	Age	$D_{OB} \pm \Delta D_{OB}$	$D_{EX} \pm \Delta D_{EX}$	$R (= D_{OB}-D_{EX}) \pm \Delta R$
ERJ	Upper Jurassic	334 ± 9	324 ± 4	10 ± 13
CYB	Bajocian-Bathonian	54 ± 6	340 ± 7	74 ± 13
CYK	Kimmeridgian	26 ± 3	324 ± 4	62 ± 7
ALJ	Aalenian	44 ± 19	340 ± 7	64 ± 26
BRJu	Upper Jurassic	354 ± 15	324 ± 4	30 ± 19
BRJm	Aalenian	23 ± 60	340 ± 7	43 ± 67
PNS	Middle Jurassic	51 ± 18	340 ± 7	71 ± 25
CM	Aalenian	21 ± 19	340 ± 7	41 ± 26
SL	Bajocian-Bathonian	117 ± 12	340 ± 7	137 ± 19
CAZ	Upper Jurassic	0 ± 8	324 ± 4	36 ± 12

D_{OB}, ΔD_{OB}: *Observed declination and corresponding confidence limit; D_{EX}, ΔD_{EX}, expected declination and confidence limit; R, ΔR, bulk rotation and corresponding error (DEMAREST, 1983).*

been computed separately for the Upper Jurassic (from STEINER *et al.*, 1985; GALBRUN *et al.*, 1990; JUÁREZ *et al.*, 1996, 1998) and for the Middle-Lower Jurassic (from SCHOTT *et al.*, 1981; GIALANELLA, 1999). At the BRJ locality the two sites investigated (BRJs and BRJm) have been considered separately because of the differences in age of the sites. The statistical confidence limit for the rotation at the BRJm site is high due to the low intensity of component B and the low number of available samples. Figure 5 illustrates the rotation of the investigated region. The palaeomagnetic results indicate a significant and systematic clockwise block rotation of all investigated sites, with the exception of ERJ sites.

When considering the data from the E-W oriented profile in the central part of the Subbetic Zone, a continuous gradient in the rotation pattern cannot be established. In contrast, a discrete pattern is observed. Two regions can be differentiated: 1) the most external part of the Subbetic (ERJ sites), which is not rotated, and 2) the remaining sites from the central part of the Subbetic that experienced homogeneous clockwise rotations (CYK, CYB, ALJ, BRJ, PNS and CM sites).

The two sites at the ERJ locality exhibit westerly declinations and the estimated rotation is not significant ($10° \pm 10°$). ERJ sites are located in the so-called "Northern External Subbetic" (GARCÍA-DUEÑAS, 1967). About 15 km to the south of ERJ sites, and close to the locality of Carcabuey, the "Southern External Subbetic" were sampled at CYB and CYK sites. These sites have experienced around 65° of dextral rotation. Palaeomagnetic data obtained in the Carcabuey region are consistent with that obtained by OGG *et al.* (1984) and STEINER *et al.* (1987). The different rotational pattern of these two localities (ERJ and CY), that are relatively close, make this region of special interest in the study of block rotation mechanisms. Unfortunately there are no structural kinematic data in this area to solve this problem yet. Nonetheless we would like to emphasise the potential interest of the area for a detailed structural study.

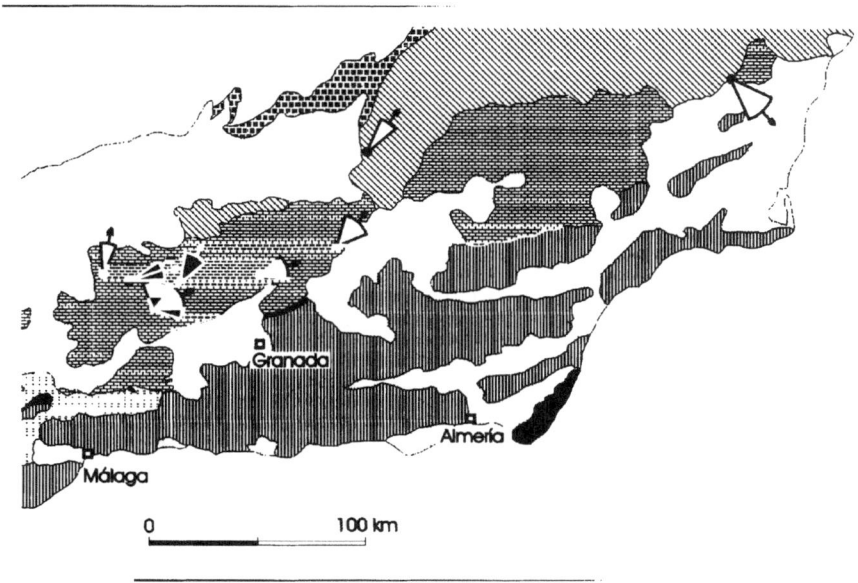

Figure 5
Sketch map showing the total rotation (R, Table 4) of each locality, with respect to geographic north, estimated from the Jurassic component. ΔR is graphically indicated on each arrow.

The remaining sites from the central part of the Subbetic (ALJ, BRJ, PNS and CM sites) are located in the Middle Subbetic paleogeographical domain. A homogeneous pattern of clockwise rotations is observed. The values of rotation range between 30° and 71°.

The SL site, located in the eastern part of the Subbetic, is strongly rotated (137° ± 14°). Our results confirm the large rotations that also have been observed in the eastern Betics by MAZAUD et al. (1986), OGG et al. (1988), OSETE et al. (1989) and ALLERTON et al. (1993).

The declination from the two CAZ sites exhibits small but significant (36° ± 9°), rotation relative to the Iberian reference direction. ALLERTON et al. (1993) reported results from one site from the Prebetic Zone (to the north of our site) that shows no important rotation. Palaeomagnetic data from the Prebetic Zone indicate that the overall rotation in the Prebetic seems to be significantly less than in the Subbetic, although differential rotations about vertical axis have also been observed.

Timing of Rotation

To investigate the timing of tectonic rotation we have used the information given by component A. This syn-folding secondary component can be used for tectonic purposes only if the proper tectonic correction is applied to the data (VILLALAÍN et al.

1996). The necessary information to calculate it is given by the incremental fold test that allows the estimation of the palaeohorizontal at the time of remagnetisation. The ALJ and BRJ sites have been considered because there are incremental fold-test results for component A in these two localities (4 sites). In addition, the positive result of the fold-test for component B at these sites allows us to exclude the possibility that the region has experienced more than one horizontal axis rotation. Table 3 gives the mean directions of component A for the two localities after 30% and 20% of unfolding. The declinations indicate no significant rotation relative to the Iberian Oligocene-Miocene direction ($D_{expected} = 4.1° \pm 8.5°$, BARBERÁ *et al.*, 1996), suggesting that this part of the Subbetic Zone has not rotated since the remagnetisation time.

The mean locality declination data of component A are displayed on Fig. 6, which also includes mean directions obtained by VILLALAÍN *et al.* (1994) from the western Subbetic. A contrasting rotational pattern is observed in these two regions: no rotations are detected in the central part of the Subbetic and significant rotations, about 40°–60°, are observed in the western Subbetic. On the basis of the analysis of directions of components A and B, VILLALAÍN *et al.* (1994, 1996) conclude that in the western Subbetic rotations took place after the remagnetisation event. These data demonstrate that, in the central part of the Subbetic, rotations occurred *before* the

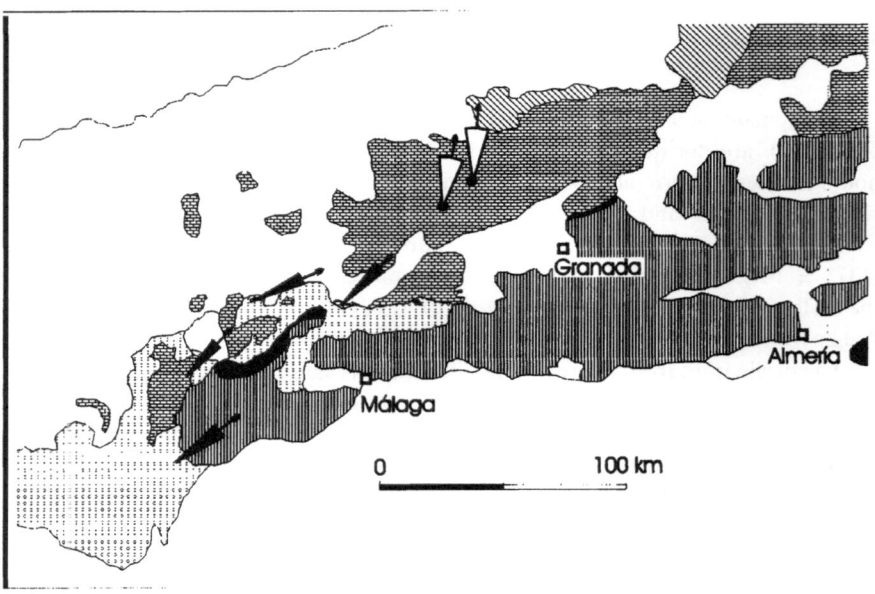

Figure 6
Geological map showing palaeodeclinations of the secondary (Neogene) component. ΔD is graphically indicated on each arrow. In white: data from ALJ and BRJ localities (this study). In grey: data from the western Subbetic (VILLALAÍN *et al.*, 1994).

acquisition of the secondary component. Consequently, if it is assumed that the remagnetisation event was synchronous in all of the Subbetic Zone, it can be concluded that the central part of the Subbetic rotated prior to the western part. Alternatively, if rotations took place in both regions at the same time, then the secondary component in the western Subbetic was acquired prior to that in the central Subbetic.

Considering that: 1) the Neogene component shows the same normal polarity in all the Subbetic Zone and 2) the Neogene is a period of mixed polarity, the hypothesis that remagnetisation was acquired over a short time-span and was synchronous in the whole Subbetic seems very likely.

Finally, it is emphasised that there is a strong potential for the use of the remagnetisation component to constrain the timing of rotational motions in the Betic Cordillera. Previous palaeomagnetic studies (PLATZMAN and LOWRIE, 1992; ALLER-TON et al., 1993, CALVO et al., 1994, 1997) have shown that only a few Tertiary sedimentary lithologies in the Betics are suitable for palaeomagnetic studies. In this situation, probably the best Tertiary palaeomagnetic data could be obtained from the remagnetisation component.

Conclusions

The magnetic behaviour of the ammonitico rosso samples from the Subbetic differs from that observed in the Prebetic Zone. Two stable components of magnetisation could be isolated by thermal cleaning in specimens from the Subbetic Zone: 1) component A, that always exhibited normal polarity and a maximum unblocking temperature of 450°C, which has been interpreted as a Neogene secondary component and 2) component B, considered as the original Jurassic magnetisation, with a maximum unblocking temperature of 575°C, showing both normal and reversed polarities of magnetisation. In contrast, in the Prebetic region, no evidence for a secondary overprint has been detected. Only the original Jurassic component could be isolated.

The spatial distribution of block rotations has been investigated (across a W-E oriented transect) in the central part of the Subbetic. The two sites located in the most external part of the region (north-western end of the profile) are not rotated. The remaining sites experienced clockwise rotations of about 60°. A single site located in eastern Subbetic is strongly rotated by 140°. The block rotations in the Prebetic Zone are significantly smaller than in the Subbetic. Two sites exhibited about 35° of dextral rotation.

A significant and widespread event took place in the whole Subbetic Zone during the Neogene. This produced partial remagnetisation of Jurassic nodular limestones. The secondary component is always of normal polarity, suggesting that remagnet-isation was acquired over a short time span ($< 10^6$ years). Incremental fold-test analyses have demonstrated that, in the central part of the Subbetic, remagnetisation was syn-folding (20% and 30% of unfolding), whereas in the western part it occurred at post-, syn- and mostly pre-folding situations (VILLALAÍN *et al.*, 1994). The presence of both components of magnetisation of Jurassic and Neogene ages in the same sample, and the use of the incremental fold-test allowed evaluation of the timing of tectonic rotations in the central and western part of the Subbetic. These results indicate that rotations in the central Subbetic took place before the remagnetisation event, whereas in the western Subbetic rotations occurred after it.

Acknowledgements

This work has been supported by the Dirección General de Investigación Científica y Tecnológica DGICYT (projects PB98-0834, PB97-0826 and BTE2002-00854). Some measurements were carried out by two of the authors (M.L.O and A.P) during visits to the Palaeomagnetic Laboratory of the ETH Zürich. They gratefully acknowledge F. Heller and W. Lowrie for their advice and generous offer of the use of Zürich facilities.

REFERENCES

ALLERTON, S. (1994), *Vertical Axis Rotation Associated with Folding and Thrusting: an Example from the Eastern Subbetics of Southern Spain*, Geology *22*, 1039–1042.
ALLERTON, S., LONERGAN, L., PLATT, J. P., PLATZMAN, E. S., and McCLELLAND, E. (1993), *Paleomagnetic Rotations in the Eastern Betic Cordillera, Southern Spain*, Earth Planet. Sci. Lett. *119*, 225–241.
ALLERTON, S., REICHERTER, K., and PLATT, J. P. (1994), *A Structural and Palaeomagnetic Study of a Section through the Eastern Subbetic, Southern Spain*, J. Geol. Soc. *151*, 659–668.
BALANYÁ, J. C., and GARCÍA-DUEÑAS, V. (1987), *Les directions structurales dans le Domaine d'Alboran de part et d'autre du Détroit de Gibraltar*, C.R. Acad. Sci. Paris. 304, II, 15, 929–933.
BANKS, C. J. and WARBURTON, J. (1991), *Mid-crustal Detachment in the Betic System of Southern Spain*, Tectonophysics *191*, 275–289.
BARBERÁ, X., CABRERA, L., GOMIS, E., and PARÉS, J. M. (1996), *Determinación del polo paleomagnético para el límite Oligoceno-Mioceno en la Cuenca del Ebro*, Geogaceta 20 (5), 1014–1016.
BÓGALO, M. F., OSETE, M. L., ANCOCHEA, E., and VILLALAÍN, J. J. (1994), *Estudio paleomagnético del volcanismo de Campos de Calatrava*, Geogaceta *15*, 117–120.
CALVO, M., OSETE, M. L., and VEGAS, R. (1994), *Palaeomagnetic Rotations in Opposite Senses in Southeastern Spain*, Geophys, Res Lett. *21*, 761–764.
CALVO, M., VEGAS, R., and OSETE, M. L. (1997), *Palaeomagnetic Results from Upper Miocene and Pliocene Rocks from the Internal Zone of the Eastern Betic Cordilleras (Southern Spain)*, Tectonophysics 277, 271–283.
CANDE, S. C. and KENT, D. V. (1995), *Revised Calibration of the Geomagnetic Polarity Timescale for the Late Cretaceous and Cenozoic*, J. Geophys. Res. *100*, B4, 6093–6095.
DEMAREST, H. H. (1983), *Error Analysis for the Determination of Tectonic Rotation from Paleomagnetic Data*, J. Geophys. Res. *88*, B5, 4321–4328.
DIJKSMAN, A. A. (1977), *Geomagnetic Reversals as Recorded in the Miocene Redbeds of Calatayud- Teruel Basin (Central Spain)*, Ph.D. Thesis, University of Utrecht.

DOBSON, J. P. and HELLER, F. (1992), *Remagnetization in Southeast China and the Collision and Suturing of the Huanan and Yangtze Blocks*, Earth Planet. Sci. Lett. *111*, 11–21.

FISHER, R. A. (1953), *Dispersion on a Sphere*, Proc. R. Soc. London A *217*, 295–305.

GALBRUN, B., BERTHOU, P.Y., MOUSSIN, C., and AZÉMA, J. (1990), *Magnétostratigraphie de la limite Jurassique-Crétacé en faciés de la plate-forme carbonatée: la coupe de Bias do Norte (Algarve, Portugal)*, Bull. Soc. Geol. Fr. *8*, VI (1), 133–143.

GARCÍA-DUEÑAS, V. (1967), *La zona Subbética al Norte de Granada*, Ph.D. Thesis, Univ. of Granada.

GARCÍA-DUEÑAS, V., BALANYÁ, J. C., and MARTÍNEZ-MARTÍNEZ, J. M. (1992), *Miocene Extensional Detachments in the Outcropping Basement of the Northern Alboran Basin (Betics) and their Tectonic Implications*, Geo-Marine Lett. *12*, 88–95.

GARCÍA-HERNÁNDEZ, M., LÓPEZ-GARRIDO, A. C., RIVAS, P., SANZ DE GALDEANO, C., and VERA, J. A. (1980), *Mesozoic Palaeogeographic Evolution of the External Zones of the Betic Cordillera*, Geologie Mijnb. *59*, 155–168.

GIALANELLA, P. R. (1999), *Analisi magnetostratigrafiche di sequenze mesozoiche affioranti in Spagna e in Italia*, Ph.D Thesis, Universitá di Napoli e Palermo.

GRADSTEIN, F. M., AGTERBERG, F. P., OGG, J. G., HARDERBOL, J., VAN VEEN, P., THIERRY, J., and HUANG, Z. (1994), *A Mesozoic Time Scale*. J. Geophys. Res. *99*, 24,051–24,074.

JIMÉNEZ, A. P. and RIVAS, P. (1979), *El Toarcense en la Zona Subbética*, Cuadernos de Geología *10*, 397–411.

JUÁREZ, M. T., OSETE, M. L., VEGAS, R., LANGEREIS, C. G., and MELÉNDEZ, G. (1996), *Palaeomagnetic study of Jurassic limestones from the Iberian Range (Spain): Tectonic implications*, (Morris, A. and Tarling, D.H., eds), *Palaeomagnetism and Tectonics of the Mediterranean Region*, Geological Society Special Publication *105*, 83–96.

JUÁREZ, M. T., LOWRIE, W., OSETE, M. L., and MELÉNDEZ, G. (1998), *Evidence of Widespread Cretaceous Remagnetisation in the Iberian Range and its Relation with the Rotation of Iberia*, Earth Planet. Sci. Lett. *160*, 729–743.

KIRKER, A. and MCCLELLAND, E. (1996), *Application of net tectonic rotations and inclination analysis to a high-resolution palaeomagnetic study in the Betic Cordillera* (Morris, A. and Tarling, D. H., eds). *Palaeomagnetism and Tectonics of the Mediterranean Region*, Geological Society Special Publication *105*, 19–32.

LINARES, A. and SANDOVAL, J. (1993), *El Aaleniense de la Cordillera Bética (Sur de España): Análisis bioestratigráfico y caracterización paleobiogeográfica*, Rev. Soc. geol. Esp. *6*, 177–206.

MAZAUD, A., GALBRUN, B., AZEMA, J., ENAY, R., FOURCADE, E., and RESPLUS, L. (1986), *Données magnétostratigraphiques sur la Jurassique Supérieur et la Berriasien du NE des Cordillères Bétique*, C. R. Acad. Sc. Paris. *302*, Série II, 18, 1165–1170.

MCCABE, C. and ELMORE, R. D. (1989), *The Occurrence and Origin of Late Paleozoic Remagnetization in the Sedimentary Rocks of North America*, Rev. Geophys. *27*, 471–494.

MCCABE, C., VAN DER VOO, R., PEACOR, D. R., SCOTESE, C. R., and FREEMAN, R. (1983), *Diagenetic Magnetite Carries Ancient yet Secondary Remanence in Some Paleozoic Sedimentary Carbonates*, Geology *11*, 221–223.

MCFADDEN, P. L. and JONES, D. L. (1981), *The Fold Test in Palaeomagnetism*, Geophys. J. Int. *67*, 53–58.

MCFADDEN, P. L. and LOWES, F. J. (1981), *The Discrimination of Mean Directions drawn from Fisher Distributions*, Geophys. J. R. Astron. Soc. *67*, 19–33.

MCFADDEN, P. L. and MCELHINNY, M. W. (1990), *Classification of the Reversal Test in Palaeomagnetism*, Geophys. J. Int. *103*, 725–729.

OGG, J. G., STEINER, M. B., OLORIZ, F., and TAVERA, J. M. (1984), *Jurassic Magnetostratigraphy, 1. Kimmeridian-Tithonian of Sierra Gorda and Carcabuey, Southern Spain*, Earth Planet. Sci. Lett. *71*, 147–162.

OGG, J. G., STEINER, M. B., COMPANY, M., and TAVERA, J. M. (1988), *Magnetostratigraphy across the Berriasian-Valanginian Stage Boundary (Early Cretaceous), at Cehegin (Murcia Province, Southern Spain)*, Earth Planet. Sci. Lett. *87*, 205–215.

OLIVER, J. (1986), *Fluids Expelled Tectonically from Orogenic Belts: Their Role in Hydrocarbon Migration and other Geologic Phenomena*, Geology *14*, 99–102.

OLÓRIZ, F. (1978), *Kimmeridgiense-Tithónico inferior en el sector central de las Cordilleras Béticaas. Zona Subbética. Paleontologia. Bioestratigrafia*, Ph.D. Thesis, Univ. Granada.

OSETE, M. L., FREEMAN, R., and VEGAS, R. (1988), *Preliminary Palaeomagnetic Results from the Subbetic Zone (Betic Cordillera, Southern Spain): Kinematic and Structural Implications*, Phys. Earth Planet. Inter. *52*, 283–300.

OSETE, M. L., FREEMAN, R., and VEGAS, R. (1989), *Palaeomagnetic evidence for block rotations and distributed deformation of the Iberian-African plate boundary* (C. Kissel and C. Laj, eds.), Palaeomagnetic Rotations and Continental Deformation. NATO ASI Series, (Kluwer Academic Publishers) 254, 381–391.

PLATT, J. P. and VISSERS, R. L. M. (1989), *Extensional Collapse of Thickened Continental Lithosphere: A Working Hypothesis for the Alboran Sea and Gibraltar Arc*, Geology 17, 540–543.

PLATT, J. P., ALLERTON, S., KIRKER, A., and PLATZMAN, E. (1994), *Origin of the Western Subbetic (South Spain): Palaeomagnetic and Structural Evidence*. J. Struct. Geol. *17*, 6, 765–775.

PLATZMAN, E. (1992), *Palaeomagnetic Rotations and the Kinematics of the Gibraltar Arc*, Geology 20, 311–314.

PLATZMAN, E. and LOWRIE, W. (1992), *Paleomagnetic Evidence for Rotation of the Iberian Peninsula and the External Betic Cordillera, Southern Spain*, Earth Planet. Sci. Lett. *108*, 45–60.

SANDOVAL, J. (1983), *Bioestratigrafía y Paleontología (Stephanocerataceae y Perisphinctaceae) del Bajocense y Bathonense de las Cordilleras Béticas*, Ph.D. Thesis, Univ. of Granada.

SCHOTT, J. J., MONTIGNY, R., and THUIZAT, R. (1981), *Paleomagnetism and Potassium-argon Age of the Messejana Dike (Portugal and Spain): Angular Limitation to the Rotation of the Iberian Peninsula since the Middle Jurassic*, Earth Planet. Sci. Lett. *53*, 457–470.

SEQUEIROS, L. (1974), *Paleobiogeografía del Calloviense y Oxfordense en el sector Central de la Zona Subbética*, Ph.D. Thesis, Univ. Granada.

STEINER, M. B., OGG, J. G., MELÉNDEZ, G., and SEQUEIROS, L. (1985), *Jurassic magnetostratigraphy. 2. Middle-Late Oxfordian of Aguilón, Iberian Cordillera, Northern Spain*, Earth Planet. Sci. Lett. *76*, 151–166.

STEINER, M. B., OGG, J. G., and SANDOVAL, J. (1987), *Jurassic Magnetostratigraphy, 3. Bathonian-Bajocian of Carcabuey Sierra Harana and Campillo de Arenas (Subbetic Cordillera, Southern Spain)*, Earth Planet. Sci. Lett. *82*, 357–372.

SUK, D., PEACOR, D. R., and VAN DER VOO, R. (1990), *Replacement of Pyrite Framboids by Magnetite in Limestones and Implications for Paleomagnetism*, Nature 345, 611–613.

SUK, D., VAN DER VOO, R., and PEACOR, D. R. (1993), *Origin of the Magnetite Responsible for Remagnetization of Early Paleozoic Limestones of New York State*, J. Geophys. Res. *98*, 419–434.

VAN FOSSEN, M. C. and KENT, D. V. (1990), *High Latitude Palaeomagnetic Poles from Middle Jurassic Plutons and Moat Volcanics in New England and the Controversy Regarding Jurassic Apparent Polar Wander for North America*, J. Geophys. Res. *95*, 17,503–17,516.

VILLALAÍN, J. J. (1995), *Estudio paleomagnético de las Béticas Occidentales y sus implicaciones tectónicas. Descripción de una reimanación regional neógena*, Ph.D. Thesis, Complutense University of Madrid.

VILLALAÍN, J. J., OSETE, M. L., VEGAS, R., GARCÍA DUEÑAS, V., and HELLER, F. (1994), *Widespread Neogene Remagnetization in Jurassic Limestones of the South-Iberian Paleomargin (Western Betics, Gibraltar Arc)*, Phys. Earth Planet. Inter. *85*, 15–33.

VILLALAÍN, J. J., OSETE, M. L., VEGAS, R., GARCÍA DUEÑAS, V., and HELLER, F. (1996), *The Neogene remagnetization in the western Betics: a brief comment on the reliability of palaeomagnetic directions*. In (Morris, A. and Tarling, D. H. eds.), Palaeomagnetism and Tectonics of the Mediterranean Region, Geological Society Special Publication *105*, 33–41.

(Received January 31, 2002, revised January 20, 2003, accepted January 30, 2003)

 To access this journal online:
http://www.birkhauser.ch